Lecture Notes in Artificial Intelligence   11680

Subseries of Lecture Notes in Computer Science

More information about this series at http://www.springer.com/series/1244

Kerstin Bach · Cindy Marling (Eds.)

Case-Based Reasoning Research and Development

27th International Conference, ICCBR 2019
Otzenhausen, Germany, September 8–12, 2019
Proceedings

 Springer

Editors
Kerstin Bach 🆔
Norwegian University of Science
and Technology
Trondheim, Norway

Cindy Marling 🆔
Ohio University
Athens, OH, USA

ISSN 0302-9743   ISSN 1611-3349   (electronic)
Lecture Notes in Artificial Intelligence
ISBN 978-3-030-29248-5   ISBN 978-3-030-29249-2   (eBook)
https://doi.org/10.1007/978-3-030-29249-2

LNCS Sublibrary: SL7 – Artificial Intelligence

This Springer imprint is published by the registered company Springer Nature Switzerland AG
The registered company address is: Gewerbestrasse 11, 6330 Cham, Switzerland

Preface

This volume contains the papers presented at the 27th International Conference on Case-Based Reasoning (ICCBR), which was held during September 8–12, 2019, at the European Academy of Otzenhausen, in Otzenhausen, Germany. ICCBR is the premier annual meeting of the Case-Based Reasoning (CBR) research community. The theme of ICCBR 2019 was Explainable AI (XAI). Explanation has been a core aspect of CBR since its very beginning, and it has been exciting to see the growing emphasis on XAI from the broader AI community.

This year marked the return of the conference to Otzenhausen, site of the first ICCBR in 1993. Previous ICCBRs, including the merged European Workshops and Conferences on CBR, were as follows: Otzenhausen, Germany (1993); Chantilly, France (1994); Sesimbra, Portugal (1995); Lausanne, Switzerland (1996); Providence, USA (1997); Dublin, Ireland (1998); Seeon Monastery, Germany (1999); Trento, Italy (2000); Vancouver, Canada (2001); Aberdeen, UK (2002); Trondheim, Norway (2003); Madrid, Spain (2004); Chicago, USA (2005); Fethiye, Turkey (2006); Belfast, UK (2007); Trier, Germany (2008); Seattle, USA (2009); Alessandria, Italy (2010); Greenwich, UK (2011); Lyon, France (2012); Saratoga Springs, USA (2013); Cork, Ireland (2014); Frankfurt, Germany (2015); Atlanta, USA (2016); Trondheim, Norway (2017); and Stockholm, Sweden (2018).

ICCBR 2019 received 46 submissions from 17 countries, spanning Europe, North America, Asia, and Oceania. Each submission was reviewed by three Program Committee members. Papers for which the reviewers did not reach consensus were referred to members of the ICCBR Advisory Council for meta-review. Of the 46 submissions, 15 (33%) were selected for oral presentation, and 11 (24%) were selected for poster presentation.

ICCBR 2019 began as the community assembled in Otzenhausen on the afternoon of September 8. The only formal activity on that day was for the Doctoral Consortium (DC) participants, who met their mentors for the first time and prepared for their upcoming presentations. The DC provides opportunities for PhD students to share and obtain mutual feedback on their research and career objectives with senior CBR researchers and peers. After the DC session, all conference attendees gathered together for a barbecue.

The first full day of the conference was devoted to workshops and the DC. There were five workshops in all: Process-Oriented Case-Based Reasoning; Case-Based Reasoning in the Health Sciences; Case-based Reasoning and Deep Learning; Explainable Knowledge in Computational Design, Media, and Teaching; and XCBR: Case-based Reasoning for the Explanation of Intelligent Systems. While many activities proceeded in parallel, the first XCBR workshop session was held as a plenary event to set the stage for the conference theme of Explainable AI. This joint XCBR workshop session featured an inspiring invited speaker, David Aha, of the Naval Research Laboratory, in Washington, DC. There was an afternoon break for a

"Walk & Talk" social program, culminating in a visit to a nearby historic Celtic Village. Then, it was back to the workshops and DC, until it was time for dinner. Dinner was followed by a showing of the entries in the Third ICCBR Video Competition.

The workshops and DC concluded on the morning of Day Two. The afternoon commenced with the invited talk, "Mapping the Challenges and Opportunities of CBR for Explainable AI," by Belén Díaz Agudo, of the Complutense University of Madrid. Next came a panel discussion, in which those who had attended the first ICCBR in Otzenhausen, in 1993, were invited to reminisce, reflect upon progress over the past 26 years, and prognosticate advances for the next 26 years. The panel was followed by the first oral presentations of the main technical track. The Conference Dinner was held in the evening.

Day Three began with the invited talk, "Some Shades of Grey! Interpretability and Explanatory Capacity of Deep Neural Networks," by Andreas Dengel, of the German Research Center for Artificial Intelligence (DFKI), in Kaiserslautern. The main technical track then continued throughout most of the day. Between the main technical track presentations and dinner, there was a Poster and Demo session. Prior to this session, the presenters of posters and demos each had three minutes to pitch their work at a plenary session, facilitating later discussion and networking.

Daniele Magazzeni, of King's College London, opened the final day of the conference with his invited talk, "Model-Based Reasoning for Explainable AI as a Service." This was followed by the final oral presentations in the main technical track. ICCBR 2019 then concluded with a Community Meeting and lunch. The community looks forward to ICCBR 2020 in Salamanca, Spain.

We gratefully acknowledge the support of the following people, without whose contributions ICCBR 2019 would not have been possible. Local chairs Ralph Bergmann and Klaus-Dieter Althoff did a fine job of arranging the splendid conference facilities and social events. Ralph Bergmann also managed the registration process and kept us all within budget. Workshop chairs Hayley Borck and Stelios Kapetanakis planned a lively workshop program of five individual workshops. To all of the organizers of these individual workshops, we also extend our heartfelt thanks. Publicity chairs Viktor Eisenstadt and Pascal Reuss managed the conference website and our social media presence. Antonio Sánchez-Ruiz and Michael Floyd chaired the DC, providing invaluable support to the next generation of CBR researchers. Brian Schack and Devi Ganesan chaired the Third ICCBR Video Competition, which, in addition to its entertainment value, aims to provide educational material for students and outreach to the general public. Our sponsorship chair, Michael Cox, raised much-needed funding to subsidize costs for student attendees.

We extend our gratitude to the members of the ICCBR Advisory Council, Agnar Aamodt, David Aha, Belén Díaz Agudo, Peter Funk, Mehmet Göker, and Ramon López de Mántaras, for their sage advice and unwavering support. We would also like to thank the Program Committee and additional reviewers, who thoughtfully assessed the submissions and did an excellent job of providing constructive feedback to the authors.

Finally, we are very grateful for the support of our sponsors, who, at the time of printing, included the *Artificial Intelligence Journal (AIJ)*, Springer, the Knexus Research Corporation, the Norwegian University of Science and Technology (NTNU), Ohio University, the University of Trier, the University of Hildesheim, and the Norwegian Open AI Lab.

September 2019   Kerstin Bach
                                                                              Cindy Marling

Organization

Program Chairs

Kerstin Bach   Norwegian University of Science and Technology,
                             Norway
Cindy Marling   Ohio University, USA

Local Chairs

Ralph Bergmann   University of Trier, Germany
Klaus-Dieter Althoff   University of Hildesheim and DFKI, Germany

Workshop Chairs

Hayley Borck   Honeywell, USA
Stelios Kapetanakis   University of Brighton, UK

Publicity Chairs

Viktor Eisenstadt   University of Hildesheim and DFKI, Germany
Pascal Reuss   University of Hildesheim and DFKI, Germany

Doctoral Consortium Chairs

Antonio Sánchez-Ruiz   Complutense University of Madrid, Spain
Michael Floyd   Knexus Research Corporation, USA

Video Competition Chairs

Brian Schack   Indiana University, USA
Devi Ganesan   Indian Institute of Technology Madras, India

Sponsorship Chair

Michael Cox   Wright State University, USA

Advisory Council

Agnar Aamodt	Norwegian University of Science and Technology, Norway
David Aha	Naval Research Laboratory, USA
Belén Díaz Agudo	Complutense University of Madrid, Spain
Peter Funk	Mälardalen University, Sweden
Mehmet H. Göker	Flexport, USA
Ramon López de Mántaras	IIIA-CSIC, Spain

Program Committee

Klaus-Dieter Althoff	University of Hildesheim and DFKI, Germany
Ralph Bergmann	University of Trier, Germany
Isabelle Bichindaritz	State University of New York at Oswego, USA
Hayley Borck	Honeywell, USA
Derek Bridge	University College Cork, Ireland
Sutanu Chakraborti	Indian Institute of Technology Madras, India
Alexandra Coman	Capital One, USA
Juan Corchado	University of Salamanca, Spain
Michael Cox	Wright State University, USA
Dustin Dannenhauer	Navatek, LLC, USA
Sarah Jane Delany	Technological University Dublin, Ireland
Michael Floyd	Knexus Research Corporation, USA
Ashok Goel	Georgia Institute of Technology, USA
Pedro González Calero	Universidad Politécnica de Madrid, Spain
Odd Erik Gundersen	Norwegian University of Science and Technology, Norway
Vahid Jalali	Indiana University, USA
Stelios Kapetanakis	University of Brighton, UK
Joseph Kendall-Morwick	Missouri Western State University, USA
Luc Lamontagne	Laval University, Canada
David Leake	Indiana University, USA
Jean Lieber	LORIA, France
Stewart Massie	Robert Gordon University, UK
Mirjam Minor	Goethe University Frankfurt, Germany
Stefania Montani	University of Piemonte Orientale, Italy
Héctor Muñoz-Avila	Lehigh University, USA
Emmanuel Nauer	LORIA, France
Santiago Ontañón	Drexel University, USA
Miltos Petridis	Middlesex University London, UK
Enric Plaza	IIIA-CSIC, Spain
Luigi Portinale	University of Piemonte Orientale, Italy
Juan Recio-Garcia	Complutense University of Madrid, Spain
Jonathan Rubin	Philips Research North America, USA
Antonio Sánchez-Ruiz	Complutense University of Madrid, Spain

Barry Smyth	University College Dublin, Ireland
Frode Sørmo	Amazon, UK
Ian Watson	University of Auckland, New Zealand
Rosina Weber	Drexel University, USA
David Wilson	UNC Charlotte, USA
Nirmalie Wiratunga	Robert Gordon University, UK

Additional Reviewers

Kareem Amin
Viktor Eisenstadt
Devi Ganesan
Miriam Herold
Patrick Klein
Lukas Malburg

Kyle Martin
Ikechukwu Nkisi-Orji
Jakob Schoenborn
Anjana Wijekoon
Christian Zeyen

Sponsors

Artificial Intelligence Journal (AIJ)
Knexus Research Corporation
Norwegian Open AI Lab
Norwegian University of Science and Technology (NTNU)
Ohio University
Springer
University of Hildesheim
University of Trier

Abstracts of Invited Papers

Mapping the Challenges and Opportunities of CBR for eXplainable AI

Belén Díaz Agudo

Department of Software Engineering and Artificial Intelligence,
Complutense University de Madrid, Madrid, Spain

Abstract. The problem of explainability in Artificial Intelligence is not new but the rise of the autonomous intelligent systems has increased the necessity to understand how an intelligent system achieves its solution, makes a prediction or a recommendation or reasons to support a decision in order to increase transparency and users' trust in these systems. The CBR research community has a great opportunity to provide general methods of self-understanding and introspection on other AI systems, not necessarily case-based. CBR provides a methodology to reuse experiences in interactive explanations and can exploit memory-based techniques to generate explanations to different AI techniques and domains of applications. This talk will review the state of the art of XCBR, the synergies with the XAI community, and will give the opportunity to review the underlying issues like confidence, transparency, justification, interfaces, personalization and evaluation of explanations. It will include a review of the lessons learnt at the XCBR workshop and the challenges and promising research lines for CBR research related to the explanation of intelligent systems.

Some Shades of Grey! Interpretability and Explanatory Capacity of Deep Neural Networks

Andreas Dengel

Deutsches Forschungszentrum für Künstliche Intelligenz GmbH (DFKI),
Kaiserslautern, Germany

Abstract. Based on the availability of data and corresponding computing capacity, more and more cognitive tasks can be transferred to computers, which independently learn to improve our understanding, increase our problem-solving capacity or simply help us to remember connections. Deep neural networks in particular clearly outperform traditional AI methods and thus find more and more areas of application where they are involved in decision-making or even make decisions independently. For many areas, such as autonomous driving or credit allocation, the use of such networks is extremely critical and risky due to their black box character, since it is difficult to interpret how or why the models come to certain results. The paper discusses and presents various approaches that attempt to understand and explain decision-making in deep neural networks.

Model-Based Reasoning for Explainable AI as a Service

Daniele Magazzeni

Department of Informatics, King's College London, UK

Abstract. As AI systems are increasingly being adopted into application solutions, the challenge of providing explanations and supporting interaction with humans is becoming crucial. Partly this is to support integrated working styles, in which humans and intelligent systems cooperate in problem-solving, but also it is a necessary step in the process of building trust as humans migrate greater responsibility to such systems. In this talk we discuss progress made in Explainable Planning, particularly in scenarios involving human-AI teaming, and we present recent advances in using model-based reasoning for designing Explainable AI as a Service.

Contents

Comparing Similarity Learning with Taxonomies and One-Mode Projection in Context of the FEATURE-TAK Framework

Oliver Berg[3], Pascal Reuss[1,2(✉)], Rotem Stram[1,3], and Klaus-Dieter Althoff[1,2]

[1] German Research Center for Artificial Intelligence, Kaiserslautern, Germany
reusspa@uni-hildesheim.de
[2] Institute of Computer Science, Intelligent Information Systems Lab,
University of Hildesheim, Hildesheim, Germany
[3] Department of Computer Science, Technical University of Kaiserslautern,
Kaiserslautern, Germany
http://www.dfki.de/
http://www.uni-hildesheim.de/

Abstract. This paper describes the learning of new similarity values for existing measures within the framework FEATURE-TAK. Maintenance of similarity measures is not easy, especially when having a semi-automated approach to relieve the knowledge engineer. Based on the extension of the vocabulary, the newly added values have to be integrated into the similarity measures with an initial similarity value to be useful. We describe the extension of the similarity measures with automated taxonomy extension and one-mode projections and present a comprehensive evaluation and comparison between the different approaches to highlight the advantages and short comings.

Keywords: Case-based reasoning · Similarity measures ·
Knowledge modeling · One-mode projection

1 Introduction

To solve occurring problems in areas like monitoring, maintenance and operation one requires knowledge of the situation and how to resolve the issue. Knowledge can be approximated through storing data of a past problem description together with context information and executed solutions, which forms a so-called case [1]. Case-based Reasoning (CBR) then tries to solve new problems through noticing similarities with previously solved problems and adapting their known solutions, as such modelling human reasoning [8]. This indicates that similarity of problem descriptions remains pivotal to finding adequate solutions. The quality of retrieved reference cases thus highly depends on the defined similarity measures. Exact similarity measures frequently rely on use

K. Bach and C. Marling (Eds.): ICCBR 2019, LNAI 11680, pp. 1–16, 2019.
https://doi.org/10.1007/978-3-030-29249-2_1

case specific variables and expert understanding of how an application operates. This dependence on expert knowledge imposes restrictions to completeness and performance of any CBR implementation. To this end, the **F**ramework for **E**xtraction, **A**nalysis, and **T**ransformation of **U**nstructu**RE**d **T**extual **A**ircraft **K**nowledge (FEATURE-TAK) is being developed, which aims at automating knowledge acquisition, specifically in the aviation domain. From problem fault descriptions in free-text form as provided by aviation maintenance personnel, it retrieves keywords, phrases and synonyms/hypernyms for specific attributes to enrich the systems vocabulary and allow for a more refined context description. As of now, FEATURE-TAK employs keywords and synonym structure for local (attribute level) similarity approximation for individual attribute values, and sensitivity analysis for global (case-level) similarity approximation for similarity of complete case descriptions. Similarity of single attributes is aggregated and weighted to build up complete case similarities [9,10].

Local similarity operates on mostly non-numeric symbol attribute descriptions (phrases from text) as opposed to numeric values. This makes attribute similarity difficult to infer in a non-binary - not only attribute value equal to or unequal to value X - manner. The approach currently employs taxonomies, as such extracting dependent symbols to then infer level-based similarity properties. This assesses similarity only for related keywords, which greatly reduces the actual number of at-all-similar relations. In its current implementation it presents major implications regarding modelling assumptions and does not generally capture all attribute dependencies as it relies on synonyms and hypernyms. The compromise is intentional though, because synonym-hypernym-connections enable taxonomies in the first place and any taxonomy related implementation of similarity will likely continue to employ synonyms/hypernyms. Global similarity on the other hand weights single attributes through sensitivity analysis based on relevance for solving cases. This allows the system to build averages used to calculate the global similarity value for comparing different cases. As this presents a primarily novel technique, it relies on mostly project-specific configuration and uses fewer well-established procedures.

Overall case-similarity should take the total case description into account. These local measures are then being weighted and combined into a final global similarity, being in return utilized to rank the solutions and return the most likely solution back to the system user. Exact local measures, weighting and ranking implementations are an act of great balance, as it is both delicate and non-deterministic which dependencies are impacting problem transference and to what degree. Therefore, a projection-based similarity procedure [12,13] is being incorporated into the framework's local similarity assessment to add comparison between not directly related keywords. This Weighted One-Mode Projection (WOMP) has proven successful in prior experiments [12] but was not tested on real-world application data yet. This paper specifically displays the outcome of properly integrating it into FEATURE-TAK, evaluating performance inside the framework and act as a starting point to incorporate further concepts.

In Sect. 2 we describe briefly the framework FEATURE-TAK, the WOMP and its integration into the framework. Section 3 describes the evaluation setup and the results of the comparison between the taxonomy similarity learning and the one mode projection learning. Section 4 closes with a discussion and outlook.

2   Weighted One Mode Projection in FEATURE-TAK

This section gives a brief overview of the FEATURE-TAK framework and then describes the Weighted One Mode Projection and its integration in the framework. This transitions into Sect. 3 with the description of the evaluation setup, the results, and their interpretation.

2.1   FEATURE-TAK

Based on established procedures, namely the myCBR toolkit [4, 6, 11] and the agent-based SEASALT architecture [3], the FEATURE-TAK framework has been developed to support knowledge engineers in querying data that is organized in structured and unstructured format and with highly domain-specific information. Technical maintenance data is frequently available in attribute-value pairs, whereas logbook entries and feedback only exist in free-text format. Thus, a hybrid representation, combining attribute-value and textual representation, is implemented by accounting for attributes specified from meta-information surrounding a case description and incorporating information entities from text; further detail regarding input- and attribute data in [9]. FEATURE-TAK processes said free-text input, applying natural language processing (NLP) techniques alongside CBR to extract keywords, phrases and synonyms to comprise attribute values of a predefined case-structure. In addition to known CBR procedures, some novel automated knowledge transformation is added. The framework consists of five layers: Data-, Agent-, NLP-, CBR- and Interface Layer. Inside the agent layer, multiple agents provide functionality in form of designated tasks based on given input data and the required CBR data structure. The tasks are regrouped and separated into sub-steps. From a given free-text input file together with provided mapping-, abbreviation and white-/black-list files the data sets are transformed into the internal representation format. This process adapts dynamically to available input information. The mapping file is in the XML format and describes which information in the initial data are to be mapped to which attributes in the case structure. Exact transformation is based on the mapping information, which results in a case structure with precisely defined 71 - possibly sparse - attributes. [9] The agents in the framework are responsible for performing the tasks within FEATURE-TAK and are implemented as follows:

- The **Preprocessing Agent** (Task 0) is a prerequisite for the subsequent stages as it prepares the input data through part-of-speech (POS) tagging and abbreviation identification.

- The **Collocation Agent** (Task 1) extracts phrases as recurring word combinations based on standard English grammar and domain-specific patterns.
- The **Keyword Agent** (Task 2) extracts keywords via stop-word elimination, lemmatization and single word abbreviation replacement, returning the word's base form.
- **The Synonym Agent** (Task 3) identifies synonyms and hypernyms considering word context and -sense, thus utilizing the POS information and provided black- and whitelists.
- The **Vocabulary Manager** (Task 4) adds phrases, keywords and synonyms/hypernyms to the CBR system's vocabulary.
- The **Similarity Manager** (Task 5) sets similarity values for concepts, extending attribute similarity measures to compare overall cases with each other, and respectively utilizes taxonomies on top of generated phrases, keywords and synonyms/hypernyms to further infer attribute value proximity.
- The **ARM Agent** (Task 6) searches for association rules in word occurrences within and across data sets, where - depending on data set size and performance constraints - either the Apriori [2] or the FP-Growth [5] algorithm with a high confidence of 0.9 being used to only allow rules found to be true most of the time.
- The **Clustering Agent** (Task 7) generates a case from each corpus of input data with an associated cluster (based on aircraft type and component, where components are specified with a unique digit called Air Transport Association (ATA) chapter) to persist it into the case bases.
- The **Sensitivity Agent** (Task 8) finally generates global similarity measurement between complete cases by incrementally approximating weight vectors over the set of attributes for each case cluster, resulting in a global relevance weight matrix [14].

As such, starting from the sparse 71-attribute-representation with only directly extracted values set, the pre-processing agent is engaged and subsequently the agents are traversed as described above. Note that tasks 0–3 are executed in strict sequence on the previous task's respective output, and only then slight branching is being undergone. From there, with a concise vocabulary description, similarity between values of a given attribute is aggregated, whereas global case similarity through the sensitivity analysis operates on subsets of data in form of generated case clusters as well as internal attribute similarity approximation [9].

2.2   Integration of the Weighted One-Mode Projection

An alternative to manually described similarity matrices and also taxonomies is the more general usage of projections of bipartite graph structures. Complex network analysis is, generally speaking, an emerging field regarding modelling relationship. In this context, bipartite graphs present a particular type of graph that consists of two distinct populations of nodes between which, but not within which, nodes are connected. This type of graph can be nicely used to model real-world systems such as for example describe personal preference [16] or depict co-authorship in scientific publishing [7]. These scenarios all have two groups of nodes interacting with each other, following the example of scientific publishing, for example "authors" and "papers". To find relations between entities within one of the two populations, a simple One-Mode Projection (OMP) or if weighted a Weighted One-Mode Projection (WOMP) can be calculated. Following the co-authorship analogy above, this would find relations or "similarity" between authors based on their collaboration in scientific publications. Regarding how the projection is calculated, Fig. 1 provides a simple example: The idea is that for each element in one population $l_a \in L$, one considers each other element $l_b \in L$ that can be reached via one or more elements $r_j \in R$ of the opposing population. Considering only reachability, one obtains a simple one-mode projection, while counting the number of common elements in R results in a weighted one-mode projection. We assume projections to be calculated on the set L for simplicity reasons, projecting R can be done by simply inverting symbols in the notation [15].

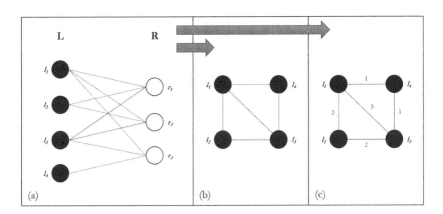

Fig. 1. Illustration of a bipartite graph and respective OMP/WOMP [12]

All of these approaches do however preserve the inherent projection problem of symmetry. Assuming symmetry in similarities cannot always be guaranteed to appy to real-world examples, as was stated in [12]. The so far discussed solutions all produce a single projected edge between any two entities l_a, l_b in population L based on common properties through connections via a number of entities of the opposing population R. The resulting edge in the projection graph is not directed, l_a is as similar to l_b as is l_b to l_a. Also, for any r_j that holds only a single connection to L, this connection cannot be properly represented in the projection graph's edge weights. Thus, the approach motivated by [12] is based on the idea introduced by [16] of nodes holding resources being distributed throughout the network. One assumes a bipartite graph with weighted edges, unweighted edges would correspond to all edges having a weight of 1.

The procedure then calculates resources to be distributed (visualized in Fig. 2). For each node $l_i \in L$ one aggregates all weights w_{ij} of edges starting in l_i and ending in all $r_j \in R$ to obtain the node's weight W_i^L as its held aggregated resources (Fig. 2(a)). Similarly, W_j^R is later calculated by aggregating all resource shares $w_{ij}^{L \to R}$ as they are propagated from $L \to R$ (Fig. 2(c)). Please note that notation is critical here: All w_{ij}, $w_{ij}^{L \to R}$ and $w_{ij}^{R \to L}$ refer to weights of the same edge connecting l_i and r_j, with the former of the three considering the original weights as of the initial weighted bipartite graph, while the latter two are being calculated by the depicted procedure.

$$W_i^L = \sum_{j=1}^{|R|} w_{ij}, \quad w_{ij}^{L \to R} = \frac{w_{ij}}{W_i^L}, \quad W_j^R = \sum_{i=1}^{|L|} w_{ij}^{L \to R}, \quad w_{ij}^{R \to L} = \frac{w_{ij}^{L \to R}}{W_j^R} \quad (1)$$

To then obtain the final projection graph, one sums up all edges connecting any pair of nodes (l_a, l_b) via nodes $r_j \in R$ of the opposing population. However, this shall result in a pair of directed weighted edges of weights w_{ab} and w_{ba}. The difference in weights is achieved by not summing up all available edge weights between l_a and l_b, but rather by following along the respective $w_{aj}^{L \to R}$ and $w_{bj}^{R \to L}$ according to the inherent direction ($a \Rightarrow b$ or $b \Rightarrow a$), in which $p_{ij} \in \{0, 1\}$ is used as indication of whether or not l_i and r_j share a connection. The resulting network is then the projection graph (not bipartite anymore) of the original (weighted) bipartite graph. This projection graph does not contain normalized edge weights. To obtain fixed weights in the interval $[0, 1]$, [12] proposes to utilize normalized weights $\hat{w}_{ab}^{L \to L}$ obtained by aggregating all incoming weights $w_{ab}^{L \to L}$ that lead into l_b and normalize by dividing though the aggregate (called $w_{bb}^{L \to L}$).

$$w_{ab}^{L \to L} = \sum_{j=1}^{|R|} p_{aj} \cdot p_{bj} \cdot (w_{aj}^{L \to R} + w_{bj}^{R \to L}), \quad \hat{w}_{ab}^{L \to L} = \frac{w_{ab}^{L \to L}}{w_{bb}^{L \to L}} \quad (2)$$

With FEATURE-TAK operating on symbol attribute descriptions retrieved from free-text descriptions from maintenance operators, exact similarity rela-

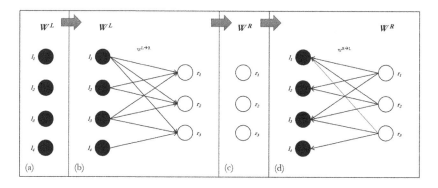

Fig. 2. Algorithmic workings of asym. WOMP with resource distribution [12]

tions between the retrieved keyword attribute values is non-trivial to infer. The goal of the current taxonomy-based similarity assessment algorithm is to position keywords alongside generated synonyms in a taxonomy structure that also builds on top of specific similarity values provided by domain experts and captures new concepts in this environment. The subsequently presented projection implementation has a different approach in that it specifically does not rely on manual similarity measures but rather derives relations between keywords by their co-occurrence in desired diagnosis results alone.

The projection operates on a bipartite graph of two populations:

– Population L, for which the one-mode projection graph is to be calculated, consists of keywords and synonyms being allowed values for a given symbol attribute description.
– Population R, which accounts for co-occurrence of to-be-projected attribute values, consists of all viable diagnoses (/ solutions) for the persisted case description.

A keyword/synonym l_i and diagnosis r_j are connected if there exists at least one case with diagnosis being r_j that also contains the keyword/synonym l_i. The bipartite graph is weighted, such that the value of an edge represents the number of cases with the respective diagnosis in which the respective keyword also occurred. As such, the final projection graph strongly connects keywords together if and only if they co-occur in many cases with the same diagnoses. Keywords and synonyms may - and probably will - occur in multiple cases with possibly multiple different diagnoses and as such will be projected with the procedure of [12] according to how a keyword's "resources" are shared across multiple diagnoses. Similarly to taxonomies, one receives separate bipartite- and projection graphs for each of FEATURE-TAK's symbolic attributes.

Regarding nodes in population R, current data does not persist proper symbolic diagnosis classes. This is resolved by pattern-matching a case's *solution-recommendation* attribute, resulting in a class label out of the set of labels of "UNKNOWN", "FIX", "REPLACE", "RESET", "DISPATCH" and "MISC".

The implemented asymmetric weighted one-mode projection algorithm is being split up into two parts, approximately depicting Eqs. 1 and 2, respectively. Firstly, resources for L- and R-vertices as well as the edges are being composed, which is done in a bipartite graph object. Inside the bipartite graph object, a composeResources() function is being called with vertex- and edge-array-lists being completed. It then assigns resources to each vertex $l_i \in L$, edge $w_{ij}^{L \to R}$, vertex $r_j \in R$ and edge $w_{ij}^{R \to L}$ in precisely this order. Note that the depicted procedure resembles the algorithm as mathematically proposed rather closely. Secondly, the actual traversal of the bipartite structure according to Eq. 2 is done to obtain the projected resource distribution relations.

Inside FEATURE-TAK's architecture, local similarity assessment is done in the *SimilarityManager* (task 5) and as such can expect to utilize all transformed results from tasks 0 to 3 and an initially set (synonym) attribute and similarity value from task 4. The current taxonomy-based similarity procedure is implemented in a single function addSynonymSimilarity inside the task directory and engaged by the respective task (5) agent object in the framework. It ultimately manipulates the taxonomy structure for the newly added synonyms (where applicable) and outputs an integer count of how many synonyms got connected. Similarly, projections would result in a secondary addProjectionSimilarity function called by the task agent.

Concerning general implementation architecture, the projection objects and graph structures are as of now disconnected from the framework's overall structure as best they can. With one of the intended benefits of a projection-based similarity computation being that of a more holistic, loosely coupled execution of similarity computation, this becomes easier to achieve with a separate implementation. This does, however, not necessarily go as well regarding integration in the object- and file formats defined in the underlying myCBR-framework. The framework is already used for internal case representations and similarity function specifications as well as persisted attributes and project-file formats. As is, the projection part of the framework does not implement interfaces of myCBR.

3   Evaluation

3.1   Similarity Matrix Computation and Modelling Assumptions

The presented projection approach is tested by loading project file information from project-files into memory. Then for each (symbolic) problem description attribute (*system, status, function, location*) taxonomy similarity values are calculated for each attribute value pair and the bipartite graph is composed based on vertices created from symbolic attribute values and calculates similarity through projection methods to compute the similarity matrix across all allowed symbolic values for each similarity function (taxonomy, simple OMP, normalized asymmetric WOMP), finally persisting similarity matrices to CSV files. This reuses components of the myCBR framework as they had been used in the framework FEATURE-TAK as well, though it omits actually executing FEATURE-TAK itself. It rather operates in an "offline" mode, not relying on having to execute all agents and waiting for feedback which itself does not provide results to similarity measures. The framework FEATURE-TAK is meant to analyze maintenance data and enrich the underlying CBR system; inspecting similarity computation assumes the CBR system to be up-to-date to allow disregarding the multi-agent system toolchain and operate on the CBR data directly. This approach is valid as long as no intermediate FEATURE-TAK-specific information and data structures are required. Specifically, synonym information would be easily extractable from these data structures.

Regarding modelling diagnosis classes, the class labels are constructed by keyword-matching according to the depicted keywords in Table 1.

The attribute "sol_recommendation" is being checked for an exact match of a part of the string-value and if the word(s) occur within the string the corresponding class label is assigned. Slight exceptions to this procedure are *UNKNOWN* (which requires not part of the value but the complete value to fit the word) and *MISC* (which is assigned if none of the other class labels can be

Table 1. Diagnosis class labels as manually assigned to case instances

Solution class label	String words
UNKNOWN	(*precisely*) "_unknown_", "none"
FIX	"fix", "de-ice"
REPLACE	"replace", "interchange", "swap", "change"
RESET	"reset", "re-power", "install new"
DISPATCH	"dispatch", "defer"
MISC	(*none of the above*)

assigned, thus acting as a default label for attribute values not captured with the generated rules).

The generated output as of the implementation of similarity computation based on respective in- and outputs consists of a total of 20 CSV files. For each of the four problem fault description attributes ("function", "location", "status", "system") as well as for each similarity computation method (taxonomy, simple- and weighted OMP, stock- and normalized asymmetric WOMP) a separate similarity matrix was generated. Note that the only formal similarity matrices are the ones of taxonomy and normalized asymmetric WOMP similarity, as no non-normalized matrix representation allows for entries in the diagonal to have similarities of 1 as they do not contain normalized values. As projections by themselves do not capture self-similarities of attributes, these needed to be added after computations have already been executed. All non-normalized quasi-similarity matrices do, however, show the intermediate relations depicted throughout the projection process, which helps quantify the projection performance on the given case instances. The taxonomy similarity matrices shall provide a baseline against which to compare projection similarity measures.

3.2   Evaluation Results

As can be seen in Table 2, which shows excerpts of the "location" attribute taxonomy similarity values, taxonomies in FEATURE-TAK produce similarity matrices which hold similar non-zero entries across rows and columns.

The location attribute was chosen as an example due to the feasible amount of 75 attributes as well as it being well populated with taxonomy similarity measures other than 1 and 0.8 (synonym relations). For distributions of (non-)zero entries and other characteristics across similarity matrices, see Table 4.

Considering the same attribute matrix but with entries calculated via projections the "location" attribute produces only two similarity values, thus Table 3 shows another excerpt of the projection similarity matrix that contains more non-zero entries. Note that the two attributes "side" and "forward": Both are approximately - but not exactly - equally similar to each other ($\text{sim}_{side \rightarrow forward} = 0.015$, $\text{sim}_{forward \rightarrow side} = 0.017$). Compare this however to the attributes "aft" and "forward": With $\text{sim}_{forward \rightarrow aft} = 0.004$, forward is much less similar to

Table 2. Excerpt of prob_fault_description_location_taxonomy.csv

Attribute value	d	d	w	D	f	w	u	u	u	u	c	i	c	b	l	l	m	c	b	m
door	1	0	0	0	0	0	0.8	0.8	0.8	0.8	0.5	0	0	0	0	0	0.8	0.5	0	0.8
deck	0	1	0	0	0	0	0	0	0	0	0	0	0	0	0	0	0	0	0	0
washstand	0	0	1	0	0	0	0	0	0	0	0	0	0	0	0	0	0	0	0	0
Door Right	0	0	0	1	0	0	0	0	0	0	0	0	0	0	0	0	0	0	0	0
fore	0	0	0	0	1	0	0	0	0	0	0	0	0	0	0	0	0	0	0	0
washbowl	0	0	0	0	0	1	0	0	0	0	0	0	0	0	0	0	0	0	0	0
upper [...] door #1	0.8	0	0	0	0	0	1	0.8	0.8	0.8	0.5	0	0	0	0	0	0.8	0.5	0	0.8
upper [...] door #2	0.8	0	0	0	0	0	0.8	1	0.8	0.8	0.5	0	0	0	0	0	0.8	0.5	0	0.8
upper [...] door #3	0.8	0	0	0	0	0	0.8	0.8	1	0.8	0.5	0	0	0	0	0	0.8	0.5	0	0.8
upper [...] door #4	0.8	0	0	0	0	0	0.8	0.8	0.8	1	0.5	0	0	0	0	0	0.8	0.5	0	0.8
cabin work station	0.5	0	0	0	0	0	0.5	0.5	0.5	0.5	1	0	0	0	0	0	0.5	0.5	0	0.5
incline	0	0	0	0	0	0	0	0	0	0	0	1	0	0	0	0	0	0	0	0
cargo	0	0	0	0	0	0	0	0	0	0	0	0	1	0	0	0	0	0	0	0
bathroom	0	0	0	0	0	0	0	0	0	0	0	0	0	1	0	0	0	0	0	0
lav	0	0	0	0	0	0	0	0	0	0	0	0	0	0	1	0	0	0	0	0
level	0	0	0	0	0	0	0	0	0	0	0	0	0	0	0	1	0	0	0	0
main [...] door #1	0.8	0	0	0	0	0	0.8	0.8	0.8	0.8	0.5	0	0	0	0	0	1	0.5	0	0.8
cookhouse	0.5	0	0	0	0	0	0.5	0.5	0.5	0.5	0.5	0	0	0	0	0	0.5	1	0	0.5
basin	0	0	0	0	0	0	0	0	0	0	0	0	0	0	0	0	0	0	1	0
main [...] door #4	0.8	0	0	0	0	0	0.8	0.8	0.8	0.8	0.5	0	0	0	0	0	0.8	0.5	0	1

aft than the other way around ($\mathrm{sim}_{aft \to forward} = 0.014$). This can be explained with forward being more closely related to other more impactful attributes (like "side" and "right"), while aft distributes its similarity over mostly shared "main deck left hand door #[1,3,4]" attributes.

Considering all the different matrix similarity representation formats, the different produced matrices show how the similarity computation proceeds across multiple stages. For illustration purposes, Fig. 3 shows original taxonomy similarity matrix, WOMP and both stock- and normalized asymmetric WOMP similarity matrices, with coloring indicating strength of similarity values. The matrices are excerpts of the problem fault description status attribute example, which is more densely populated with also different WOMP values (see Table 4 for more detail). All matrices apply on the same attribute pairs.

Figure 3 illustrates that distinct different relations exist between taxonomy- and projection measures. While the taxonomy relies on nodes modeled by domain expert and taxonomy similarities, the projection builds up over multiple stages from the symmetric WOMP of the bipartite graph, which is then traversed asymmetrically and normalized. Note that the asymmetric (non-normalized) WOMP is not computed from the simpler WOMP directly, but the properties as of the symmetric WOMP carry over into the asymmetric representation.

Table 3. Projected normalized asymmetric similarity values

Attribute value	m	m	m	s	m	m	s	w	u	u	f	u	c	r	s	s	a
main [...] #1	1	0	0.015	0	0	0.014	0	0	0	0	0.009	0	0	0	0	0	0.012
main [...] #2	0	1	0	0	0	0	0	0	0	0	0	0	0	0	0	0	0
main [...] #3	0.020	0	1	0	0	0.020	0	0	0	0	0.014	0	0	0	0	0	0.017
side	0	0	0	1	0	0	0	0	0	0	0.015	0	0	0.016	0	0	0
middle	0	0	0	0	1	0	0	0	0	0	0	0	0	0	0	0	0
main [...] #4	0.020	0	0.020	0	0	1	0	0	0	0	0.014	0	0	0	0	0	0.017
side of meat	0	0	0	0	0	0	1	0	0	0	0	0	0	0	0	0	0
washbasin	0	0	0	0	0	0	0	1	0	0	0	0	0	0	0	0	0
upper [...] #2	0	0	0	0	0	0	0	0	1	0	0	0	0	0	0	0	0
upper [...] #3	0	0	0	0	0	0	0	0	0	1	0	0	0	0	0	0	0
forward	0.004	0	0.005	0.017	0	0.005	0	0	0	0	1	0	0	0.015	0	0	0.004
upper [...] #4	0	0	0	0	0	0	0	0	0	0	0	1	0	0	0	0	0
crew rest [...]	0	0	0	0	0	0	0	0	0	0	0	0	1	0	0	0	0
right	0	0	0	0.019	0	0	0	0	0	0	0.015	0	0	1	0	0	0
sanitary	0	0	0	0	0	0	0	0	0	0	0	0	0	0	1	0	0
storey	0	0	0	0	0	0	0	0	0	0	0	0	0	0	0	1	0
aft	0.020	0	0.020	0	0	0.020	0	0	0	0	0.014	0	0	0	0	0	1

The displayed taxonomy excerpt contains 1.0 entries on the diagonal indicating self-similarity, but also in other non-diagonal entries where attributes have been connected via parent nodes of value 1.0 (e.g. "problem" and "inactive"). The 0.8 values indicate similarity between synonyms (e.g. "inactive" and "stuck"). The entries of value 0.3 are connected via weaker parent nodes, resulting in smaller similarity (e.g. "inactive" and "unserviceable").

On the projection side of things, one can observe that more entries are set to 0 (attributes not similar). Few entries have values (number of common neighbors in R-population) greater than 1. The continued asymmetric projection still captures higher values for entries with values greater than 1 in the previous weighted projection, where the asymmetry and weighting across complete attribute similarity values introduce further deviation. The final normalized asymmetric WOMP then contains entries generally being very small in similarity value. Relations from the previous non-normalized asymmetric WOMP largely persist, but are heavily scaled down, which gets further amplified for larger similarity matrices. Manually inserting self-similarity properties through 1.0 entries in the diagonal appear disproportional to other similarity values. As Fig. 3 shows, though formats of both taxonomy- and projection similarity build on fundamentally similar ideas, they yield different values through different ways of reasoning. This leaves open the question of how well these different measures are comparable not from a conceptual standpoint but from an application and data-oriented one.

To additionally compare not only representations across the same attribute, but also overall shape and how many actual similarity relations are contained

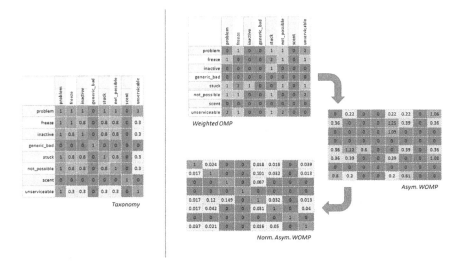

Fig. 3. Comparing similarity matrix representations of the "status" attribute

in the presented similarity matrices, Table 4 shows an overview of how many entries of a particular type of similarity value are in different similarity matrices across multiple attributes. For all four problem fault description attributes, the taxonomy and asymmetric projection matrices are covered as well as the intermediate symmetric weighted projection. The numbers for the asymmetric projection holds for both the normalized and not normalized matrices, thus it is only listed once. The symmetric projection is included to give an intuitive understanding of how many occurrences of attributes overlap in the resulted projection, to which the simple non-weighted projection does not provide additional insight, thus it is excluded from the overview.

Inspected matrix properties are the number of total entries and non-zero (as similarity is always positive this depicts all entries of $sim_{a \to b} > 0$) as well as non self-similar values (thus excluding entries d on the diagonal). For taxonomies it is also interesting how many non-intrinsically related similarity values - in the sense of values not specifically assigned by domain experts directly or the framework FEATURE-TAK implicitly through synonym structures - exist, thus only entries of similarity 0.8 are excluded further from the measure to exclude synonym relations. This does also neglect valid taxonomies just happening to return a similarity of 0.8 without the prior use of synonym constructs, but after manually inspecting the taxonomy similarity table and operation of FEATURE-TAK, this possibility was estimated to be unlikely. Note that domain experts specify intermediate nodes, which as such not necessarily depict attribute values expected to occur in real cases, and as such they needed not be inspected separately from formally calculated similarity values. For the intermediate symmetric weighted projection, an important measure is that of similarity relations depicted through values greater than 1, because these correspond to multiple

Table 4. Similarity matrices statistics across attributes

Attribute	Taxonomy	N. Asym. WOMP	Symmetric WOMP
Function (65 × 65, 4225 entries)	>0: 577 >0, ¬d: 512 >0, ¬d, ¬0.8: 0	>0: 133 >0, ¬d: 68	>0: 68 >0, >1: 6
Location (75 × 75, 5625 entries)	>0: 967 >0, ¬d: 892 >0, ¬d, ¬0.8: 498	>0: 177 >0, ¬d: 102	>0: 102 >0, >1: 0
Status (137 × 137, 18769 entries)	>0: 4151 >0, ¬d: 4014 >0, ¬d, ¬0.8: 3082	>0: 669 >0, ¬d: 532	>0: 532 >0, >1: 66
System (979 × 979, 635209 entries)	>0: 22827 >0, ¬d: 22030 >0, ¬d, ¬0.8: 21386	>0: 22373 >0, ¬d: 21576	>0: 21576 >0, >1: 62

common neighbors across attributes pairs, which in return allows to draw conclusions towards respective influences on the distribution of similarity in the normalized asymmetric projection values.

As seen in the left-most column of Table 4, the different attributes are very different in size with problem-fault-description-system being by far the largest of the four and the others being approximately of the same size with just - status being noticeably larger. As can be seen in the second and third column for taxonomy- and (normalized) asymmetric WOMP respectively, the attributes are more or less sparse depending on the type of values considered. Taxonomy values are largely zero with function, location, status and system having 14%, 17%, 22% and 4% non-zero entries, respectively. Compare this to 3%, 3%, 4% and 4% non-zero values for the asymmetric projection, which is considerably more sparse (except for the largest "system"-matrix). Though "status" is denser than "location" still being denser then "function" with taxonomies, this gets reduced to an even 3–4% across all attributes for projections. This drop is even more severe considering not only non-zero values but also values not equal to 0.8. Here, taxonomies maintain 0%, 9%, 16% and 3% similarity values with "location" keeping many-, while "status" and "system" keep most values. Problem fault description "function" only contains values of self- or synonym-similarity.

Considering weighted projections, both "function" and "status" hold a fair share of values larger than one, which allows for a more varying degree of similarity in the asymmetric projections, as opposed to larger matrices such as with "system" with fewer overlapping neighbors. In order to compare similarity measures of projections to their established taxonomy counterparts, a Mean Squared Error (MSE) estimation summing up the squared difference in values for all entries in both matrices was intended to be done, comparing the overall deviation in values between the taxonomy similarity matrix and the normalized asymmetric WOMP similarity matrix of each attribute separately. This however resulted in incredibly large deviations, because those similarity matrices do not overlap in most - if not all - non-zero values. Upon closer inspection as to why the values mostly do not overlap, it was found that the synonym structure as well as keywords related over said synonyms frequently happen to be conceptually related, without being used interchangeably in cases with similar solutions. What this means is that the taxonomies capture (pre-)defined classes of concepts, while projections rather capture concepts of different keywords used in similar situations in a scenario more closely related to the solution procedure of the case (as they get connected over solution classes).

4   Discussion and Outlook

The evaluation addresses inadequate symmetric self-similarity and small similarity values for sparse data as major issues. Regarding symmetric self-similarity, projections do not yield similarities of 1 for attributes to themselves. After having normalized all other non-self similarities, the similarities of attributes to themselves can be set to 1 manually, but they are not a result of the projection procedure itself, because the projection accounts only for co-occurrences of attribute values. The procedure of traversing the right-population has as consequence that any traversal is related to the resource being distributed to different solution classes, thus even adjusting for traversing all edges of the bipartite graph from a left-node to itself results in related twisted aggregated values due to W^R being typically non-zero, which is correlated to $w^{R \to L}$. While taxonomy similarity has attribute self similarity as inherent property, projections do not. This technically violates the fundamental notion of similarity, and requires manual adaptation if used on its own. Furthermore, as could be seen in Fig. 3, the different intermediate similarity matrices of symmetric-, asymmetric non-normalized- and asymmetric normalized WOMP have very different scaling across similarity values. When compared with similarity values in the taxonomy similarity matrix being nicely distributed over the interval $[0, 1]$, the values of the final normalized asymmetric WOMP are incredibly small and - not counting self-similarity - rarely exceed a similarity of 0.1. This trait, however, is not shared across the intermediate representations, where the non-normalized matrix contains what seems like more adequate values, which can be larger than 1. Note that the basic normalization idea of [12] still applies, but it seems to not scale well in application, especially with a small number of connections between the populations of the bipartite graph as a result of a low number of cases.

A solution to this could be a logarithmic function, which maps normalized projection-based similarity values to a range incorporating also similarities closer to 1. Compared to a baseline of no mapping, this logarithmic mapping would ensure that for small projection similarities their output similarity value in the similarity matrix would be larger - e.g. attain a value of 1 for a value of 1 in the normalized projection similarity and setting larger values to exactly 1. The steepness of the initial logarithmic incline for very small values and how quickly the maximum similarity should be attained depends on the number of input cases and distribution of attributes and attribute co-occurrences within the attribute descriptions. A further study inspecting different parameters and their influence would be worthwhile.

With projections being more of an extension to - rather than a substitution of - taxonomy relations, under the given analysis and computations it seems that a combined application of projections as well as taxonomies is more likely to succeed as compared to utilizing projections on their own. How exactly both procedures get interconnected in the application architecture of FEATURE-TAK and how nicely the different components extend already implemented functionality was sketched but can at this point only be estimated. The idea of combining projections with taxonomies works intuitively from an ideological point of view - with both incorporating expert knowledge in a network structure, while resulting in different sets of similarity relations - where the combination procedure requires further planning.

Possible future work includes incorporating projections more closely into FEATURE-TAK's workflow and implementing projection similarity measures in myCBR as the underlying CBR framework. The maintenance procedure for continuous operation of FEATURE-TAK allowing scaling across longer framework runs is considerably easier for taxonomies than for projections, but trade-offs in complexity can be expected by batch-updating the bipartite graph.

References

1. Aamodt, A., Plaza, E.: Case-based reasoning: foundational issues, methodological variations and system approaches. AI Commun. **7**(1), 39–59 (1994). http://www.idi.ntnu.no/emner/it3794/lectures/papers/Aamodt_1994_Case.pdf
2. Agrawal, R., Srikant, R.: Fast algorithms for mining association rules. In: Proceedings of the 20th International Conference on Very Large Data Bases, VLDB, pp. 487–499 (1994)
3. Bach, K.: Knowledge acquisition for case-based reasoning systems. Ph.D. thesis, University of Hildesheim (2013). Dr. Hut Verlag München
4. Bach, K., Althoff, K.-D.: Developing case-based reasoning applications using myCBR 3. In: Agudo, B.D., Watson, I. (eds.) ICCBR 2012. LNCS (LNAI), vol. 7466, pp. 17–31. Springer, Heidelberg (2012). https://doi.org/10.1007/978-3-642-32986-9_4
5. Borgelt, C.: An implementation of the FP-growth algorithm. In: Proceedings of the 1st International Workshop on Open Source Data Mining: Frequent Pattern Mining Implementations, pp. 1–5. ACM (2005)
6. Hundt, A., Reuss, P., Sauer, C.S., Roth-Berhofer, T.: Knowledge modeling and maintenance in myCBR3. In: Proceedings of the FGWM Workshop at Learning, Knowledge, Adaptation Conference 2014. Springer, Heidelberg (2014)
7. Lambiotte, R., Ausloos, M.: N-body decomposition of bipartite author networks. Phys. Rev. E **72**, 066117 (2005)
8. de Mantaras, R.L.: Case-based reasoning. In: Advanced Course on Artificial Intelligence, pp. 127–145. Springer, Heidelberg (1999)

9. Reuss, P., Stram, R., Althoff, K.-D., Henkel, W., Henning, F.: Knowledge engineering for decision support on diagnosis and maintenance in the aircraft domain. In: Nalepa, G.J., Baumeister, J. (eds.) Synergies Between Knowledge Engineering and Software Engineering. AISC, vol. 626, pp. 173–196. Springer, Cham (2018). https://doi.org/10.1007/978-3-319-64161-4_9

10. Reuss, P., et al.: FEATURE-TAK - framework for extraction, analysis, and transformation of unstructured textual aircraft knowledge. In: Goel, A., Díaz-Agudo, M.B., Roth-Berghofer, T. (eds.) ICCBR 2016. LNCS (LNAI), vol. 9969, pp. 327–341. Springer, Cham (2016). https://doi.org/10.1007/978-3-319-47096-2_22

11. Stahl, A., Roth-Berghofer, T.R.: Rapid prototyping of CBR applications with the open source tool myCBR. In: Althoff, K.-D., Bergmann, R., Minor, M., Hanft, A. (eds.) ECCBR 2008. LNCS (LNAI), vol. 5239, pp. 615–629. Springer, Heidelberg (2008). https://doi.org/10.1007/978-3-540-85502-6_42

12. Stram, R., Reuss, P., Althoff, K.-D.: Weighted one mode projection of a bipartite graph as a local similarity measure. In: Aha, D.W., Lieber, J. (eds.) ICCBR 2017. LNCS (LNAI), vol. 10339, pp. 375–389. Springer, Cham (2017). https://doi.org/10.1007/978-3-319-61030-6_26

13. Stram, R., Reuss, P., Althoff, K.-D.: Dynamic case bases and the asymmetrical weighted one-mode projection. In: Cox, M.T., Funk, P., Begum, S. (eds.) ICCBR 2018. LNCS (LNAI), vol. 11156, pp. 385–398. Springer, Cham (2018). https://doi.org/10.1007/978-3-030-01081-2_26

14. Stram, R., Reuss, P., Althoff, K.-D., Henkel, W., Fischer, D.: Relevance matrix generation using sensitivity analysis in a case-based reasoning environment. In: Goel, A., Díaz-Agudo, M.B., Roth-Berghofer, T. (eds.) ICCBR 2016. LNCS (LNAI), vol. 9969, pp. 402–412. Springer, Cham (2016). https://doi.org/10.1007/978-3-319-47096-2_27

15. Xu, K., Wang, F., Gu, L.: Behaviour analysis of internet traffic via bipartite graphs and one-mode projecions. IEEE/ACM Trans. Netw. (TON) **22**, 931–942 (2014)

16. Zhou, T., Ren, J., Medo, M., Zhang, Y.C.: Bipartite network projection and personal recommendation. Phys. Rev. E **76**, 046115 (2007)

An Algorithm Independent Case-Based Explanation Approach for Recommender Systems Using Interaction Graphs

Marta Caro-Martinez$^{(\boxtimes)}$ ⓘ, Juan A. Recio-Garcia ⓘ,
and Guillermo Jimenez-Diaz ⓘ

Department of Software Engineering and Artificial Intelligence,
Universidad Complutense de Madrid, Madrid, Spain
{martcaro,jareciog,gjimenez}@ucm.es

Abstract. Explanations in recommender systems are essential to improve user confidence in the recommendation. Traditionally, recommendation algorithms are based on ratings or additional information about the item features or the user profile. But some of these approaches are implemented as black boxes where this information is not available to provide the explanations. In this work, we propose a case-based approach to support this kind of black-box recommenders in order to find explanatory examples. It is a knowledge-light approach that only requires the information extracted from the interactions between users and items. As these interaction graphs can be analyzed through social network analysis, we propose the use of link prediction techniques to find the most suitable explanatory cases for a recommended item.

Keywords: Explanations · Interaction graphs · Recommender systems

1 Introduction

The Internet is a tool where people can access to a huge amount of information, to find or consume products. Amazon or eBay are examples of popular online shops where users can get new products of all types. Netflix or Spotify are examples of platforms where users consume products online –TV series and movies or music, respectively. However, the supply of products is so large that users have difficulties to find the most suitable products to them. This is the origin of recommender systems, which help users with this task. Therefore, the Internet is a medium to increase the development of recommender systems technology [2].

Sometimes, users do not trust in recommender systems because they do not know how they work. Recommender systems usually work as a black box system

Supported by the UCM (Research Group 921330), the Spanish Committee of Economy and Competitiveness (TIN2017-87330-R) and the funding provided by Banco Santander in UCM (CT42/18-CT43/18).

K. Bach and C. Marling (Eds.): ICCBR 2019, LNAI 11680, pp. 17–32, 2019.
https://doi.org/10.1007/978-3-030-29249-2_2

from the user's perspective. Due to this suspicion, some users reject the recommendations provided by these systems. Therefore, explanations in recommender systems are necessary to improve users' trust. If a user understands why an item was recommended to her thanks to an explanation, this user will consider this item as a promising candidate for being consumed [29].

Case-based explanations focus primarily on similar cases employed to perform the recommendation. They provide users with the cases that are most similar to the new case that needs to be explained. These similar cases are considered as cases for comparison [30]. From the point of view of recommender systems, it is an item-based style explanation, since it uses items to justify a recommendation [25]. The main advantage of this approach is that it allows users to assess the quality of the recommendation by comparing items that ideally should be related to the user's criteria.

Our proposal in this work is to provide explanatory cases using the interaction graphs that we employed to perform recommendations in previous works. These works [5,17,18] proposed a novel approach to make recommendations when we lack information about the users, items, ratings or other additional information. The classic techniques that recommender systems implement usually work with user preferences and ratings. For example, collaborative filtering uses the ratings to find similar users or similar items and produce the recommendations [1,9,28]. In the case of content-based approaches, systems just take into account item descriptions [2]. But, in certain scenarios, this information is not available and traditional techniques cannot be applied. The proposed graph-based methods have advantages over these traditional techniques since they just require information about which user interacted with which item. We represent these interactions between users and items into two graphs: an user-based graph and an item-based graph. We employ these graphs in combination with *link prediction* techniques as a way to make recommendations. Link prediction is a technique from social network analysis that predicts the existence of links that will appear in a graph over time or links that have disappeared [12,21,22,32]. We can see a recommendation as a link prediction problem [7,8,33]. We represent users and items as nodes in a graph and a link is formed when a user interacts with an item. If we can predict that a link will appear, we can provide this item as a suitable recommendation for this user.

In this paper, we use link prediction techniques to find explanatory items that are related to the recommended item according to the interaction graph. This way, we are able to provide an explanation using these related cases, taking into account the previous interactions carried out by this user with these similar items. Therefore, explanations are personalized to the user because they are based on her previous interactions and the recommendations performed by the system, which are used as cases for explanation.

One of the main advantages of the proposed technique is that it is completely independent of the recommendation algorithm and we are able to provide explanations assuming that the recommender system is a black box. Figure 1 provides an overview of the proposed approach. It only requires the set of user-item

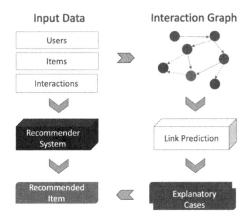

Fig. 1. General overview of the case-based explanation approach using interaction graphs

interactions as input, which is the minimum information required by any recommender system. This way, we can provide explanation examples for recommendations where the ratings, item or user features are not required at all, and we do not need to know the underlying recommender algorithm. Moreover, we can provide explanations with a positive or negative perspective, showing examples of why an item is interesting to the user or not.

The following sections describe our approach and its evaluation in depth. Section 2 relates a revision of works about explanation systems for recommendations. Section 3 describes our CBR approach and the link prediction metrics that we evaluated. Next, Sect. 4 contains the evaluation of our approach, where we demonstrate the effectiveness of our proposal in comparison with a content-based approach. We analyze the dataset and we describe the methodology carried out. Finally, in Sect. 5 we conclude and present the future work.

2 Related Work

The interest in recommender systems has grown in the last years due to the growth of online shops and entertainment platforms. These systems must help users to find interesting products and, at the same time, must retain users in order to continue using them. Explanations are necessary because users should trust in recommender systems to have a good experience in these platforms. So, recent research in recommender systems has been focused on explanations.

In the research of related work about explanations, we have found several surveys and classifications of explanations in recommender systems [3,6,11,14, 24,25,29,30], which stress the importance of explanations and describe the main features of explanations systems. In this section, we are going to describe some of these approaches, taking into account the most relevant ones.

The work in Herlocker *et al.* [15] is one of the pioneers in explanations for recommender systems. In this work, they propose many different types of visualizations to explain recommendations in MovieLens, like visualizations with tables, ratings or histograms. The work in [13] makes several proposals of explanations for MovieLens and compare them with the previous work. These proposals include new ways of visualizing explanations, with charts and personalized tag clouds. One of the most complete approaches is the work in [20], where they describe a mobile application that recommends and provides explanations about clothing. It is a flawless example because it includes text explanations with templates, visual explanations, different points (positive or negative) of view and possibility of user interaction. Another interesting approach is the work in [31], where an innovative explanation system is introduced. It shows personalized tags with a positive or negative perspective that is represented with different colors and sizes. The work in [27] is an explanation system based on matrix factorization to provide transparency to users. In [26], authors proposed an explanation system that personalizes explanations for recommendations using social elements. The target of these explanations are groups of users and they try to satisfy all of its members with text and visual approaches. In [19], we have one example of a very recent explanation approach. It describes an explanation method for hybrid recommender systems that includes personalization with different styles: user-based, item-based, content-based, social-based and item popularity. Another explanation method published recently is the work in [16], where the explainability is improved combining factorization results with templates. In the work of Musto *et al.* [23], an algorithm-independent and domain-independent framework that generates explanations based on natural language is presented. It uses Linked Open Data (LOD) to create these explanations.

Link prediction techniques, which are one of the pillars of our current work, are described in several paper reviews [12,21,22,32]. Moreover, there are some approaches that use these methods on graphs to make recommendations. The work in [7] describes a recommender system based on interaction graphs to make recommendations based on collaborative filtering. In [33], personal recommendations are presented using bipartite network projection. The work in [8] uses link prediction techniques on an user-item graph to make recommendations in User-Generated Content systems (UGCs) as Youtube or Flickr. Regarding explanations, there are few approaches in the field. For example, the work described in [4] proposes a model based on graphs and link prediction techniques to predict new links in the social network. They make recommendations and provide an explanation thanks to the reason that describes why the link was created. However, there is still more work to do with these techniques and other unexplored ones in the field of explanations for recommendations.

There are two types of link prediction techniques [32]: learning-based and similarity-based methods. *Learning-based* methods use machine learning techniques to establish if two non-connected nodes have probabilities to be linked in the future, whereas *similarity-based* methods use similarities between two nodes to determinate if this pair of nodes should be linked. In our work we will use similarity-based methods because we think that they are the most suitable approach for a case-based reasoning system.

3   Explanations Based on Interaction Graphs

Generally speaking, recommendation systems have information about the interactions that are carried out in the platform. This information can be represented as a tuple $R = (t, u, i, x)$, where t is the timestamp when the interaction happened, and u is the user that interacted with item i. Finally, x represents additional information associated with the interaction. In many cases, x is the rating that u gives to i, but there are other types of interactions, such as if u watched the movie i, read the book i, or listened to the song i. With the information about the interactions, we can build an adjacency matrix $A = \{A_{ui}\}$, where an element A_{ui} is equal to 1 if user u has interacted with i.

Taking into account the adjacency matrix, we can represent the interactions as a non-weighted bipartite graph. In this graph, nodes represent users belonging to the user set U or items that belong to the item set I. Next, we can apply a network projection to convert the original graph into a non-bipartite graph. From our interaction graph we can generate both an item-based graph, and a user-based graph. Figure 2 shows how a bipartite network projection works. In our case, we are interested in the item-based graph because it will allow us to collect similar items to the recommended one in order to find explanatory cases for the user. The item-based graph is formed by nodes that belong to I. Links are created between two nodes when at least one user has interacted with both items. Additionally, the original non-weighted graph is transformed into a weighted graph to avoid losing information. The weight of a link represents the number of common different users that have interacted with both items.

The main idea of our explanation method is to use the item-based graph in order to find the most relevant explanatory items using *link prediction* techniques. Therefore, we can consider two nodes as pairwise relevant if the probability of linking these nodes is high.

If we consider the items that a concrete user has interacted with as its personal explanatory case-base, we can also consider the recommended item from any recommendation algorithm as the query of an explanatory CBR system. This way, using the item-based graph and link prediction techniques, we can find the most relevant explanatory cases for that recommendation according to the previous user interactions. Next section describes and formalizes the approach.

3.1   The Case-Based Explanation System

The proposed explanatory CBR approach does not take into account the recommendation algorithm. It only requires information about the user-item interactions and, therefore, it is algorithm independent.

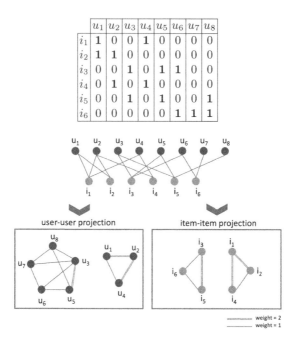

	u_1	u_2	u_3	u_4	u_5	u_6	u_7	u_8
i_1	1	0	0	1	0	0	0	0
i_2	1	1	0	0	0	0	0	0
i_3	0	0	1	0	1	1	0	0
i_4	0	1	0	1	0	0	0	0
i_5	0	0	1	0	1	0	0	1
i_6	0	0	0	0	0	1	1	1

Fig. 2. Example of bipartite network projection from the adjacency matrix (on top).

The main knowledge container of this CBR system is the item-based graph specified before. It will serve as the case base containing the explanatory examples. The graph is defined from the adjacency matrix A as $G = \langle I, L \rangle$, where items I are the nodes and the links $L = \{(i, j, w)\}$ represent the interactions of users with two different items i and j, where the weight w is the number of common different users that have interacted with both items. This graph is considered as the global container of cases from which we can define the personal case base for every user u. A case base for user u is composed by any previous item i that u has interacted with:

$$CB_u = \{i \in I_u\} \tag{1}$$
$$I_u = \{i \in I\} : A_{ui} \neq 0 \tag{2}$$

Now, given a recommended item i_r for a target user u, we retrieve all the cases in the personal case base CB_u and we compute the link prediction from any item $i \in CB_u$ to i_r. The link prediction represents the probability of creating a new link from i to i_r, and therefore, the probability of choosing the recommended item i_r because user u has already interacted with i. Thus, to explain the recommendation i_r we will present to the user the previous cases i that obtain the highest link prediction[1].

---

[1] We are assuming link prediction metrics as similarity measures, although in our approach they are not normalized to [0,1] because we only need the resulting score to rank and compare items.

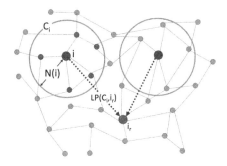

Fig. 3. Global scheme of the CBR explanation system [33]

The link prediction techniques that we use in our approach are the similarity-based methods. They allow us to get a score that denotes the likeness between two items according to the interactions carried out by users with the items. Similarity-based metrics, at the same time, can be classified into four groups [32]: *Node-based* metrics take into account the node properties to find the similarities; *Neighbour-based* metrics use the information about the node's neighbourhood, $N(i)$, that is defined as any other item j that a user has also interacted with; *Path-based* metrics take into account the paths that join two nodes; and, *random walk-based* metrics that make social simulations through transition probabilities between nodes and neighbours.

In this work we have focused on node-based and neighbour-based techniques. Therefore, we have to reformulate the definition of the user case base to include the node's neighbourhood $N(i)$ as follows:

$$CB_u = \{C_i : i \in I_u\} \tag{3}$$

$$C_i = \langle i\,, N(i)\rangle \tag{4}$$

$$N(i) = \{j \in I : (i,j) \in L\} \tag{5}$$

This way, the cases C_i obtained by the link prediction function may explain the recommendation to the user not only through the corresponding item i, but also by presenting the associated neighborhood of items $N(i)$.

Finally, we only need to define the link prediction function as follows:

$$lp_u(C_i, i_r) : [CB_u \times I] \to \mathbb{R} \tag{6}$$

Figure 3 illustrates the structure of the case base CB_u and the main schema of the proposed explanation system. In the next section, we discuss the different link prediction techniques that we could apply to relate the items in our graph.

3.2   Link Prediction Similarity Measures

To compute the score between a pair of nodes (i, j) that expresses the probability of linking from i to j, we have made a variation of the classic link prediction metrics. The metrics used in this work are the same that we used in our previous graph-based recommender approaches [5,17,18] and are detailed in [21,32]. Most of the metrics can be defined into two versions: weighted and unweighted. Moreover, most of the metrics are not normalized: they do not lie in the range [0,1]. This is not a problem in our approach because we only need the score to rank similar items.

Before discussing the link prediction metrics, we need to introduce a specific notation to clarify the following discussion.

– $N(i)$ represents the neighbours of node i.
– $|N(i)|$ represents the number of neighbours (or *node degree*) of node i.
– L_{ij} represents the weight w of the link between nodes i and j.
– $W(i) = \sum L_{ix} : x \neq i \in I$ represents the *weighted node degree* of node i, which means the sum of the weights in the links directly connected with node i.

Next, we detail the link prediction metrics considered for our explanation system:

Edge Weight (EW). This metric measures the similarity between two nodes as the weight of the link between them. $L_{ij} = 0$ represents node i and node j are not connected. An unweighted version of this metric exists ($L_{ij} = 1$ if the link exists; 0 otherwise) but we have not used it because it is too simple.

$$EW(i,j) = L_{ij}$$

Common Neighbours (CN). Using this metric, the similarity between two nodes is the number of neighbours they have in common. The rationale behind this metric is that the greater the intersection of the neighbour sets of any two nodes is, the greater the chance of a future association between them. Weighted Common Neigbours (WCN) is the weighted version of this metric.

$$CN(i,j) = |N(i) \cap N(j)|$$

$$WCN(i,j) = \sum_{z \in N(i) \cap N(j)} L_{iz} + L_{jz}$$

Jaccard Neighbours (JN). This metric is an improvement of $CN(i,j)$ as it measures the number of common neighbours of i and j compared with the number of total neighbours of both nodes. It does not have a weighted metric version.

$$JN(i,j) = \frac{|N(i) \cap N(j)|}{|N(i) \cup N(j)|}$$

Adar/Adamic (AA). This metric also measures the intersection of neighbour-sets of two nodes in the graph, but emphasizing in the smaller overlap. The weighted version of this metric is Weighted Adar/Adamic (WAA).

$$AA(i,j) = \sum\nolimits_{z \in N(i) \cap N(j)} \frac{1}{log|N(z)|}$$

$$WAA(i,j) = \sum\nolimits_{z \in N(i) \cap N(j)} \frac{L_{iz} + L_{jz}}{log(1 + W(z))}$$

Preferential Attachment (PA). It is based on the consideration that nodes create links, with higher probability, with those nodes that already have a larger number of links. The probability of creating an link between nodes i and j is computed as the product of the degree of the nodes i and j, so the higher the degree of both nodes, the higher is the probability of linking. This metric has the drawback of leading to high probability values for highly connected nodes to the detriment of the less connected ones in the network. Weighted Preferential Attachment (WPA) is the weighted version of the previous one. It is an improvement of PA, where the link weights are taken into account when computing the degree of nodes i and j.

$$PA(i,j) = |N(i)| \cdot |N(j)|$$

$$WPA(i,j) = W(i) \cdot W(j)$$

Once we have described the link prediction techniques that we could apply to find explanatory cases, next section presents a comparative evaluation of their performance.

4   Evaluation

We have designed our evaluation to be applied in a Matrix Factorization recommender system. Our graph-based explanation approach addresses the matrix factorization system as a black box system. It generates the explanations finding the most similar cases to the recommended one in order to show to the user an explanation like: "This product X was recommended to you because it is similar to the product Y, which you have liked". We have chosen a Matrix Factorization system because we need to use the rating predicted by it to evaluate our approach.

One of the main advantages of our method is its decoupling from the recommendation algorithm and its low requirements regarding input data. It does not require information about the features of users or items unlike standard content-based strategies. Content-based strategies are the obvious alternative for finding explanatory cases for a recommended item, and therefore, our baseline to compare with.

In order to compare graph-based and content-based approaches we can measure the quality of the explanatory cases obtained in both approaches. Our hypothesis is that our graph-based approach is able to find explanatory cases with a relevance to the target user similar to the cases obtained by content-based strategies, which are much more knowledge demanding.

The comparison of the explanatory cases obtained by both graph-based and content-based approaches is performed by calculating the error between the rating predicted for the recommended item and the actual rating of the explanatory cases. Here, we assume that the user should rate the explanatory cases with a similar rating value than the predicted for the recommended item.

In Sects. 4.1 and 4.2, we describe the details of the dataset and the evaluation process, respectively. In Sect. 4.3, we discuss the results.

4.1   Data

To perform this evaluation, we use the 100K MovieLens dataset[2]. It includes 100K ratings in tuple form $R = (t, u, i, x)$, where u is the user, i is the movie that u has watched, x is the rating that u has provided to i and t is the timestamp when u rated i.

From this dataset we are able to build the adjacency matrix to create the interaction graph. It is important to note that this interaction graph does not require any further information about users, item's features, or even ratings. In this adjacency matrix, $A_{ui} = 1$ if user u has rated the movie i.

To implement the content-based approach, we need to enrich the representation of the items from the MovieLens dataset in order to include their features. To do that, we decide to use the IMDB dataset[3], which provides a knowledge-rich description of the movies, like genres, directors or stars. However, not all of the movies in MovieLens dataset are in IMDB dataset. For this reason, we have to filter the dataset in order to get just the movies that appear in both datasets. The resulting dataset, denoted as A, is split into the training set A_t with the 90% of the ratings, and the evaluation set A_e with the 10% of the ratings.

Before performing the comparative evaluation of both graph-based and content-based approaches, we have analyzed the datasets. We have followed the model proposed in [10] to make this descriptive analysis. The analysis of the dataset A_e (Table 1) revealed that it has a bias because the number of items for each rating value is clearly unbalanced.

In order to evaluate the impact of this bias in both approaches, we will perform an additional evaluation using stratified evaluation sets B_e, generated from the previous A_e. In each stratified evaluation set, the number of items for each rating value is equally distributed. This way, each B_e will contains 35 items for each rating value as it is the lowest value for 2.5 stars rating.

---

[2] https://grouplens.org/datasets/movielens/100k/.
[3] https://www.imdb.com/.

Table 1. Descriptive analysis of the datasets used for the evaluation. ML is the original Movielens dataset.

Metric	ML	A	A_t	A_e	B_e
# Ratings	100,000	11,477	10,330	1,147	280
# Items	1,682	164	164	145	109
# Users	943	587	584	394	134
Density	0.06	0.12	0.11	0.02	0.02
Items					
Maximum # ratings per item	583	329	305	30	10
Median # ratings per item	27	43.5	39	5	2
Average # ratings per item	59.45	69.98	62.99	7.91	2.57
Minimum # ratings per item	1	1	1	1	1
Users					
Maximum # ratings per user	737	128	113	15	11
Median # ratings per user	65	12	11	2	1
Average # ratings per user	106.05	19.55	17.69	2.91	2.09
Minimum # ratings per user	20	1	1	1	1
Ratings					
% Ratings ≥ 4	55.38	52.54	52.66	51.44	37.50
% Ratings < 4	44.62	47.46	47.34	48.56	62.50

4.2   Experimental Setup

In order to evaluate our case-based explanation approach, we have implemented a recommender system based on Matrix Factorization and we use cross validation with the training and evaluation sets previously described. For each recommended item i_r for a target user u in the evaluation set, we obtain a set with the most suitable explanatory cases through any of the *link prediction* techniques that we detailed before: CN, EW, AA, JN, PA, WCN, WAA, WPA. Analogously, we obtain a set of explanatory cases using a content-based approach, that selects the most similar items according to the list of binary movie features. The similarity is computed using three different algorithms: Euclidean, Cosine and Jaccard.

To compare both sets of explanatory cases, we compute the Root Mean Square Error (RMSE) between the predicted rating for i_r and the rating of the explanatory cases.

Additionally, we analyzed the impact of the number of explanatory cases presented to the user. To do that, the size of the set of explanatory cases is controlled with a k parameter, which ranges from 1 to 10. In this case, we will remove from the evaluation the users who did not rate enough items to complete the set of explanatory cases with size k. For example, if a target user u_t watched 6 movies, the set with the 7 top-most similar cases recovered to

Table 2. Results from evaluation with original evaluation set (A_e).

k	1	2	3	4	5	6	7	8	9	10
AA	1.095	0.865	0.792	0.769	0.747	0.723	0.702	0.686	0.678	0.664
CN	1.035	0.821	0.750	0.713	0.683	0.666	0.649	0.640	0.636	0.635
EW	1.087	0.878	0.797	0.754	0.734	0.706	0.679	0.666	0.651	0.639
JN	**0.961**	**0.734**	**0.658**	**0.624**	**0.599**	**0.573**	**0.562**	**0.549**	**0.540**	**0.534**
PA	1.126	0.897	0.806	0.789	0.782	0.762	0.741	0.728	0.718	0.706
WAA	1.113	0.968	0.908	0.855	0.822	0.794	0.775	0.747	0.723	0.705
WCN	1.113	0.968	0.907	0.855	0.821	0.796	0.778	0.749	0.723	0.704
WPA	1.115	0.970	0.910	0.852	0.821	0.793	0.779	0.747	0.723	0.703
Cosine	0.973	1.036	1.064	1.078	1.087	1.100	1.101	1.104	1.108	1.111
Euclidean	0.966	1.032	1.063	1.079	1.092	1.092	1.099	1.100	1.102	1.105
Jaccard	0.974	1.037	1.064	1.078	1.087	1.099	1.101	1.104	1.109	1.111

explain a recommendation would be incomplete. Therefore, we will not use u_t to compute the RMSE for $k \geq 7$.

We run two different experiments, using the original evaluation set A_e and with stratified evaluation set B_e. Because B_e is created randomly, choosing 35 items from A_e for each rating value, we repeated the evaluation process using the stratified dataset 100 times, using 100 different evaluation sets B_e created randomly.

4.3   Results

Table 2 reports the results using the original evaluation set A_e. Each column represents the average RMSE between the recommended items and the set of the corresponding k explanatory cases. The best result for each k is emphasized.

From these results we can remark that our item graph-based approaches always get better results than the content-based ones. The difference is more noticeable as long as k grows.

When comparing only the item graph-based approaches, the best results are achieved using the Jaccard Neighbours (JN) similarity measure, which represents the rate of common neighbours between two items.

It is worth noting that the average RMSE with the set of explanatory cases is reduced as long as the number of cases k grows, in contrast with the content based approaches.

Table 3 shows the average results from evaluations performed using stratified evaluation sets B_e. Taking into account this evaluation we can conclude that, again, item graph-based approaches always get better results than content-based algorithms. The difference is also more noticeable when k grows. This fact reveals that the proposed approaches are not affected by the bias in the dataset.

As in the previous experiment, the best results are achieved by the approach that uses JN. The average RMSE with the set of explanatory cases is improved as long as the number of cases k grows.

Table 3. Average results from the 100-fold evaluation with stratified evaluation sets (B_e).

k	1	2	3	4	5	6	7	8	9	10
AA	1,164	0,947	0,861	0,839	0,817	0,796	0,774	0,764	0,754	0,738
CN	1,084	0,893	0,811	0,761	0,739	0,721	0,696	0,695	0,695	0,698
EW	1,142	0,938	0,863	0,827	0,802	0,766	0,743	0,736	0,714	0,702
JN	**1,004**	**0,748**	**0,673**	**0,643**	**0,629**	**0,597**	**0,584**	**0,573**	**0,566**	**0,566**
PA	1,182	1,000	0,902	0,883	0,873	0,852	0,835	0,819	0,809	0,793
WAA	1,156	1,061	1,002	0,944	0,922	0,894	0,874	0,838	0,810	0,790
WCN	1,149	1,054	0,994	0,937	0,917	0,890	0,873	0,837	0,807	0,787
WPA	1,152	1,065	1,011	0,940	0,921	0,889	0,873	0,837	0,812	0,789
Cosine	1,117	1,130	1,125	1,125	1,125	1,120	1,129	1,125	1,121	1,121
Euclidean	1,090	1,105	1,102	1,110	1,105	1,106	1,106	1,111	1,112	1,112
Jaccard	1,052	1,052	1,054	1,067	1,084	1,097	1,100	1,099	1,102	1,100

We would like to highlight that the weighted versions of the link prediction similarity metrics employed in our approach generally perform worse than the non-weighted versions, in contrast with the results obtained in a previous work [5]. In that work, the item-based approach was employed to recommend items to a target user and the weighted metrics performed the best results. However, the domain and datasets were completely different (online judges), so we should repeat these experiments in different domains.

We can conclude that in all of the experiments carried out, our graph-based approaches are better than the content-based methods in order to choose the best cases employed to explain the recommendation performed by a recommender system. The exception occurs when we try to explain the recommendation with only one case ($k = 1$). Comparing the results with the evaluation datasets A_e and B_e, we conclude that the one with stratified evaluation set is worse than the one that uses original evaluation set. This makes sense, because we have removed the bias.

5   Conclusions and Future Work

The trust on recommender systems can be enhanced by providing explanations to the users. In this work we have detailed a case-based explanation approach for recommender systems that relies on the interaction graph model that can be inferred for every recommendation system. This approach uses a set of items previously consumed by the user as an explanation for the recommended item.

We highlight two main advantages of our approach. On the one hand, it is independent of the algorithm employed by the recommendation system. On the other hand, it does not need any additional information to provide an explanation because it is based on the past experiences of the users, modelled in the interaction graph.

Our approach has been evaluated against a knowledge-rich content-based approach and the results revealed that the sets of the explanatory cases selected by our approach are, in general, more accurate with the preferences of the user who is being recommended.

In this approach we use an item-based interaction graph. However, in our previous works using this interaction models, we also employed user-based approach as a method to perform a recommendation. We plan to adapt this user-based interaction graphs in order to generate explanations based on similar users and their common tastes.

Finally, our approach has demonstrated a good performance in our offline experimental evaluation. However, it should be tested by real users, who should evaluate if the explanations provided are useful and if they enhance their experience with the recommendation system. This is one of the most important lines of our future work.

References

1. Adomavicius, G., Tuzhilin, A.: Toward the next generation of recommender systems: a survey of the state-of-the-art and possible extensions. IEEE Trans. Knowl. Data Eng. **6**, 734–749 (2005)
2. Aggarwal, C.C.: Recommender Systems. Springer, Cham (2016). https://doi.org/10.1007/978-3-319-29659-3
3. Al-Taie, M.Z., Kadry, S.: Visualization of explanations in recommender systems. J. Adv. Manag. Sci. **2**(2), 140–144 (2014)
4. Barbieri, N., Bonchi, F., Manco, G.: Who to follow and why: link prediction with explanations. In: 20th ACM SIGKDD International Conference on Knowledge Discovery and Data Mining, pp. 1266–1275. ACM (2014)
5. Caro-Martinez, M., Jimenez-Diaz, G.: Similar users or similar items? Comparing similarity-based approaches for recommender systems in online judges. In: Aha, D.W., Lieber, J. (eds.) ICCBR 2017. LNCS (LNAI), vol. 10339, pp. 92–107. Springer, Cham (2017). https://doi.org/10.1007/978-3-319-61030-6_7
6. Caro-Martinez, M., Jimenez-Diaz, G., Recio-Garcia, J.A.: A theoretical model of explanations in recommender systems. In: ICCBR 2018, p. 52 (2018)
7. Chen, H., Li, X., Huang, Z.: Link prediction approach to collaborative filtering. In: Proceedings of the 5th ACM/IEEE-CS Joint Conference on Digital Libraries, JCDL 2005, pp. 141–142. IEEE (2005)
8. Chiluka, N., Andrade, N., Pouwelse, J.: A link prediction approach to recommendations in large-scale user-generated content systems. In: Clough, P., Foley, C., Gurrin, C., Jones, G.J.F., Kraaij, W., Lee, H., Mudoch, V. (eds.) ECIR 2011. LNCS, vol. 6611, pp. 189–200. Springer, Heidelberg (2011). https://doi.org/10.1007/978-3-642-20161-5_19
9. Desrosiers, C., Karypis, G.: A comprehensive survey of neighborhood-based recommendation methods. In: Ricci, F., Rokach, L., Shapira, B., Kantor, P.B. (eds.) Recommender Systems Handbook, pp. 107–144. Springer, Boston (2011). https://doi.org/10.1007/978-0-387-85820-3_4
10. Dooms, S., Bellogín, A., Pessemier, T.D., Martens, L.: A framework for dataset benchmarking and its application to a new movie rating dataset. ACM Trans. Intell. Syst. Technol. (TIST) **7**(3), 41 (2016)

11. Friedrich, G., Zanker, M.: A taxonomy for generating explanations in recommender systems. AI Mag. **32**(3), 90–98 (2011)
12. Furht, B.: Handbook of Social Network Technologies and Applications. Springer, Boston (2010). https://doi.org/10.1007/978-1-4419-7142-5
13. Gedikli, F., Jannach, D., Ge, M.: How should I explain? A comparison of different explanation types for recommender systems. Int. J. Hum Comput Stud. **72**(4), 367–382 (2014)
14. He, C., Parra, D., Verbert, K.: Interactive recommender systems: a survey of the state of the art and future research challenges and opportunities. Expert Syst. Appl. **56**, 9–27 (2016)
15. Herlocker, J.L., Konstan, J.A., Riedl, J.: Explaining collaborative filtering recommendations. In: Proceedings of the 2000 ACM Conference on Computer Supported Cooperative Work, pp. 241–250. ACM (2000)
16. Hong, M., Akerkar, R., Jung, J.J.: Improving explainability of recommendation system by multi-sided tensor factorization. Cybern. Syst. **50**(2), 97–117 (2019)
17. Jimenez-Diaz, G., Gómez-Martín, P.P., Gómez-Martín, M.A., Sánchez-Ruiz, A.A.: Similarity metrics from social network analysis for content recommender systems. AI Commun. **30**(3–4), 223–234 (2017)
18. Jimenez-Diaz, G., Gómez Martín, P.P., Gómez Martín, M.A., Sánchez-Ruiz, A.A.: Similarity metrics from social network analysis for content recommender systems. In: Goel, A., Díaz-Agudo, M.B., Roth-Berghofer, T. (eds.) ICCBR 2016. LNCS (LNAI), vol. 9969, pp. 203–217. Springer, Cham (2016). https://doi.org/10.1007/978-3-319-47096-2_14
19. Kouki, P., Schaffer, J., Pujara, J., O'Donovan, J., Getoor, L.: Personalized explanations for hybrid recommender systems. In: Proceedings of the 24th International Conference on Intelligent User Interfaces, pp. 379–390. ACM (2019)
20. Lamche, B., Adıgüzel, U., Wörndl, W.: Interactive explanations in mobile shopping recommender systems. In: Joint Workshop on Interfaces and Human Decision Making in Recommender Systems, p. 14 (2014)
21. Liben-Nowell, D., Kleinberg, J.: The link-prediction problem for social networks. J. Am. Soc. Inform. Sci. Technol. **58**(7), 1019–1031 (2007)
22. Lü, L., Zhou, T.: Link prediction in complex networks: a survey. Phys. A **390**(6), 1150–1170 (2011)
23. Musto, C., Narducci, F., Lops, P., de Gemmis, M., Semeraro, G.: Linked open data-based explanations for transparent recommender systems. Int. J. Hum Comput Stud. **121**, 93–107 (2019)
24. Nunes, I., Jannach, D.: A systematic review and taxonomy of explanations in decision support and recommender systems. User Model. User-Adap. Inter. **27**(3–5), 393–444 (2017)
25. Papadimitriou, A., Symeonidis, P., Manolopoulos, Y.: A generalized taxonomy of explanations styles for traditional and social recommender systems. Data Min. Knowl. Disc. **24**(3), 555–583 (2012)
26. Quijano-Sanchez, L., Sauer, C., Recio-Garcia, J.A., Diaz-Agudo, B.: Make it personal: a social explanation system applied to group recommendations. Expert Syst. Appl. **76**, 36–48 (2017)
27. Rastegarpanah, B., Crovella, M., Gummadi, K.P.: Exploring explanations for matrix factorization recommender systems (2017)
28. Schafer, J.B., Frankowski, D., Herlocker, J., Sen, S.: Collaborative filtering recommender systems. In: Brusilovsky, P., Kobsa, A., Nejdl, W. (eds.) The Adaptive Web. LNCS, vol. 4321, pp. 291–324. Springer, Heidelberg (2007). https://doi.org/10.1007/978-3-540-72079-9_9

29. Tintarev, N., Masthoff, J.: Evaluating the effectiveness of explanations for recommender systems. User Model. User-Adap. Inter. **22**(4–5), 399–439 (2012)
30. Tintarev, N., Masthoff, J.: Explaining recommendations: design and evaluation. In: Ricci, F., Rokach, L., Shapira, B. (eds.) Recommender Systems Handbook, pp. 353–382. Springer, Boston, MA (2015). https://doi.org/10.1007/978-1-4899-7637-6_10
31. Vig, J., Sen, S., Riedl, J.: Tagsplanations: explaining recommendations using tags. In: Proceedings of the 14th International Conference on Intelligent User Interfaces, pp. 47–56. ACM (2009)
32. Wang, P., Xu, B., Wu, Y., Zhou, X.: Link prediction in social networks: the state-of-the-art. Sci. China Inf. Sci. **58**(1), 1–38 (2015)
33. Zhou, T., Ren, J., Medo, M., Zhang, Y.C.: Bipartite network projection and personal recommendation. Phys. Rev. E **76**(4), 046115 (2007)

Explanation of Recommenders Using Formal Concept Analysis

Belen Diaz-Agudo[✉], Marta Caro-Martinez, Juan A. Recio-Garcia,
Jose Jorro-Aragoneses, and Guillermo Jimenez-Diaz

Department of Software Engineering and Artificial Intelligence,
Universidad Complutense de Madrid, Madrid, Spain
{belend,martcaro,jareciog,jljorro,gjimenez}@ucm.es

Abstract. Formal Concept Analysis is a mathematical approach which enables formalisation of concepts as basic units of human thinking and analysing data in the object-attribute form. In this paper, we propose the use of FCA as a general resource for explanations and apply it to explain the results of recommender systems. Our method is reusable and applicable to different domains. We define different types of explanations by travelling the lattice structure and analyse how the lattice metrics can be used to characterise the different types of user profiles.

Keywords: Explanations · Explainable artificial intelligence · Formal concept analysis · Recommender systems

1 Introduction

Explainable artificial intelligence and case based explanations have become active areas of research in the last few years [13,21,26]. In recommender systems, explanations are essential to improve user trust and persuasion [12,19] and there are different approaches that have been reviewed elsewhere [6,25].

The term *explanation* can be interpreted in two different ways: in AI in general and in recommender systems in particular [1,4,24]. First interpretation refers to *transparency* and deals with explanations as part of the reasoning process itself and with the goal of understanding how the reasoning process works. The other interpretation deals with *justification* or attempting to make a certain reasoning process, or its result, understandable to the user. Recommendations resulting from *content-based strategies* are more comprehensible for users, because they are based on the explicit user profiles. The *content filtering* approach creates a profile for each user or product to characterise its contents and recommends a similar product that matches the user profile. Most of the content-based recommenders typically generate case-based explanations presenting the

Supported by the UCM (Research Group 921330), the Spanish Committee of Economy and Competitiveness (TIN2017-87330-R) and the fundings provided by Banco Santander in UCM (CT17/17-CT17/18) and (CT42/18-CT43/18).

K. Bach and C. Marling (Eds.): ICCBR 2019, LNAI 11680, pp. 33–48, 2019.
https://doi.org/10.1007/978-3-030-29249-2_3

items that are most similar to a new user profile, and the similarity and dis-similarity knowledge between the user profile preferences (query) and the item set and the user's experiences (likes or dislikes) [10,16,18,20,28]. This is related to the view of CBR systems as self explainable systems because having cases as precedents of similar problem solving experiences are by themselves useful pieces of knowledge to explain the system outputs [15]. Most of the approaches using explanatory cases do not explain how the system has reached its solu-tion. However, the usefulness of explanatory cases as a support or justification of the results has been demonstrated [24]. Note also that many authors [9,22] agree that displaying the best case is not always sufficient explanation. That is especially true when the goal of the explanation is to provide transparency and when the solution is not a simple reuse of a similar experience but it emerges from complex retrieval and adaptation processes. In this situation specific expla-nation knowledge, apart from the cases, is required [2,17,22], and explanations may make explicit details about the features that the query has in common with the retrieved cases, why some similarities and differences are more relevant than others for the solution and why the system performs a certain adaptation. When recommender systems use the *collaborative filtering* approach, the system does not use an explicit user profile because the recommendation algorithm relies only on user ratings. Collaborative filtering identifies new user-item associations and predicts users preferences as a linear, weighted combination of other user preferences. Collaborative filtering is more flexible and generally more accurate than content-based techniques [3]. However, it suffers from the self explanatory capability as it lacks from explicit profiles. Other related algorithms like matrix factorization also suffers the same problem.

In this paper, we propose a general approach for explanations in recommender systems using Formal Concept Analysis (FCA). We study how the use of FCA helps in finding the knowledge structure of a recommender system and how this knowledge is useful as the explanation knowledge in the system. Note that the sense of the term explanation here refers to *justification*, as it attempts to make the result of a recommender systems understandable to the user. We propose different approaches that vary in the way we apply FCA and travel the lattice. We propose building the *user profile lattice* with the user personal best rated items. This lattice can be used itself to explain the user profile, and the diversity of her preferences, and let her refine her ratings or understand why a certain item has been recommended. Besides, we also explore how the dependencies between attributes and the maximal groups of items are useful as explanation knowledge in different ways: *item-style*, *property-style* and *dependency style*. We first review the related work in Sect. 2. Section 3 introduces the basics of FCA. Section 4 describes the explanation algorithm and Sect. 5 evaluates the structural properties of profile lattices. Section 6 concludes the paper and outlines the future work.

2   Related Work

Case based explanations have become an active area of research in the last few years [21]. Even if most of the approaches use explanatory cases that do not explain how the AI system has reached its solution (i.e. transparency), the usefulness of explanatory cases as a support or justification of the results of a twin black-box AI system has been demonstrated [24]. Applying explanations in recommendation systems is an important area of research in this type of systems. The main problem with recommendation systems is that users do not know why a product has been recommended to them. Recommender systems that use explanations improve user confidence in those recommendations [25]. In addition, users consume more products that are the result of the recommendations that are explained to them [12].

In previous work [8] we have proposed the use of FCA to help knowledge acquisition and refinement and to help the CBR processes. We studied how the use of FCA can support the task of discovering knowledge embedded in a case base. FCA application provides an internal sight of the case base conceptual structure and allows finding regularity patterns among the cases. Moreover, FCA lattice supports classification based retrieval processes and extracts dependence rules between the attributes describing the cases, that is useful to guide the query formulation process. Given a query, the concept lattice allows accessing all the cases that share properties with the query at the same time so that they are grouped under the same concept. In [7] we used FCA to elicit knowledge from the case based including dependencies between attributes. In the proposed general explanation framework we also explore these dependencies and the maximal groups of items are useful as explanation knowledge. Since its origin in early 80s [27] FCA has became a popular human-centred tool for knowledge representation, data analysis and knowledge discovery with numerous applications. Ontology engineering and big data and their analysis attracted the attention of some researchers using FCA to find the pattern structure and its visualisation [23]. We are not aware of any work using FCA to generate explanations.

3   Formal Concept Analysis

FCA is a mathematical approach to data analysis based on the lattice theory of Birkhoff [5]. It provides a way to identify maximal groupings of objects with shared properties, and enables formalisation of concepts as basic units of human thinking and analysing data in the object-attribute form. This is a clear characteristic of recommender systems where there are items described by properties. Even for collaborative filtering approaches based on collecting ratings there are object-attribute knowledge about the items.

FCA application provides with a conceptual hierarchy, because it extracts the *formal concepts* and the hierarchical relations among them, where related items are clustered according to their shared properties. The lower in the graph, the more characteristics can be said about the items; i.e. the more general concepts are higher up than the more specific ones. In this paper, we propose using

Table 1. Sample set G with 6 movies

Movie Id.	Movie title	Director	Genre	Actors	Year
223	Clerks	Kevin Smith	Comedy	Jason Mewes Jeff Anderson	1994
231	Dumb & Dumber	Peter Farrelly	Comedy	Lauren Holly Teri Garr	1994
235	Ed Wood	Tim Burton	Biography, Comedy Drama	Johnny Deep Martin Landau	1994
110	Braveheart	Mel Gibson	Biography, Drama, History, War	Mhairi Calvey James Robinson	1995
151	Rob Roy	Michael Caton-Jones	Adventure, Biography	Liam Neeson Eric Stoltz	1995
1	Toy Story	John Lasseter	Adventure, Animation, Comedy, Family, Fantasy	Tom Hanks Jim Varney	1995

this conceptual structure as the knowledge base of an explanation framework for recommender systems. In the proposed general explanation framework we also explore dependencies between attributes and metrics on the lattices as explanation knowledge (see Sect. 4.2).

We first briefly review the basics of the FCA technique. See [8] for a description of our previous work on FCA to elicit knowledge from CBR systems. We refer the interested reader to [11,23] for a complete description on FCA and its applications.

A *formal context* is defined as a triple $\langle G, M, I \rangle$ where there are two sets G (of objects) and M (of attributes), and a binary (incidence) relation $I \subseteq G \times M$, expressing which attributes describe each object (or which objects are described using an attribute), i.e., $(g, m) \in I$ if the object g carries the attribute m, or m is a descriptor of the object g. With a general perspective, a concept represents a group of objects and is described by using *attributes* (its intent) and *objects* (its extent). The extent covers all objects belonging to the concept while the intent comprises all attributes (properties) shared by all those objects. With $A \subseteq G$ and $B \subseteq M$ the following operator (*prime*) is defined as:

$$A\prime = \{m \in M | (\forall g \in A)(g, m) \in I\} \qquad B\prime = \{g \in G | (\forall m \in B)(g, m) \in I\}$$

A pair (A,B) where $A \subseteq G$ and $B \subseteq M$, is said to be a *formal concept* of the context $\langle G, M, I \rangle$ if $A' = B$ and $B' = A$. A and B are called the *extent* and the *intent* of the concept, respectively.

It can also be observed that, for a concept (A, B), $A'' = A$ and $B'' = B$, which means that *all objects of the extent of a formal concept, have all the attributes of the intent of the concept, and that there is no other object in the set G having all the attributes of (the intent of) the concept.*

id	John Lasseter	Kevin Smith	Mel Gibson	Michael Caton-Jones	Peter Farrelly	Tim Burton	Adventure	Animation	Biography	Comedy	Drama	Family	Fantasy	History	War	Eric Stoltz	James Robinson	Jason Mewes	Jeff Anderson	Jim Varney	Johnny Depp	Lauren Holly	Liam Neeson	Martin Landau	Mhairi Calvey	Teri Garr	Tom Hanks	1994	1995
	DIRECTOR						GENRE									ACTORS												YEAR	
223	X									X						X	X											X	
231			X							X											X				X			X	
235					X		X	X	X												X		X					X	
110		X								X		X		X	X	X									X				X
151				X			X			X						X								X					X
1	X						X	X		X		X	X								X					X			X

Fig. 1. Example of applying FCA: cross table

Example. We illustrate how to apply FCA to a set G of objects containing 6 movies. The set is described in Table 1, where the selected attributes are the columns in the table. The binary (incidence) relation $I \subseteq G \text{x} M$ is represented by the cross table in Fig. 1. Figure 2 shows the Hasse diagram of the concept lattice resulting of the FCA application[1]. Each node represents a *formal concept* of the context, and the ascending paths of line segments represent the subconcept-superconcept relationship. The lattice contains exactly the same information that the cross table (Fig. 1), so the incidence relation I can always be reconstructed from the lattice.

In Fig. 2 the attributes from the intent are inside grey box labels and the objects from the extent are inside white box labels. A lattice node is labelled with the attribute $m \in M$ if it is the upper node having m in its intent; and a lattice node is labelled with the object $g \in G$ if it is the lower node having g in its extent. Using this reduced labelling, each label (attribute or object name) is used exactly once in the diagram. If a node C is labelled by the attribute m and the object g then all the concepts more general than C (above C in the graph) have the object g in their extents, and all the concepts more specific than C (below C in the graph) have the attribute m in their intents. This way, the intent of a concept in a Hasse diagram in Fig. 2 can be obtained as the union of the attributes in its grey-boxed label and attributes in the grey-boxed labels of the concepts above it in the lattice. Conversely, the extent of a concept is

---

[1] *Conexp tool* (https://sourceforge.net/projects/conexp/).

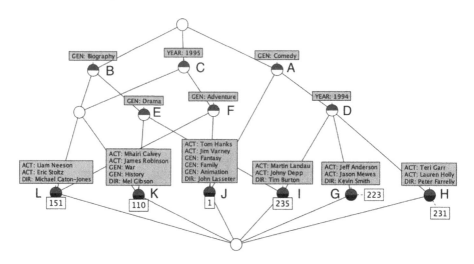

Fig. 2. Example of applying FCA: Hasse diagram of the Lattice.

Table 2. Dependency rules examples extracted from lattice in Fig. 2

GEN: Drama → *GEN: Biography*
GEN: Adventure → *YEAR: 1995*
YEAR: 1994 → *GEN: Comedy*
ACT: Liam Neeson, ACT: Eric Stoltz, DIR: Michael CatonJones → *GEN: Biography, YEAR: 1995*
ACT: Mhairi Calvey, ACT: James Robinson, GEN: War, GEN: History, DIR: Mel Gibson → *GEN: Drama, YEAR: 1995*
ACT: Martin Landau, ACT: Johny Depp, DIR: Tim Burton → *GEN: Drama, YEAR: 1994*

obtained as the union of the objects in its white-boxed label and objects in the white-boxed labels of the concepts below it in the lattice.

Besides the hierarchical conceptual clustering of the objects, FCA captures knowledge about the co-appearance or associations between attributes. A *dependence rule* [8] between two attribute sets (written $M1 \rightarrow M2$, where $M1, M2 \subseteq M$) means that any object having all attributes in $M1$ has also all attributes in $M2$. We can read the dependence rules in the graph as follows:

- Each line between nodes labelled with attributes means a dependence rule between the attributes from the lower node to the upper one.
- When there are several attributes in the same label it means that there is a co-occurrence of all these attributes for all the objects in the sample and we can infer co-dependence rules.

For example, in the lattice of Fig. 2 we find a dependence between nodes E and B meaning that, in the set of items used to build the lattice, all the Drama movies are also Biographies (*GEN: Drama* → *GEN: Biography*). Table 2 shows some of the dependency rules.

The operation $classify(i)$ returns the formal concept C in the lattice that recognises *all* the attributes that describe i. For example, given a movie m

with known attributes [*GEN: Comedy, ACT: Tom Hanks*], m is classified in the concept A because it is the lower concept where all the given attributes are fulfilled. The concept J does not recognise the individual as an instance because, even if it has the attribute [*ACT: Tom Hanks*], there are other required attributes [*GEN: Fantasy, GEN: Family, GEN: Animation,..*] that m should fulfil.

4   FCA-Based Explanation Algorithm

A collaborative filtering approach recommends items based on users' past behaviour and ratings. However, it lacks from an explicit aggregated model of the user preferences and the capability of explaining their results. We propose building the *user profile lattice* with the user personal best rated items. This lattice can be used as a model to explain the user profile and the diversity of her preferences. This explanation allows the self-comprehension of the user profile based on her ratings (Sect. 4.1) or understanding why a certain movie has been recommended (Sect. 4.2). The explanation lattice is computed for each user and it is reused for different recommendations.

Our approach to generate explanations is general and applicable to different recommendation domains. It uses a set of the user ratings and the item properties to generate the FCA lattice based on the best rated items, and generates personalised explanations for each particular user profile. The *FCA-based explanation algorithm* (see Algorithm 1) allows us to organise the knowledge on the user preferences and obtain the vocabulary to explain the user profile and, according to this profile, why an item has been recommended using either the *item-style* explanation, which includes the similar items rated by the user; the *property-style* explanation, which describes the properties from the formal concepts; and the *dependency-style* explanation, that includes the description of the association rules elicited by the FCA. The general process runs as follow:

Step 1 Selection of the M_u (attributes) and G_u (items) sets used to build the lattice (details of selection strategies sel_g and sel_m in Sect. 5.2).
Step 2 Apply FCA and evaluate and refine the resulting lattice.
Step 3 Choose between explaining the user profile (details in Sect. 4.1) or explain a specific recommendation (*rec*), so we classify *rec* to generate more specific explanations (details in Sect. 4.2).
Step 4 Explanations are generated from textual templates filled with the corresponding elements obtained while travelling the lattice *item-style, property-style, dependency-style.*

The different styles of explanations can be combined. In the *property-style* explanation we show the intent of the formal concepts that explains the properties that make these items be grouped together. In the *item-style* explanation we show the extent of the formal concepts. It shows items that are somehow similar according to the maximal groups. Association (or dependency) rules are very interesting pieces of knowledge, difficult to see at first sight from the items, and very useful for explanations.

Algorithm 1. Travelling the lattice to build Explanations

Input: G_u, M_u, I_u, rec, sel_g, sel_m
Output: *Expl-item, Expl-property, Expl-dependency*
1 $G_u' = sel_g(G_u)$
2 $M_u' = sel_m(M_u)$
3 $Ret = FCA(G_u', M_u', I_u)$
4 $C_r \leftarrow Ret.classify(rec) \parallel TOP$
5 *Expl-item* $\leftarrow \{traverseLevels(C_r.extent)$ $\}$
6 *Expl-property* $\leftarrow \{traverseLevels(C_r.intent)$ $)$ $\}$
7 *Expl-dependency* $\leftarrow$ {obtainRules(Ret) }

Table 3. Example of selected best rated items for user u

Movie Id.	Movie title	Director	Genre	Actors	Year	Rating
594	Snow White and the Seven Dwarfs	William Cottrell	Animation, Family, Fantasy, Musical	Adriana Caselotti Lucille La Verne	1937	5.0
596	Pinocchio	Norman Ferguson	Animation, Family, Fantasy, Musical	Mel Blanc Cliff Edwards	1940	4.5
588	Aladdin	Ron Clements	Adventure, Animation, Comedy, Family, Fantasy, Musical, Romance	Robin Williams Scott Weinger	1992	5.0
364	The Lion King	Roger Allers	Adventure, Animation, Drama, Family, Musical	Matthew Broderick Niketa Calame	1994	5.0
317	The Santa Clause	John Pasquin	Comedy, Drama, Family, Fantasy	Judge Reinhold Peter Boyle	1994	3.5
34	Babe	Chris Noonan	Comedy, Drama, Family	Miriam Margolyes Roscoe Lee Browne	1995	4.0
158	Casper	Brad Silberling	Comedy, Family, Fantasy	Eric Idle Cathy Moriarty	1995	3.0
48	Pocahontas	Mike Gabriel	Adventure, Animation, Drama, Family, History, Musical, Romance	Christian BaleIrene Bedard	1995	5.0

We have generated a user profile using the items in Table 3. The corresponding lattice is shown in Fig. 3 and it will be used as running example in the following subsections.

4.1 Explanation of the User Profile

The user profile and the corresponding explanations are personalised for each user because the item set G_u' used to generate the lattice are selected from the user ratings. In the example, Table 3 represents the selected best rated items and attributes for a certain user G_u' and M_u'. We apply FCA and generate the corresponding lattice (Fig. 3). According to the general Algorithm 1 described in Sect. 4 we propose a property-style explanation that travels the lattice, level by level, from top to bottom. Note that the label [*GEN: Family, LAN: English*] in the TOP concept means that all the movies in G_u' have these properties

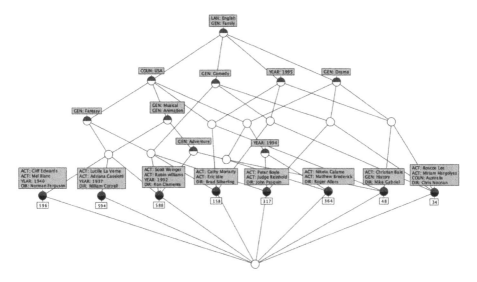

Fig. 3. User profile lattice Ret $= \mathrm{FCA}(G'_u, M'_u, I_u)$ with G'_u M'_u and I_u in Table 3

Table 4. Example *Property-style* explanations for the user profile in Fig. 3

Level 0	*"All your high scored movies share GEN: Family LAN: English"*
Level 1	*"Most of your high scored movies are GEN: Drama or GEN: Comedy"*
	"Would you like to see them?" (item-style)
	"Your movies in GEN: Drama *are:* 590, 364, 317, 34, 48" (Extent of concept)
	"Your movies in GEN: Comedy *are:* 588, 317, 34, 158" (Extent of concept)
	"Most of your high scored movies are COUN: USA"
	"Would you like to see them?" (item-style)
	"Most of your high scored movies are YEAR: 1995"
	"Would you like to see them?" (item-style)
Level 2	*"Some of your high scored movies are* GEN: Musical, *GEN: Animation or GEN: Fantasy"*

and, therefore, all the concepts below TOP inherit the property. The *property-style* explanation can be combined with *item-style* explanations by retrieving the extent of a certain concept. In the example, the user chooses when to see the specific items. The textual explanation generated for this user profile will employ different templates for each level in the lattice of Fig. 3, as described in Table 4.

In addition to *property-style* explanations we can also use complementary *dependency-style* explanations, which provide association between attributes in G'_u. Dependency-style explanations are very useful to help the user to understand her profile and refining the scoring if the user disagrees with any of the extracted rules. Note that transitivity could be applied between rules and that some of the explanations are redundant with respect to the *property-style* explanations on the same lattice. Table 5 shows some examples of dependency rules between attributes in the lattice of Fig. 3 and the corresponding textual explanations.

Table 5. *Dependency-style* explanations generated for the user profile in Fig. 3

GEN: Comedy → *GEN: Family, LAN: English*
"All your high scored movies that have GEN: Drama have also GEN: Family and LAN: English"
GEN: Animation, GEN: Musical → *COUN: USA*
"All your high scored movies that have GEN: Animation and GEN: Musical have also COUN: USA"
GEN: Adventure → *GEN: Animation, GEN: Musical*
"All your high scored movies that have GEN: Adventure have also GEN: Animation and GEN: Musical"
GEN: Romance → *GEN: Adventure*
"All your high scored movies that have GEN: Romance have also GEN: Adventure"
GEN: Drama → *GEN: Family*
"All your high scored movies that have GEN: Drama have also GEN: Family"
. . .

Note that the textual templates for explaining dependency rules fill the gaps using the rule (Left/Right) Hand Sides (L/R)HS): *"All your high scored movies that have"* LHS *"have also"* RHS.

4.2   Explaining a Recommendation

Besides explaining user profiles, the FCA lattice can be used to explain a recommendation, i.e, why a particular item has been recommended over others. Step 3 in Algorithm 1 classifies an item rec in the lattice by its properties. Then, we build an *item-style* or *property-style* explanation using the concept C_r that recognises the object rec.

As we described in Sect. 3, rec is classified in the lattice $-classify(rec)-$ if any of lattice concepts recognises all rec properties. In [7] we proposed a classification based retrieval method, where a partially defined query is classified in the FCA lattice that organise the case base. We proposed a query completion process based and the use of dependency rules that helps to complete the query towards similar cases. We cannot apply this method here because rec is not a partially defined query, but a complete individual with all its properties. Note that the classification process fails if none of the formal concepts recognises all the properties in rec.

An easy approach would be rebuilding the FCA lattice using $\{G_u \cup r\}$ and generating the explanations using Algorithm 1. Depending on the size of the formal context and the optimisation of the FCA implementation that could become inefficient. As an alternative, we propose using each property separately, completing with dependency rules when possible and generating partial explanations. For example, if the recommender systems recommends item 223, whose properties are $[GEN:\ Comedy,\ YEAR:\ 1995,\ (...)]$, we can generate the explanations detailed in Table 6 using the properties one by one.

5   Evaluation

In order to evaluate the possibilities of the FCA lattice to provide explanations we have analysed several descriptive metrics. These metrics let us conclude the

Table 6. Explanations of a recommendation based on properties of the proposed item.

"This movie is recommended to you because of the GEN: Comedy property" (property-style)

"The recommended movie shares the property GEN: Comedy with 588, 317, 34, 158" (Extent of concept)
(item-style)

"The recommended movie shares the property YEAR: 1995 with 34, 48, 158" (Extent of concept)
(item-style)

---

feasibility of the method for a real dataset and discuss the potential quality of the explanations according to them. Concretely, we have combined two popular datasets. The first of them is the MovieLens dataset [14]. This dataset contains 100,000 ratings made by users in the MovieLens recommendation system. The second dataset contains the features of 5000 movies extracted from IMDB[2]. The features of the movies used in the evaluation are: genres, directors, actors, screenwriters and the decade in which they were released. The metrics defined to analyse the lattice are:

Num. Nodes (N) represents the number of nodes in the lattice. It includes *top* and *bottom* nodes. We can measure the number of nodes with respect to the number of items used to generate the lattice. It allows us to measure the proportion between the nodes in the lattice and the number of items used to generate it.

Level Width (LW) is the highest number of nodes in a lattice level. It represents the maximum width of the lattice.

Depth (D) measures the length of the longest path from *bottom* to *top* of the lattice. Width and depth allow to study the distribution and the diversity/homogeneity of the attributes in the items of the user's profile.

Branch Factor (BF) measures the average number of children for the nodes in the lattice.

First, we have analysed the global behaviour of the lattices when varying the number of movies used to create them. Later, we have studied the behaviour of the lattices when using two different approaches for selecting the movies. Next, we discuss both evaluations and their impact in the quality of the explanations.

5.1   Global Behaviour of the FCA Lattices

This first evaluation aims to explore the features of the lattices with respect to the number of movies used to create them. This way we can figure out the optimal number of movies required to create the lattice that provides the explanations and analyse the global behaviour of lattices regarding diversity, homogeneity and complexity. To perform this evaluation we have chosen randomly 100 users and generated their lattices with a fixed number of movies (from 2 to 75). These movies are the best rated by the user. Then for each number of movies we average the results of the 100 users.

---

[2] https://www.imdb.com/.

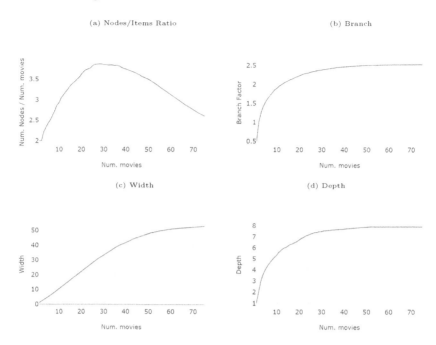

Fig. 4. Lattice properties analysis w.r.t. the number of movies $|G'_u|$ (x-axis)

Figure 4 summarises our findings. The first chart (Fig. 4(a)) shows the relationship between the number of nodes of the lattices and the items used to create them. As expected, the lattice grows as the number of items increases. However, it saturates around $|G'_u| \sim 20$ and starts decreasing. This way, we can conclude that it is the optimal size to compute the user's lattice.

Then, Figs. 4(c) and (d) describe the behaviour of the width and depth metrics. The decrease of the lattice size as the number of items is greater than approximately 20, makes this ratio to slightly slow down after that value in the case of the width metric. The depth metric stabilises completely for large lattices. As this metric is associated to homogeneity of the user preferences, results clearly denote that larger profiles are more heterogeneous because they became wider but depth value is stabilised. If we analyse the width/depth ratio we can clearly obtain that conclusion. For example, lattices built with $|G'_u| = 20$ items are on average 3 times wider.

Finally, Fig. 4(b) shows the branch factor. Here we can a observe a behaviour similar to the depth ratio. Increasing the number of items does not imply a linear raise of the complexity of the lattices, i.e., the number of intermediate nodes representing shared properties does not raise proportionally.

Once we have analysed the global behaviour of the lattices we can evaluate the item selection strategies used to create them.

Table 7. Averaged results of the selection strategies.

Metric	sel_g^{best}	sel_g^{20}
Num. Nodes/Num. Movies	3.711	3.611
Depth/Num. Movies	0.270	0.336
Width/Num. Movies	1.078	1.101
Branch factor/Num. Movies	0.087	0.111

5.2   Item Selection Strategies

Before applying FCA we select the item's attributes and the subset of movies from the user's profile that will be analysed. It is the first step of the general process outlined in Sect. 4 and represented as the sel_m and sel_g functions in Algorithm 1.

The selection of the item's attributes (sel_m) is a simple process that depends on the domain. In our example we can discard the *id* and *title* attributes as they cannot be used to classify the movies in the lattice.

However, there are several alternatives for selecting the movies used to compute the FCA lattice (sel_g strategy). As we have concluded from the previous evaluation, 20 is approximately the optimal number of movies to generate the FCA lattice. Therefore, we could select the 20 movies with the maximum score, select them randomly, use a stratified selection according to the rating values, etc. On the other hand, we could ignore this fixed number of items and implement an alternative with all the items with a high score. In order to evaluate the impact of this selection method we have computed the previous metrics for every user but selecting items according to the following strategies:

sel_g^{best}: Select all the items with score $>= 4$ stars.
sel_g^{20}: Select the 20 items with the maximum scores.

From sel_g^{best} and sel_g^{20}, we are able to compute two lattices, respectively, for each user. Our goal to compare the properties of both lattices using the previous metrics. However, we cannot compare them directly as the number of items used to generate the lattices $((|G_u'|))$ is different. Therefore, we will normalise every metric (nodes, wide, depth and branch factor) according to the number of movies. Again, we have randomly chosen 100 users that have watched more than 20 movies. The mean number of movies rated by these 100 users is 40.20.

To summarise the achieved results, we have averaged every metric for the whole set of users. Results are shown in Table 7. Analysing this table we can conclude that both approaches got very similar results. To analyse the non-aggregated results we have generated the graphs in Fig. 5. Here, we compare the difference obtained by every metric, once has been normalised, between the pair of lattices generated for each user. Therefore, every column represents sel_g^{20} − sel_g^{best}. Columns have been ordered according to the difference value to be able to compare which is the the winning strategy. Beginning with the number of

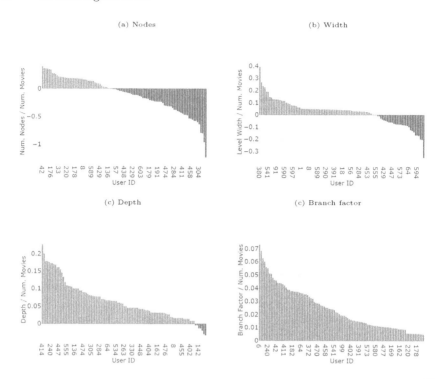

Fig. 5. Differences in the evaluation of every user using both selection strategies (sel_g^{20} − sel_g^{best}). Every metric has been previously normalised w.r.t. the number of movies.

nodes shown in Fig. 5(a), it is slightly balanced to the sel_g^{best} method (there are more negative values). It means that this selection strategy provides items that generate larger lattices. Regarding the width of the lattices, Fig. 5(b), it is quite balanced. However, Fig. 5(c) shows that the sel_g^{20} method obtains deeper lattices, associated to homogeneous profiles. It makes sense as this strategies retrieves the best rated items for the user, and they are usually quite similar. Finally, we cannot conclude any significant result regarding the branch factor because the magnitude of the differences is too low.

6   Conclusions and Future Work

In this paper, we have proposed the use of Formal Concept Analysis to help finding the knowledge structure of the user profiles in recommender systems and how to use this knowledge as explanation knowledge for the system. We have proposed a FCA-based explanation approach to organise the knowledge about users' preferences in a lattice that helps to obtain the vocabulary to explain why an item is recommended using other either similar items (*item-style explanations*) that the user has rated or similar properties (*property-style explanations*) with the best rated items for that user. Additionally, the *user profile lattice* itself can be employed as a form of scrutability of the recommender system, where the user understands which profile is inferred by the system according to her ratings.

We have analysed a set of user profile lattices according to their structural properties and the next step is to evaluate the explanations with real users. Besides, the study of lattice metrics and properties opens a line of future work where we plan to study further the different types of user profiles in terms of their rating behaviours, both in quantity and distribution of the rating values. Our hypothesis is that the number of ratings and the distribution of the rating values would affect the structural properties of the user profile lattice. For this reason we should test with different methods to refine the way the set of items (G'_u) employed to generate the user profile are chosen. Similarly, we would like to test different ways to select the set of properties (M'_u) that will be used to generate the lattice and, therefore, the explanations.

References

1. Aamodt, A.: Explanation-driven case-based reasoning. In: Wess, S., Althoff, K.-D., Richter, M.M. (eds.) EWCBR 1993. LNCS, vol. 837, pp. 274–288. Springer, Heidelberg (1994). https://doi.org/10.1007/3-540-58330-0_93
2. Armengol, E., Plaza, E.: Using symbolic descriptions to explain similarity on CBR. In: Proceedings of the 8th International Conference of the ACIA, Alguer, Italy, pp. 239–246 (2005)
3. Bell, R., Koren, Y., Volinsky, C.: Matrix factorization techniques for recommender systems. Computer **42**(08), 30–37 (2009)
4. Biran, O., Cotton, C.V.: Explanation and justification in machine learning: a survey. In: IJCAI 2017 Workshop on Explainable AI (XAI) (2017)
5. Birkhoff, G.: Lattice Theory, vol. 25. American Mathematical Society, New York (1940)
6. Caro-Martinez, M., Jimenez-Diaz, G., Recio-Garcia, J.A.: A theoretical model of explanations in recommender systems. In: ICCBR 2018, p. 52 (2018)
7. Díaz-Agudo, B., González-Calero, P.A.: Classification based retrieval using formal concept analysis. In: Aha, D.W., Watson, I. (eds.) ICCBR 2001. LNCS (LNAI), vol. 2080, pp. 173–188. Springer, Heidelberg (2001). https://doi.org/10.1007/3-540-44593-5_13
8. Díaz-Agudo, B., González-Calero, P.A.: Formal concept analysis as a support technique for CBR. Knowl.-Based Syst. **14**(3–4), 163–171 (2001)
9. Doyle, D., Cunningham, P., Bridge, D.G., Rahman, Y.: Explanation oriented retrieval. In: Funk, P., González Calero, P.A. (eds.) ECCBR 2004. LNCS (LNAI), vol. 3155, pp. 157–168. Springer, Heidelberg (2004). https://doi.org/10.1007/978-3-540-28631-8_13
10. Doyle, D., Tsymbal, A., Cunningham, P.: A review of explanation and explanation in case-based reasoning (2019)
11. Ganter, B., Wille, R.: Formal Concept Analysis: Mathematical Foundations. Springer, Heidelberg (2012)
12. Gedikli, F., Jannach, D., Ge, M.: How should I explain? A comparison of different explanation types for recommender systems. Int. J. Hum Comput Stud. **72**(4), 367–382 (2014)

13. Gunning, D.: Darpa's explainable artificial intelligence (XAI) program. In: 24th IUI, Marina del Ray, CA, USA (2019)
14. Harper, F.M., Konstan, J.A.: The movielens datasets: History and context. ACM Trans. Interact. Intell. Syst. **5**(4), 19:1–19:19 (2015)
15. Leake, D.B., McSherry, D.: Introduction to the special issue on explanation in case-based reasoning. Artif. Intell. Rev. **24**(2), 103–108 (2005)
16. Lee, O., Jung, J.J.: Explainable movie recommendation systems by using story-based similarity. In: 23rd ACM IUI, Tokyo, Japan (2018)
17. Maximini, R., Freßmann, A., Schaaf, M.: Explanation service for complex CBR applications. In: Funk, P., González Calero, P.A. (eds.) ECCBR 2004. LNCS (LNAI), vol. 3155, pp. 302–316. Springer, Heidelberg (2004). https://doi.org/10.1007/978-3-540-28631-8_23
18. McSherry, D.: Explanation in recommender systems. Artif. Intell. Rev. **24**(2), 179–197 (2005)
19. Muhammad, K., Lawlor, A., Smyth, B.: A multi-domain analysis of explanation-based recommendation using user-generated reviews. In: 31st FLAIRS, Melbourne, USA, pp. 474–477 (2018)
20. Pu, P., Chen, L.: Trust-inspiring explanation interfaces for recommender systems. Knowl.-Based Syst. **20**(6), 542–556 (2007)
21. Schank, R.C., Kass, A., Riesbeck, C.K.: Inside Case-Based Explanation. Psychology Press (2014)
22. Roth-Berghofer, T.R.: Explanations and case-based reasoning: foundational issues. In: Funk, P., González Calero, P.A. (eds.) ECCBR 2004. LNCS (LNAI), vol. 3155, pp. 389–403. Springer, Heidelberg (2004). https://doi.org/10.1007/978-3-540-28631-8_29
23. Singh, P.K., Aswani Kumar, C., Gani, A.: A comprehensive survey on formal concept analysis, its research trends and applications. Int. J. Appl. Math. Comput. Sci. **26**(2), 495–516 (2016)
24. Sørmo, F., Cassens, J., Aamodt, A.: Explanation in case-based reasoning-perspectives and goals. Artif. Intell. Rev. **24**(2), 109–143 (2005)
25. Tintarev, N., Masthoff, J.: Explaining recommendations: design and evaluation. In: Ricci, F., Rokach, L., Shapira, B. (eds.) Recommender Systems Handbook, pp. 353–382. Springer, Boston, MA (2015). https://doi.org/10.1007/978-1-4899-7637-6_10
26. Weber, R.O., Johs, A.J., Li, J., Huang, K.: Investigating textual case-based XAI. In: Cox, M.T., Funk, P., Begum, S. (eds.) ICCBR 2018. LNCS (LNAI), vol. 11156, pp. 431–447. Springer, Cham (2018). https://doi.org/10.1007/978-3-030-01081-2_29
27. Wille, R.: Restructuring lattice theory: an approach based on hierarchies of concepts. In: Rival, I. (ed.) Ordered Sets, pp. 445–470. Springer, Dordrecht (1982). https://doi.org/10.1007/978-94-009-7798-3_15
28. Zanker, M., Ninaus, D.: Knowledgeable explanations for recommender systems, vol. 1, pp. 657–660 (2010)

FLEA-CBR – A Flexible Alternative to the Classic 4R Cycle of Case-Based Reasoning

Viktor Eisenstadt[1(✉)], Christoph Langenhan[2], and Klaus-Dieter Althoff[1,3]

[1] Institute of Computer Science, University of Hildesheim,
Samelsonplatz 1, 31141 Hildesheim, Germany
ayzensht@uni-hildesheim.de, klaus-dieter.althoff@dfki.de
[2] Chair of Architectural Informatics, Technical University of Munich,
Arcisstrasse 21, 80333 Munich, Germany
langenhan@tum.de
[3] German Research Center for Artificial Intelligence (DFKI),
Trippstadter Strasse 122, 67663 Kaiserslautern, Germany

Abstract. This paper introduces FLEA-CBR, an alternative approach for composition of case-based reasoning (CBR) processes. FLEA-CBR extends the original 4R *(Retrieve, Reuse, Revise, Retain)* CBR cycle with a flexible order of execution of its main steps. Additionally, a number of combinatorial features for a more comprehensive and enhanced composition can be used. FLEA is an acronym for *Find, Learn, Explain, Adapt* and was initially created to solve the restrictiveness issues of case-based design (CBD) where many existing approaches consist of the retrieval phase only. However, the methodology can be transferred to other CBR domains too, as its flexibility allows for convenient adaptation to the given requirements and constraints. The main advantages of FLEA-CBR over the classic 4R cycle are the ability to combine and activate the main steps in desired or arbitrary order and the use of the explainability feature together with each of the steps as well as a standalone component, providing a deep integration of Explainable AI (XAI) into the CBR cycle. Besides the CBR methods, the methodology was also conceptualized to make use of the currently popular machine learning methods, such as recurrent and convolutional neural networks (RNN, ConvNet) or general adversarial nets (GAN), for all of its steps. It is also compatible with different case representations, such as graph- or attribute-based. Being a template for a distributed software architecture, FLEA-CBR relies on the autonomy of implemented components, making the methodology more stable and suitable for use in modern container-based environments. Along with the detailed description of the methodology, this paper also provides two examples of its usage: for the domain of CBR-based creativity and library service optimization.

Keywords: CBR cycle · Distributed CBR · Explainable AI ·
Adaptation · Case-based design · Artificial neural network ·
Software architecture

© Springer Nature Switzerland AG 2019
K. Bach and C. Marling (Eds.): ICCBR 2019, LNAI 11680, pp. 49–63, 2019.
https://doi.org/10.1007/978-3-030-29249-2_4

1   Introduction

In case-based reasoning (CBR), the historical development of the research area established the 4R cycle [2], that consists of the consecutive steps *Retrieve, Reuse, Revise,* and *Retain,* as the main underlying structure for implemented and theoretical approaches that explicitly identify themselves or were designated by researchers as CBR-based. Quickly after its introduction in 1994, the 4R cycle became a widely accepted synonym for case-based reasoning and is referenced nowadays in nearly every CBR-related research work.

Generally, two types of such CBR research works exist nowadays: *one-step-specific* and *generative application.* The works of the first type deal with only one of the R-steps (where Retrieve and Reuse make up the majority) and are rarely created for one specific domain, providing improvement of a universally applicable method. In contrast to the first type, the generative application of the CBR cycle is usually conceptualized for a specific domain (often including the related domains as well), taking the CBR cycle as a whole and using all of the R-steps to build a CBR-based solution for a specific problem of the domain. Sometimes this type is combined with other AI methods providing the so-called *hybrid* approaches. Both types have in common that they do not try to modify the complete CBR cycle and adapt it to own purposes, instead the 4R cycle is used "as is". These modifications, however, exist and make up the third, very rare, type of CBR-related research: the *explicit modifications* of the classic CBR cycle. The reasons for development of such modifications are different, but can mostly be narrowed to the (partial) incompatibility of the application domain in question to the original order of execution of 4R and/or to the restrictions of the domain that allow for execution of a subset of the four R-steps only.

In this work, we present *FLEA-CBR,* a methodology for modification of the original CBR cycle, whose aim is to bring more flexibility to the cycle structure and to offer an alternative underlying approach for systems where the four R-steps may not or cannot be applied subsequently in the original order. FLEA is an acronym for *Find, Learn, Explain,* and *Adapt.* These four components represent the main functionalities of the methodology. Main features of FLEA are the possibility of standalone usage of the components as well as the arbitrary combination of them and the deep integration of Explainable AI (XAI) into the cycle making it an equivalent step among the already existing ones. FLEA was conceptualized for the domain of case-based design (CBD), however, its flexibility allows for application to other domains as well. The only requirement is that the case representation types of the domain(s) are compatible with FLEA-CBR.

This paper is structured as follows: first, we present work related to the research topic of this paper, i.e., the modifications of the 4R CBR cycle. We show differences between hybrid CBR approaches and cycle modifications and present an overview of selected modifications. In the following Sect. 3, FLEA-CBR's core components and features are presented in detail. In Sect. 4, existing and theoretical example usages of FLEA-CBR are presented for the domains of CBR-based creativity and library service optimization respectively. Finally, an outlook to the future of the methodology concludes this work in Sect. 5.

2 Related Work

The explicit modifications of the original order of execution of the four R-steps have a long tradition in CBR, being present after the 4R cycle started to gain its popularity. However, it is important to mention that this rare type of CBR-related research should not be confused with the hybrid CBR approaches, i.e., those that combine CBR with other machine learning techniques or related methods. The examples of such hybrid approaches can be found in combination with deep learning [6], genetic algorithm optimization [21], or support vectors [11]. Additionally, an overview of such combinations [28] provides an entry point to start research into these approaches. The hybrid approaches mostly do not change or adapt the cycle itself, whereas the CBR cycle modifications explicitly edit the 4R cycle: for example, an additional step can be appended or prepended to the cycle, or one or more of the existing steps can be replaced by other steps, combined into a single step, or managed by an additional non-CBR module, such as an artificial neural network (ANN). The following Table 1 provides a comparison of selected CBR cycle modifications that either explicitly identify themselves as a 'modification' or provide enough evidence to count as such, i.e., edit the cycle as described above. The approaches that were proposed as modified CBR cycles, but in fact are hybrid systems, were not considered. Besides the descriptions of the respective approaches and the corresponding cycle modifications, this comparison also provides information on suitability of the approach for universal use and if it implements explainability features.

Table 1. An overview of selected CBR cycle modifications. *U* stands for the universal type of use, *DB* for domain-bound use. XAI stands for explainability/explainable AI.

Description	Publication	Modification	U/DB	XAI
CBR-based recipe recommender system	Skjold and Øynes [36]	Second Retrieve in ephemeral case base betw. Reuse and Revise	DB	No
Recipe generation with CBR and deep learning	Grace et al. [16]	Dual-cycle CBR	DB	No
CBR real-time planning for industrial systems	Navarro et al. [25]	Starts with learning (Revise + Retain)	DB	No
Decision support with neocortex imitation	Hohimer et al. [18]	Exec. order: *Reason*, Retain, *Review*, Revise	U	No
Tagging + retrieval of similar code passages	Roth-Berghofer and Bahls [34]	*Explain + Customize* replace Reuse + Revise	DB	Yes
Reorganization of the case bases	Finnie and Sun [14]	*Repartition* betw. Reuse and Revise	U	No
Maintenance of CBR systems	Reinartz et al. [29]	*Review* and *Restore* after Retain	U	No
Oceanography forecasts with CBR and ANN	Lees and Corchado [23]	ANN governs Reuse and Revise	DB	No

Interpreting the Table 1, we can conclude that certain similarities between hybrid approaches and cycle modifications exist, in some cases it is also hard to tell if the given approach is of hybrid type or an explicit modification. Both types are mostly used for a specific domain or task. The universal approaches contained in Table 1, such as the neocortex imitation [18], the intermediate reorganization of case bases [14], or the 6R cycle [29] that adds the maintenance steps Review and Restore to 4R, build the minority. Another example is the real-time intelligent decision support [12], which, however, was not included in the comparison, as it makes the human operator responsible for Retain and provides no adaptation, restricting the CBR process to the retrieval phase only.

The obvious problem of cycle modifications is that no methodology exists that is universal as well as flexible in terms of execution order. Other problem is the lack of XAI features. The only exception is the approach [34] that, however, provides explanations only for results of the retrieval phase.

3   FLEA-CBR

This section describes the FLEA-CBR methodology in detail and compares it to the 4R cycle in terms of flexibility and universality of use. First, the problem of non-flexibility of the 4R cycle will be described, after that an overview and each of the R-steps will be compared to its approximate counterpart of FLEA-CBR.

3.1   Problem Description

During the last decades, the 4R cycle (see Fig. 1) became an ubiquitous part of case-based approaches that mostly improve one of the R-steps or apply the cycle to the given domain to produce a complete CBR-based solution to the given problem space (see also Sect. 1). However, one of the main problems of the 4R cycle is that it requires the consecutive execution of the R-steps in order to work properly and to be used for the problem space as a suitable solution approach. This problem also precludes the 4R cycle from being more universal and forces the applications to follow the only available order (Retrieve → Reuse → Revise → Retain) and to be non-flexible in terms of selection of the currently required step. This issue is also valid for the contemporary alternatives to the 4R cycle, such as the cycles presented by Hunt [20] or in Leake's work [22]. Changing the order of execution might result in reassessment of the complete approach and in the subsequent decision of the developers of the system to use methods other than case-based. A number of modified execution orders and the nevertheless existing problems and issues were already listed in Sect. 2.

A major technical flexibility problem of the 4R cycle and the majority of its modifications is that the retrieval step is required in all application cases and all other steps depend on its results. While many problems of many domains can be solved with this order and restrictions (for example, in the classic domains of CBR, such as mechanical engineering tools diagnosis or medical applications [19]), other domains, especially those that make use of creativity (for example, architecture or game development), may require other, more complex and custom orders that may not look reasonable to humans but are completely understandable and executable by the machines.

Another main structure-related issue is that, while being simulated from human thinking and reasoning abilities, the 4R cycle and its contemporary alternative cycles also follow the human methods of experience-based decision making. This is reasonable if the computer should make decisions that the humans can relate to, but computers are generally more flexible in that regard and should not be restricted to one order only and should be able to decide autonomously which order is the most suitable for the current problem space. Therefore, especially case-based reasoning should provide more high-level features, for example to mix the steps and allow for repeating of some, if necessary.

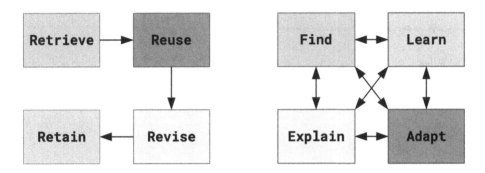

Fig. 1. FLEA-CBR in direct comparison with the 4R cycle.

3.2   Overview and Background

To overcome the issues described above and to make it possible for affected domains to make more use of CBR, we developed the methodology FLEA-CBR that introduces flexibility to the usage of the CBR features and allows for *mixing*, *repeating*, *sequencing*, or *cycling* (further referred as FLEA's *core features*) of the steps *Find, Learn, Explain,* and *Adapt* (further referred as FLEA's *core components*). FLEA-CBR was created in research context of MetisCBR [7], a framework for distributed case-based decision support during the early conceptual phases in architecture, that can find similar spatial configurations and contextually explain the retrieval results, suggest the next design step, and evolve the current design state to show how it can look in the future.

In the following sections the four core features of FLEA-CBR are described first, after that the four core components are presented. In Fig. 1, a high-level overview of FLEA-CBR and its direct comparison to the 4R cycle are available. Figure 2 demonstrates examples of core feature usages. Figures 3 and 4 contain feature usages for real domains.

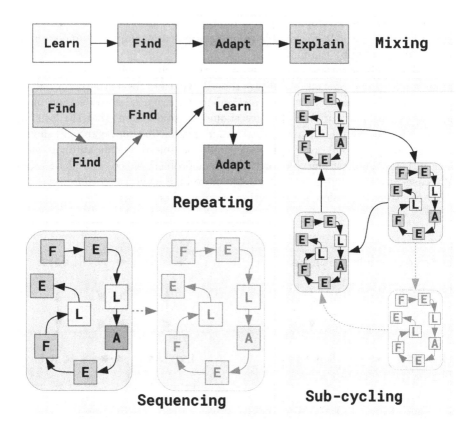

Fig. 2. Core features of FLEA-CBR.

3.3 Core Features

Mixing. This feature is crucial for FLEA-CBR's flexibility advantage as it allows for breaking the original 4R order and mix the mains steps in the order that is necessary for the domain or just for trying out and find the best option. Mixing can be done by a human operator as well as by a machine, e.g., a 'parent' CBR cycle. In Sect. 4, all examples are mixed and demonstrate how this action provides the FLEA-CBR implementation with the flexibility advantage.

Repeating. The goal of this functionality is to allow for continuous re-execution of a step with a predefined or arbitrary count of executions, e.g., to try other configuration options or different methods within one cycle run. This feature makes the CBR process with FLEA more fine-grained and exact, but also versatile in terms of selection of the most suitable method. This core feature can be executed with specific high-level frameworks, such as Keras [9] for deep learning.

Sequencing. This feature introduces the possibility of summarizing of (repeated) steps into sequences that can be used for similar domains or tasks. Unlike repeating, the sequences are more symbolic, i.e., do not contain information about exact methods or approaches applied, and can consist of different steps. This gives the system designers freedom to decide which methods are the most suitable, but preclude them from changing the execution order, transferring only the structure of the sequence. Along with Mixing, Sequencing bears a close resemblance to the original 4R cycle.

Sub-cycling. This feature introduces FLEA-CBR sub-cycles that can be used to encapsulate certain sequences and use the final outcome only in the parent cycle. Sub-cycles' goal is to increase the likelihood of possibly useful outcomes that, however, might be helpful or reused *later* in the parent process. That is, the sub-cycles can be skipped or executed asynchronously. Furthermore, sub-cycles can also be configured by the parent cycle. Technically, sub-cycles were added to the FLEA-CBR features in order to add a higher compatibility to the modern microservices-based software architectures. They can be executed in a dedicated container with methods that totally differ from other sub-cycles in terms of execution time or system resources usage.

3.4   Find

FLEA-CBR's *Find* step is the approximate equivalent of the *Retrieve* phase of the 4R cycle. Its main task is to search in the given collection of the domain cases (the case base) for entries with the highest similarity to the received problem description (query). However, generally, Find differs from Retrieve in the way that it can receive queries or requests from other FLEA-CBR modules or sub-cycles and forward them to modules other than Reuse in the format suitable for this module or sub-cycle.

Beyond the retrieval of the most similar cases, Find can also be used for looking for contextual connections between a subset of the data collection entries in order to provide the user or a FLEA-CBR application with information about how these entries, i.e., cases, are related to each other. In theory, it is also possible to integrate Find as a simple full text or image search engine to perform a common precision-recall-based information retrieval process, however, the modules that receive the results should be able to work with them, e.g., infer similarity values when necessary. All the retrieval methods described above can be combined and/or executed consecutively using the Repeating feature.

Context of Usage. When used as an intermediate step between other steps or modules, it is advisable to provide the FLEA-CBR sequence that contains Find with information on the execution context, e.g., the domain, in order to select the most suitable retrieval or matching method. This information can be received from the Learn or Explain component. Similarly to Retrieve, Find can also be used as an entry point to the case-based reasoning process.

3.5   Learn

The *Learn* step of the methodology was developed to enrich the FLEA-CBR-based approaches and applications with abilities of making reasonable conclusions based on features of the cases and predicting contextual and relational connections between them, but also to implement the retaining techniques that govern how and why the cases should be saved in the application's case collections. Therefore, the Learn step can also be seen as an approximate further development of 4R cycle's Retain and a high-level interface for gathering and combining the learning strategies and case preservation methods. To accomplish these tasks, Learn is conceptualized to make use of the modern deep learning methods, such as recurrent neural networks based on LSTM [17] or GRU [8], convolutional neural networks, or Generative Adversarial Networks (GAN) [15]. However, the established CBR learning methods such as Instance-Based Learning [4] can be used as well. A combination by using a subset of these technologies is possible too, e.g., in the Repeating feature (see Sect. 3.2).

Context of Usage. Learn can be used in context of classification of cases into different categories for later saving in this category's case base (if the application or approach make use of distributed architecture with multiple case bases). It can also be used in the CBRs usual context of judging if the case is save-worthy or not. On application's entry point level Learn can separate the initial case set into different clusters based on separations conducted for similar case sets. Intermediately, Learn can function as a bridge between other modules (i.e., activated each time after Find, Explain, or Adapt) to draw conclusions from the results of the respective step and inform the subsequent module about its findings in order to influence its behavior and produce a more reasonable outcome.

3.6   Explain

The current development of the research area of artificial intelligence emphasizes the increasing requirement of the AI systems to include a component or module that implements an algorithm that is able to explain the behavior and decisions of the system during the current operation. Based on this increasing need, a research area of Explainable AI (XAI) has emerged and resulted in a number of corresponding research initiatives, such as the well-known DARPA Explainable AI program and a multitude of thematic workshops, such as IJCAI XAI workshops [5] or the XCI workshop [27]. One of the most recent research works, a comprehensive guide to XAI in machine learning applications by Molnar [24], shows the future directions of XAI and provides instructions and best practices for its implementation. Furthermore, a survey on XAI [3] can be examined.

In case-based reasoning, the requirement for proper explanations was examined before the 4R cycle was first introduced [1]. However, the development of the XAI area in CBR concentrated mostly on simplistic explanations (i.e., the why- and how-explanations), as pointed out in foundational issues of explanations in CBR [33]. The cognitive explanations, that extend their explanation space beyond the simplistic questions and whose task is, inter alia, to provide answers to the questions of relations between results of operations (e.g., retrieval) or contextual connections between concepts, were rarely implemented. This problem extends to the most systems and approaches from other AI domains. Being very complex systems (which especially relates to the decision support systems), they are not always able to explain their behavior to the user. The reasons for this are either the non-transparent reasoning process, which could be implemented on purpose this way (e.g., in a black-box API), or a system design that does not allow for explanation interfaces to be implemented and/or connected. The latter problem is more common: many systems were constructed without having in mind the importance of an explicit explanation module, i.e., conceptualized and implemented before XAI emphasized the importance of such functionalities.

With FLEA-CBR's *Explain*, we intend to provide the systems that use the methodology as their underlying structure, with a high-level approach that can combine, select, and apply different types of explanations in order to compose an algorithm that can give insights into the behavior of this system. *As a standalone component, Explain is the main structural difference between 4R and FLEA-CBR.* Its inclusion in the methodology as a particular step of the reasoning process is based on the XAI importance described above. The main goal of Explain is to build connections that can provide a structure for exchange of explainable data between the modules and sequences of a FLEA-CBR implementation and the user. In particular, we differentiate between the *internal* and *external* types of explanations that can be handled and produced by Explain:

- *Internal* – This type of explanations is intended for internal use between the components. The data transferred within these explanations does not have to be human-readable, as its goal is to provide the receiving module with a foundation for its reasoning process.
- *External* – The explanation data of this type is delivered to the user providing her with the corresponding information. It is up to the system designer as well as depends on the domain how the explanations will be presented. Many of the systems use textual explanations (e.g., Roth-Berghofer and Bahls [34], see Sect. 2), however, graphical or audiovisual explanations can be produced too. The most important fact to have in mind is that XAI is a matter of user experience and system usability too, therefore, explanations should be designed in a way that is most familiar to the target group.

Both explanation types can be *final* as well as *temporal*. The final explanations provide a finite state where the expression or data cannot be modified anymore and represents the final output of the *Explain* module implementation. This type can be used for static modules that do not update their state with new results, i.e., do not run iteratively to improve the results, e.g., use only the Mixing feature (see Sect. 3.3). In contrast to the final ones, the temporal explanations can be changed over time and are more suitable for systems that are able to run iteratively or apply a number of concurrent processes. Hence, the temporal type allows for continuous update of explanations making the systems more dynamic and their mode of operation more asynchronous. Such approaches would most likely use either Sequencing or Repeating features for their systems.

Context of Usage. The necessity to use Explain can be found at any point in the runtime of the system where a component needs to send or receive data required for justification or transparency of its actions. In structural CBR systems, for example, local similarities on the attribute level can be used to produce a detailed transparency report that makes clear how the overall similarity value for the case has been calculated. This report can be filled iteratively with temporal explanations constructed for different similarity measures. In textual CBR systems, Explain can be used for labeling of the most important text passages providing, for example, the Learn module with information about those passages in order to learn the class of the text or transfer the labeling information to the current domain. As a standalone component, Explain can also be used separately, e.g., to explain differences between selected cases.

3.7   Adapt

Adaptation of selected cases, being a core feature of case-based reasoning, has also found its way to FLEA-CBR. Similarly to Find and Learn, the step *Adapt* inherits the original task of its approximate 4R equivalent Reuse, i.e., the modification of cases according to the adaptation rules, and extends it with additional operations to catch up with the currents trends in AI. Adapt unifies methods for both transformation-based and generative adaptation [37] together with the data augmentation, completion, and generation approaches.

Additionally, Adapt is responsible for automatic revision of transformed cases, i.e., it imitates the Revise step of the 4R cycle in order to rate the helpfulness of adaptation and to decide if the transformation has a potential to solve the given problem. For this special case, a combination of Adapt and Explain seems reasonable: with Explain's abilities to construct a human-readable expression an explanation can be delivered to the user informing her about possible solution transformation and application.

Context of Usage. Besides the classic usages of Reuse named above, Adapt can be used for the currently very frequently executed tasks of data augmentation and data generation. Both tasks are required in applications where the available amount of data is small and can be extended and/or enhanced with augmentation or generation to produce the amount required for application to work properly or to be able to conduct experiments with a prototype of the approach.

4   Example Usages

The following examples intend to demonstrate how the dynamic structure of FLEA-CBR would (in theory and practice) outperform the static structure of 4R on problem spaces that are suitable for solving with the CBR methods.

4.1   CBR and Creativity

The first example and at the same time the showcase of FLEA-CBR is the problem space of the *CBR and creativity* domain. Among the CBR community, this notation normally subsumes the application of CBR to the human creativity-related tasks, such as architecture [31] and computer-aided design (CAD) [26] (both are usually summarized as case-based design, CBD), cooking [10], knitting [30] or similar domains. The common problem space of all these sub-domains of CBR-based creativity can be narrowed to the following characteristics:

1. *The complexity of cases* – Mostly, the cases available in the case bases of the creativity domain applications and approaches have a complex, very often nested, attribute-value-based structure. Knowledge described with these cases is comprehensive and many cases differ only in small details. An example of such case is a CAD model that consists of many parts combined together or a floor plan with many differently shaped rooms. As a result of such complexity, the 4R cycle may take a very long time to finish its process.
2. *Adaptation is not available* – The use of CBR knowledge containers (case bases, adaptation knowledge, similarity measures, vocabulary) [32] qualifies an approach as CBR-based. However, quite often the adaptation knowledge is not available in CBR creativity applications due to the lack of the adaptation feature. Such approaches are often restricted to retrieval only leaving the user without creative solution recommendations or an adaptation to the styles of other designers (style transfer).

FLEA-CBR's ability to run tasks in parallel and partially can solve such problems without being modified or adapted specifically, the only important fact to have in mind is that a suitable configuration of FLEA-CBR is required.

For example, using a configuration with the Sub-cycling feature, the FLEA-CBR-based application can divide the execution of the whole cycle into separate sub-cycles in order to provide each of these sub-cycles with a part or a layer of the complex case. The sub-cycles would run concurrently and asynchronously and speed up the process, without waiting for the previous process to finish. For each part or layer a modification of the case into other representation can be provided as well using it as input for an adaptation approach that can work with this type of data. More specifically, the Generative Adversarial Nets (GAN) [15], as an artificial neural network type that was specifically created for generation of objects that would appear real to a human and also accepted as such, can be used as an adaptation approach for creativity cases, provided that it receives input data in the proper format. As adaptation knowledge the data transferred with Learn from other domains and accepted generated objects can be used. Additionally, Learn can suggest the proper next actions of the current creativity process. In Fig. 3, an example implementation of FLEA-CBR for the creativity domain is shown. This example is based on the only existing and evaluated [13,35] implementation of FLEA-CBR: namely, in the framework MetisCBR for support of floor plan design process (see also Sect. 3.2). The figure demonstrates the current state and the planned further development of the framework.

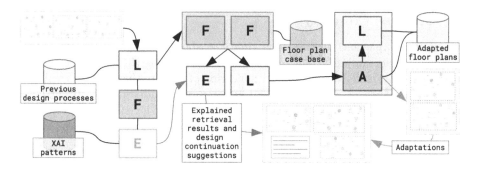

Fig. 3. FLEA-CBR implementation in MetisCBR [7]. The modules that are currently in early stage of development are represented with a slight transparency. The dark gray group with Learn and Adapt, that produces possible design evolutions with GAN, represents the Sub-cycling feature. The double Find group represents Repeating.

4.2 Library Service Optimization

The second example demonstrates how the optimization of the scientific library service process can be achieved with a FLEA-CBR-based software architecture. For demonstration purposes, an everyday process of recording of bibliographic items, e.g., scientific monographies, and the following item borrowing process from the library's stock will be used. We do not assume a specific case representation, but the case itself, i.e., the bibliographic item to be recorded, can be incomplete or in foreign language. The general tasks of the planned system in this particular case can be summarized as follows:

1. *Automatic translation of the item* – If the item is not in the language(s) of library's country then it should be translated for correct keyword extraction and, if required, transliterated to be readable by the library staff and users.
2. *Looking for similar items in case bases of items* – This task subsumes the search for similar items and the subsequent completion of the given item, if some information is missing.
3. *Entry and search in the knowledge graph* – Usually, a catalogue search would be used instead, however, taking the current scholarly information processing trends into account, we replace it with a knowledge graph.

Additionally, the XAI features, that provide the results of the tasks named above with human-understandable justifications to receive confirmations or corrections from the user and to proceed with the next step, can also be used.

The task of the FLEA-CBR implementation is to guide the human operator during the item recording process: suggest the best way of transliteration based on the item's language, fill up the missing information based on similar previous cases, insert information into the knowledge graph, suggest catalogue categories and a signature, and track the item's status (borrowed, not returned etc.). Figure 4 demonstrates the possible FLEA-CBR configuration for this use case.

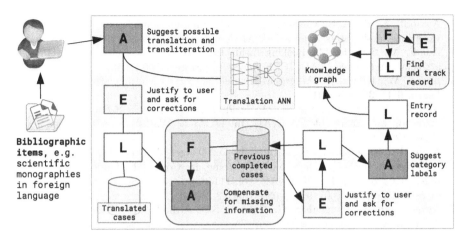

Fig. 4. FLEA-CBR configuration example for a library service optimization. The pink group represents the Sequencing feature where information about incomplete parts can be transferred from other librarian records categories (e.g., periodical publications). The green group represents Sub-cycling, i.e., searching and status tracking of the record. (Color figure online)

5 Conclusion and Future Work

In this paper, we presented FLEA-CBR, an alternative approach to the well-established 4R (Retrieve, Reuse, Revise, Retain) cycle. FLEA-CBR's core components Find, Learn, Explain, and Adapt can be executed in order desired or suitable for the current domain or task, unlike the 4R steps that require the execution in the predetermined order, starting with Retrieve. FLEA-CBR's goal is to solve this flexibility issue, that the 4R itself and its contemporary alternatives and modifications possess. Furthermore, the FLEA methodology makes the XAI component an equivalent and ubiquitous part of the CBR cycle. Additionally, FLEA-CBR offers a number of features that extend the flexibility of the cycle: Mixing, Sub-cycling, Sequencing, and Repeating. The suitability of FLEA-CBR for use as the underlying structure for CBR approaches was demonstrated in an evaluated existing approach for CBR-based creativity and a theoretical application for the library service domain. Future plans for FLEA-CBR include, for example, an implementation of a symbolic programming framework for prototyping of applications that are based on the methodology and conducting of new experiments for existing implementations to further confirm the methodology's suitability for use as the underlying software architecture.

References

1. Aamodt, A.: Explanation-driven case-based reasoning. In: Wess, S., Althoff, K.-D., Richter, M.M. (eds.) EWCBR 1993. LNCS, vol. 837, pp. 274–288. Springer, Heidelberg (1994). https://doi.org/10.1007/3-540-58330-0_93

2. Aamodt, A., Plaza, E.: Case-based reasoning: foundational issues, methodological variations, and system approaches. AI Commun. **7**(1), 39–59 (1994)
3. Adadi, A., Berrada, M.: Peeking inside the black-box: a survey on explainable artificial intelligence (XAI). IEEE Access **6**, 52138–52160 (2018)
4. Aha, D.W., Kibler, D., Albert, M.K.: Instance-based learning algorithms. Mach. Learn. **6**(1), 37–66 (1991)
5. Aha, D., Darrell, T., Pazzani, M., Reid, D., Sammut, C., Stone, P.: IJCAI-17 workshop on explainable AI (XAI). In: IJCAI-17 Workshop on Explainable AI (XAI) (2017)
6. Amin, K., Kapetanakis, S., Althoff, K.-D., Dengel, A., Petridis, M.: Answering with cases: a CBR approach to deep learning. In: Cox, M.T., Funk, P., Begum, S. (eds.) ICCBR 2018. LNCS (LNAI), vol. 11156, pp. 15–27. Springer, Cham (2018). https://doi.org/10.1007/978-3-030-01081-2_2
7. Ayzenshtadt, V., Langenhan, C., Bukhari, S.S., Althoff, K.D., Petzold, F., Dengel, A.: Thinking with containers: a multi-agent retrieval approach for the case-based semantic search of architectural designs. In: Filipe, J., van den Herik, J. (eds.) 8th International Conference on Agents and Artificial Intelligence (ICAART-2016), Rome, Italy, 24–26 February. SCITEPRESS (2016)
8. Cho, K., et al.: Learning phrase representations using RNN encoder-decoder for statistical machine translation. arXiv preprint arXiv:1406.1078 (2014)
9. Chollet, F., et al.: Keras: Deep learning library for theano and TensorFlow (2015). https://keras.io
10. Cordier, A., et al.: Taaable: a case-based system for personalized cooking. In: Montani, S., Jain, L. (eds.) Successful Case-Based Reasoning Applications-2, pp. 121–162. Springer, Heidelberg (2014). https://doi.org/10.1007/978-3-642-38736-4_7
11. De Paz, J.F., Bajo, J., González, A., Rodríguez, S., Corchado, J.M.: Combining case-based reasoning systems and support vector regression to evaluate the atmosphere-ocean interaction. Knowl. Inf. Syst. **30**(1), 155–177 (2012)
12. Eremeev, A.P., Vagin, V.N.: Common sense reasoning in diagnostic systems. In: Efficient Decision Support Systems-Practice and Challenges From Current to Future. IntechOpen (2011)
13. Espinoza-Stapelfeld, C., Eisenstadt, V., Althoff, K.-D.: Comparative quantitative evaluation of distributed methods for explanation generation and validation of floor plan recommendations. In: van den Herik, J., Rocha, A.P. (eds.) ICAART 2018. LNCS (LNAI), vol. 11352, pp. 46–63. Springer, Cham (2019). https://doi.org/10.1007/978-3-030-05453-3_3
14. Finnie, G., Sun, Z.: R5 model for case-based reasoning. Knowl.-Based Syst. **16**(1), 59–65 (2003)
15. Goodfellow, I., et al.: Generative adversarial nets. In: Advances in Neural Information Processing Systems, pp. 2672–2680 (2014)
16. Grace, K., Maher, M.L., Wilson, D.C., Najjar, N.A.: Combining CBR and deep learning to generate surprising recipe designs. In: Goel, A., Díaz-Agudo, M.B., Roth-Berghofer, T. (eds.) ICCBR 2016. LNCS (LNAI), vol. 9969, pp. 154–169. Springer, Cham (2016). https://doi.org/10.1007/978-3-319-47096-2_11
17. Hochreiter, S., Schmidhuber, J.: Long short-term memory. Neural Comput. **9**(8), 1735–1780 (1997)
18. Hohimer, R., Greitzer, F.L., Noonan, C.F., Strasburg, J.D.: Champion: intelligent hierarchical reasoning agents for enhanced decision support. In: STIDS, pp. 36–43 (2011)
19. Holt, A., Bichindaritz, I., Schmidt, R., Perner, P.: Medical applications in case-based reasoning. Knowl. Eng. Rev. **20**(3), 289–292 (2005)

20. Hunt, J.: Evolutionary case based design. In: Watson, I.D. (ed.) UK CBR 1995. LNCS, vol. 1020, pp. 17–31. Springer, Heidelberg (1995). https://doi.org/10.1007/3-540-60654-8_19
21. Kim, S., Shim, J.H.: Combining case-based reasoning with genetic algorithm optimization for preliminary cost estimation in construction industry. Can. J. Civ. Eng. **41**(1), 65–73 (2013)
22. Leake, D.B.: Case-Based Reasoning: Experiences, Lessons and Future Directions. MIT Press, Cambridge (1996)
23. Lees, B., Corchado, J.: Neural network support in a hybrid case-based forecasting system. In: Case-Based Reasoning Integrations, pp. 85–90 (1998)
24. Molnar, C.: Interpretable Machine Learning (2019). https://christophm.github.io/interpretable-ml-book/
25. Navarro, M., De Paz, J.F., Julián, V., Rodríguez, S., Bajo, J., Corchado, J.M.: Temporal bounded reasoning in a dynamic case based planning agent for industrial environments. Expert Syst. Appl. **39**(9), 7887–7894 (2012)
26. Ociepka, P., Herbuś, K.: Application of the CBR method for adding the process of cutting tools and parameters selection. In: IOP Conference Series: Materials Science and Engineering, vol. 145, p. 022029. IOP Publishing (2016)
27. Pereira-Fariña, M., Reed, C.: Proceedings of the 1st workshop on explainable computational intelligence (XCI 2017). In: Proceedings of the 1st Workshop on Explainable Computational Intelligence (XCI 2017) (2017)
28. Prentzas, J., Hatzilygeroudis, I.: Combinations of case-based reasoning with other intelligent methods. Int. J. Hybrid Intell. Syst. **6**(4), 189–209 (2009)
29. Reinartz, T., Iglezakis, I., Roth-Berghofer, T.: Review and restore for case-base maintenance. Comput. Intell. **17**(2), 214–234 (2001)
30. Richards, P., Ekárt, A.: Supporting knitwear design using case-based reasoning. In: Proceedings of the 19th CIRP Design Conference-Competitive Design. Cranfield University Press (2009)
31. Richter, K.: Augmenting Designers' Memory: Case-based Reasoning in Architecture. Logos-Verlag, Berlin (2011)
32. Richter, M.M.: Knowledge containers. In: Readings in Case-Based Reasoning. Morgan Kaufmann Publishers, San Francisco (2003)
33. Roth-Berghofer, T.R.: Explanations and case-based reasoning: foundational issues. In: Funk, P., González Calero, P.A. (eds.) ECCBR 2004. LNCS (LNAI), vol. 3155, pp. 389–403. Springer, Heidelberg (2004). https://doi.org/10.1007/978-3-540-28631-8_29
34. Roth-Berghofer, T.R., Bahls, D.: Code tagging and similarity-based retrieval with myCBR. In: Bramer, M., Petridis, M., Coenen, F. (eds.) International Conference on Innovative Techniques and Applications of Artificial Intelligence, pp. 19–32. Springer, London (2008). https://doi.org/10.1007/978-1-84882-171-2_2
35. Sabri, Q.U., Bayer, J., Ayzenshtadt, V., Bukhari, S.S., Althoff, K.D., Dengel, A.: Semantic pattern-based retrieval of architectural floor plans with case-based and graph-based searching techniques and their evaluation and visualization. In: 6th International Conference on Pattern Recognition Applications and Methods (ICPRAM 2017), Porto, Portugal, 24–26 February 2017 (2017)
36. Skjold, K., Øynes, M.S.: Case-based reasoning and computational creativity in a recipe recommender system. Master's thesis, NTNU (2017)
37. Wilke, W., Bergmann, R.: Techniques and knowledge used for adaptation during case-based problem solving. In: Pasqual del Pobil, A., Mira, J., Ali, M. (eds.) IEA/AIE 1998. LNCS, vol. 1416, pp. 497–506. Springer, Heidelberg (1998). https://doi.org/10.1007/3-540-64574-8_435

Lazy Learned Screening for Efficient Recruitment

Erik Espenakk[1], Magnus Johan Knalstad[1], and Anders Kofod-Petersen[1,2(✉)]

[1] Department of Computer Science,
Norwegian University of Science and Technology, 7491 Trondheim, Norway
erik.espenakk@seaonics.com, magnus@knalstad.no
[2] The Alexandra Institute, Njalsgade 76, 3. sal, 2300 København S, Denmark
anders@alexandra.dk
https://www.ntnu.edu/idi
https://alexandra.dk/uk

Abstract. The transition from traditional paper based systems for recruitment to the internet has resulted in companies in getting a lot more applications. A majority of these applications are often unstructured documents sent over email. This results in a lot of work sorting through the applicants. Due to this, a number of systems have been implemented in an effort to make the screening phase more efficient. The main problems consist of extracting information from resumes and ranking the candidates for positions based on their relevance.

We develop a system that can learn how to rank candidates for a position based on knowledge obtained from earlier screening phases. This Candidate Ranking System (CRS) is based on Case-based Reasoning, combined with semantic data models. The systems performance is evaluated in conjunction with a large international Job company and a software company in an actual recruitment process.

Keywords: Candidate ranking · Human resources · Recruitment

1 Introduction

Traditionally recruitment has been separated into to the following three main steps: (*i*) Sourcing, (*ii*) Screening, and (*iii*) Selection. The sourcing step is where the candidate list gets filled with potential candidates. The sourcing step has traditionally been solved in a passive way, by using ads in papers and more recently over the internet. However exceptions include the use of recruitment agencies as some of these tend to actively seek out candidates via their online persona. Screening is the second step of the recruitment process. This is also the step that we want to focus on in this study.

The screening step is where you have a group of candidates for a job and you want to sort them from most suitable to least suitable. In the area of screening there are several assessment suites and tests that can be performed, each of

K. Bach and C. Marling (Eds.): ICCBR 2019, LNAI 11680, pp. 64–78, 2019.
https://doi.org/10.1007/978-3-030-29249-2_5

these claiming that they can tell something about how the candidate will perform. We will however mostly stick to a slightly earlier phase of the screening, before resources are spent requesting tests. In the job market today, many of the job-applications come in as emails containing CV, cover letter and an application text. Some companies offering attractive positions can get hundreds of applications, some of which are not even relevant. This is where we would like to put our focus, using Case-based Reasoning to do some pre-screening in order to reduce the human resource usage.

2 Related Work

A body of work already exist on the screening process. In addition to different approaches to this there are also existing semantic resources available to help recruiters and automatic systems. This section covers the relevant existing body of work, as well as relevant semantic resources.

2.1 Existing Approaches to Screening

Kessler et al. has over the course of several studies developed the "E-Gen" system, a Natural Language Processing (NLP) and Information Retrieval (IR) system [12–15]. The system uses Support Vector Machines (SVM) to analyse the content of a candidate's email.

Faliagka et al. [10] uses analytical hierarchy process (AHP) to rank candidates based on information from Linkedin and text mining on personal blogs. In [8] they take this further, comparing a set of learning to rank methods in their ability to predict relevance for candidates compared to a human recruiter. In [6] they add a taxonomy to distinguish between certain ICT skills.

Gil et al. [17] presented a solution based on machine learning. Their solution is able to approximate a function for determining the distance between two resumes or a resume and a job description. They define the distance as a set of replacement, deletion and insertion costs on the attributes in a resume. The attributes are identified using the *International Standard of Occupations* (ISCO) taxonomy. Machine learning is used to train a model of these (replacement, deletion, and insertion costs), by comparing a resulting list with a list created by a human recruiter.

Kmail et al. [16] uses NLP and semantic resources to extract and relate candidate concepts from job descriptions and resumes. Another approach is introduced in [18]. It introduces an "Emotional Aptitude Evaluation Module". The idea is to measure the applicants' emotional intelligence based on their twitter posts.

A CBR approach is presented in [21]. In their approach they represent jobs and job seekers as a set of attributes (Gender, Age, Race, State/geographic location, Qualification, Grade point average, etc.). A feature vector representation approach with feature similarity is used. The system was tested by a selection of employers and they were asked to fill out a questionnaire. The results from the questionnaire showed that the employers think that they could be more effective using the proposed system.

In the study by Salazar et al. [20], a CBR system using an ontology as representation is proposed. The study explores the use of the HR-XML standard for representing CVs. In their system, the CBR is implemented as an agent in a multiagent environment. They use a small part of HR-XML to represent a candidate's CV and job offers. This limits their systems ability to do detailed comparisons.

Siting et al. [22] survey different recommendation-system approaches applied in the job domain. These range from systems using collaborative recommendation that recommends jobs that similar profiles have liked, to bilateral recommendation systems that are able to provide recommendation to both job seeker and employer. The systems covered in this study are mostly approaches used in the early job portals in combination with information retrieval techniques.

2.2   Existing Semantic Resources

There are several existing ontologies defining information relevant for the job market. Among these is the *European Skills, Competences, Qualifications and Occupations* (ESCO) ontology has been specifically developed to aid in developing suggestion systems, job search algorithms and job matching algorithms. This ontology includes modules that contain elements such as occupations, knowledge, skills, competencies and qualifications. These modules are combined with the hierarchy specified in the ISCO to form a useful ontology that can be used for classification purposes.

The data model of ESCO is structured into three main pillars: (*i*) The occupations pillar; (*ii*) The knowledge, skills and competencies pillar; and (*iii*) The qualifications pillar. The three pillars are interlinked. Skills can be both attributed to an occupation as required skills, and to a qualification. This makes it easy to query for skills required for a certain occupation, and also makes it easy to query for occupations that require a certain skill.

The occupations pillar should not be confused with jobs, jobs are not covered by ESCO. An occupation is a grouping of jobs that require the same type of skills and involve similar tasks, while a job is a set of tasks and duties meant to be executed by one person [19]. Each occupation is linked to their own set of metadata as well as as an ISCO-08 code. The ISCO-08 code can be used as a hierarchical structure for the occupations pillar.

The ISCO standard divides all occupations into ten major groups. In the current version of the ESCO classification [19], ISCO provides the top four levels for the occupations pillar. ESCO occupations are located at level five and lower.

The skills pillar consists of both knowledge, skills and competencies. There are in total 13,492 skill concepts in ECSO v1. The ESCO classification uses the following definitions for skill, knowledge and competence, taken from [4]: *Knowledge*) The body of facts, principles, theories and practices that is related to a field of work or study. Knowledge is described as theoretical and/or factual, and is the outcome of the assimilation of information through learning; *Skill*)

The ability to apply knowledge and use know-how to complete tasks and solve problems. Skills are described as cognitive (involving the use of logical, intuitive and creative thinking) or practical (involving manual dexterity and the use of methods, materials, tools and instruments); and *Competence*) The proven ability to use knowledge, skills and personal, social and/or methodological abilities in work or study situations, and professional and personal development.

Like for the competence pillar, each skill concept in the skill pillar contains a set of useful metadata that both describe the skill and its context. The data includes the type of skill and a relation to a broader skill. The relationship between knowledge, skills and competences has been captured to a certain degree.

In addition to the meta-data mentioned above, skills are assigned a reusability level. The reusability level can be either *transversal, cross-sector*, or *sector-specific*. Transversal skills are relevant to a broad range of occupations and are often referred to as *core skills*. *Cross-sector* skills are relevant across several economic sectors, but not as general as the transversal. A *sector-specific* skill is used in a specific sector, but can be used in a many occupations within that sector.

The skills hierarchy is divided into five major categories: thinking; language; application of knowledge; social interaction; attitudes and values. Dividing the skills into this hierarchy makes it easier to use ESCO in certain situations, such as in a CV creation situation.

The qualifications pillar is intended to become a comprehensive listing of all the qualifications that are relevant for the European labour market. ESCO includes qualifications both directly and indirectly. The indirect-inclusion is based on data gathered from EU countries' national qualification databases. Some occupations can have mandatory skills on a national level or skills that are optional.

3   Design and Implementation

The implementation presented here approaches the screening problem by implementing a CBR system for ranking candidates applying for jobs. The section details the case representation, the similarity functions, and gives a brief overview of the implemented CBR cycle. MyCBR is used for implementation[1].

3.1   Case Representation

The case-base is represented uses an object-oriented representation. The chosen representation has full support in the MyCBR Framework and is based on two main types of relations (*is-a, part-of*).

The representation is divided into *concepts, attributes* and *relations*. In our representation we have six concepts: *Skill, Language, Education, Occupation, Candidate*, and *Job*. Each of these concepts have their own set of attributes. Figure 1 depicts the case representation.

---

[1] http://www.mycbr-project.org/index.html.

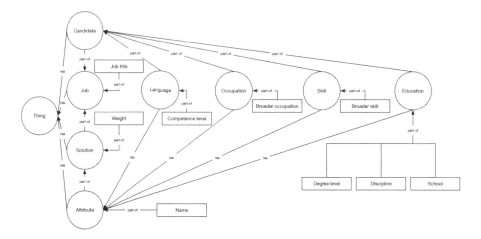

Fig. 1. A diagram of the case representation

The concepts used to build the case representation reflect the main part of the information gathered from candidates linked to the information made available by the ESCO Ontology described above. As an example, each instance of the Skill concept will have been reported in as part of our data collection and it will have a link to a ESCO skill resource URI, if the skill exist in the ESCO Ontology.

3.2 Similarity Functions

The case representation consists of several concepts, each of these concepts has a set of attributes. These attributes are of varying importance when it comes to identifying the concept itself. Due to this, each concept has its own global similarity measure, also called an amalgamation function. Each of these amalgamation functions will be used to collect the similarity of each of the concept's attributes into a global similarity measure.

In Fig. 2 the amalgamation function of the Skill concept is depicted. The purpose of this function is to combine the two local similarity function of Skills attributes. Skill contains two string attributes, each of these use a cosine string similarity function. Cosine similarity was chosen to limit the variables involved in the systems performance, allowing focus on the gains from lazy learning (CBR). Techniques such as NLP would potentially improve the system overall. From the `SkillAmalgamation` function you can see that the skill attributes are weighted differently, with name being twice as important as `broaderSkill`. From ESCO we have that each Skill can have a broader-skill, this broader-skill property is shared by many similar skills, some more similar than others, therefore the broaderSkill attribute gets a lower weight.

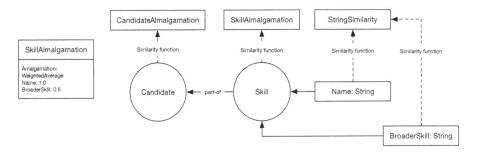

Fig. 2. Amalgamation function combining local similarity measures using weighted average

Local Similarities. The concepts in the case representation (see Fig. 1) each have their own set of attributes. The attributes consist of the types String, Symbol, Integer and using a concept as an attribute. The attributes with Integer as type is part of the solution concept. We do not include the solution itself into the similarity measure during retrieval. The other attribute types are String, Symbol and Concept, String attributes use a string similarity function based on cosine similarity with trigrams. Concepts used as attributes will convey their local similarity measures through their amalgamation function.

For some of our attributes we setup and configure unique symbolic similarity function. This is true for the **degreeLevel** attribute of the education concept, a screendump of this symbolic function can be seen in Fig. 3.

We structured the symbolic function for the degree level to prefer job queries looking for lower degrees when a similar level is not available. In the event that a company is looking for a Phd to fill a vacancy they will generally have a slightly different approach compared to when looking for a Master's candidate. For instance: grades and academic accomplishments will usually not be used to differ between Phd level candidates, while this can still be important when looking at candidates at the Master's level. We decided to reflect this in our symbolic function by making it asymmetric and weighting BSc and MSc closer compared to the Phd level.

When modeling the symbolic similarity function of the level of language competency it decided to give solutions with a higher competency criteria a slight bias, should the queried level not exist. This is reflected in the symbolic function seen in Fig. 4.

Symmetry ○ symmetric ● asymmetric

	Bachelor	Doctor	Master
Bachelor	1.0	0.4	0.7
Doctor	0.0	1.0	0.3
Master	0.5	0.7	1.0

Fig. 3. A screendump of the MyCBR Workbench modeling the degree level symbolic function

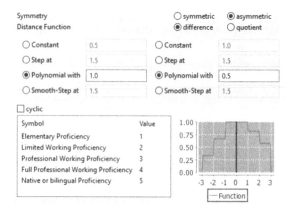

Fig. 4. A screendump of the MyCBR Workbench modeling the language level symbolic function

Table 1. Amalgamation function Education concept

Attribute	Weight
Disipline	1.0
Degree level	0.5
School	0.25

Table 2. Amalgamation function Skill concept

Attribute	Weight
Name	1.0
BroaderSkill	0.5

Table 3. Amalgamation function Language concept

Attribute	Weight
Name	1.0
LanguageLevel	0.75

Table 4. Amalgamation function Occupation concept

Attribute	Weight
Name	1.0
BroaderOccupation	0.5

Global Similarities. All the concepts in our representation that has more than one attribute requires some form of global similarity measure. The global similarity measure is used to combine the local similarities, in this project weighted average has been used with adequate results. Each of the concepts will have their own set of tweaked weights, this weight is based on the current knowledge of the authors as well as some testing in MyCBR Workbench.

The Education concept has three attributes, (degree level, discipline, and school). Discipline was found to be the most important attribute, with degree level coming in second. School could be important for some companies, therefore it is left in the model. Table 1 displays a table of the weights in the education amalgamation function.

Table 2 describe the skill amalgamation function. The `BroaderSkill` attribute has only one third of the total weight. This way a similar skill name will have much higher priority than one that only matches a broader skill, and the instance where both name and broader match have the highest similarity. Both the skill name and the broader skill use our default string similarity function.

Finally, Tables 3 and 4 describe the amalgamation functions for the Language concept and Occupation concept, respectively.

3.3   The CBR Cycle

This implementation follows the traditional four step CBR process of: Retrieve, Reuse, Revise, and Retain [1].

Retrieval is carried out using the similarities described above. It is implemented using the MyCBR library.

Reuse is carried out as a process that resembles direct reuse but with some simple adaption rules.

When the adaptation starts the first step involves stripping the case for superfluous attributes. As an example, if we query for a job with the skills "Programming" and "User interface design", if the closest case also contains a entry for Language "English, Native or bilingual proficiency" then the adaption will remove this superfluous language attribute. This leaves the two skills and their respective weights.

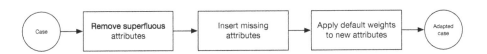

Fig. 5. Diagram of the adaption behaviour

The adaptation mechanism deals with models, it takes as input the model used as the query and the model retrieved. It follows the behaviour in Fig. 5. If the case being adapted has a low similarity, below 50% then the adaptor attempts to query ESCO for suggestions. This is only possible if the job query has an ESCO occupation as part of the experience job criteria.

Revision. The revision phase of the CBR system has been separated out from the core implementation and is implemented client-side. The revision process itself has borrowed some ideas from the Conversational CBR retrieval/query building phase. We hypothesised that this is a good solution considering that in a sense our cases are "filter settings/ranking strategies" and to improve/edit a ranking strategy requires some visual feedback in how the list of ranked candidates change depending on the strategy used. Therefore a "Conversational" approach is implemented to guide the users into creating or editing a case, which is the strategy/weights used to rank the candidates for the relevant job position.

Retention. In the initial training and experimentation phase of this project, the size of the case base is not expected to be issue. Thus, all revised and validated cases that are not already 98% equal to an already existing case are being retained.

4   Test and Evaluation

The evaluation plan includes training the system, testing it on real world recruitment data, and comparing the results with the results of a human expert. This method of evaluation has been used with other studies [6,7,9,11,18], all of these focusing on ranking/screening candidates in recruitment. Mentioned by [16], the weakness of this is that candidates selected by the human expert does not necessarily represent the best list. To explore this further, an additional informal experiment has been designed, inspired by the Turing test [23].

The Turing test inspired experiment is simple, human recruiter is given a job description and some lists of candidates. They are then asked to rank these lists based on the suitability of their proposed candidates and also with a focus on the list's ranking. Participants in a Turing test does not know if they are chatting with an AI, in our test the recruiters do not know which of the lists are created by an algorithm.

The experiments are all carried out in cooperation with a large international company working with job listing and career information, denoted *job company*; and a national software company as part of their recruitment process, denoted *business*. The collection module was integrated into the platforms of two companies. Combined a total of 476 candidates were collected using the collection module, distributed in two separate databases.

4.1   Setup

The ranking system is evaluated against data collected from real and simulated recruitment scenarios. The system compares against the information retrieval ranking algorithm Okapi BM25, as suggested in [11]. The Okapi BM25 is a good baseline for determining the performance of learning techniques as this is one of the common bag of words based information retrieval techniques. Okapi BM25 is readily available, provides consistent results and has been used as a baseline in previous studies.

In a study by Arnulf et al. [2], it was found that the layout of CVs had a measurable degree of influence on the decision of recruiter. They found that for the same candidate, when their resume was structured in a formal way the candidate was twice as employable as compared to when their resume had a more creative layout. Because our system does have access to the same layout information we have decided to remove resumes that do not follow the formal layout described in [2].

We have also decided to remove information about gender, age and ethnicity to limit the amount variables. The field study [3] found indications of a gender bias in resume evaluation. Findings from the study [5] indicates the existence of an ethnicity bias in the screening of resumes.

In order to have expertise evaluation, we contacted and chose a variety of recruiters that had experience with screening and scoring candidates. However, there's not a single answer key for each scoring. We therefore asked each of the evaluators to write down which list they found most relevant. In addition, they should also write a comment to justify their selection. The last step we asked them for was to give us their subjectively opinion of how many of the candidates from each list they found to be relevant for the job position.

4.2   Experiment 1

A set of 40 jobs where taken from *job company's* platform. These jobs were selected within categories matching the most common education fields among the candidates collected through the *job company*. From the set of 40 jobs, a subset of 20 was chosen at random. The training module was then used by a recruiter to create job queries and revise these according to the criteria found in the 20 job ads.

The data-set used consists of four job positions selected by the authors from the list of 20 jobs not used in training the system. For each job position we picked 30 applicants out of the 476 candidates harvested through the *job company*. Resulting in four *screening packages*, each containing a job ad bundled with 30 applicants.

The four *screening packages* were distributed to four recruiters, each recruiter belonging to a different company. From each of these *screening packages* a list was produced by respective recruiter ranking the top 5 of the 30 applicants in regards to the bundled job ad.

The same *screening packages* were also evaluated by our CRS, BM25 and Random. Producing a total of four top 5 applicant lists per each of the four jobs. A fifth list was added, this list was the top 5 applicants produced by the recruiter and rearranged by the CRS. The resulting set of 4 jobs each bundled with 5 different lists of top 5 screened applicants formed our *evaluation packages*.

The *evaluation packages* were sent out to the 4 companies participating in the experiment. Each company were sent 3 of the 4 *evaluation packages*, the excluded package containing the applicant list produced by a recruiter in the company. Each of the Evaluation packages were bundled with a link to an online form, enabling several recruiters from each company to participate.

An evaluation consisted of ranking the 5 lists of screened candidates from best to worst, taking into consideration both the quality of the candidates and the internal ranking in the list itself.

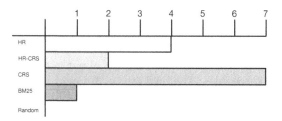

Fig. 6. Results of the different ranking approaches

4.3   Experiment 2

Our *business* contact agreed to facilitate a real world recruitment scenario for us. To this end they gave us access to the applicants for a position that they were hiring for, the job description and the shortlist of top 25 candidates that resulted after screening the 170 applicants. The job in question is a "Business Intelligence Consultant" position.

The candidate ranking system was trained using a six month old recruitment scenario in which the *business* hired a Data Analyst. The authors manually filled in the resumes into the data collection module. Four months later the *business* was hiring again, this time for a Business Intelligence Consultant. The authors manually filled in the candidates CVs into the data collection module.

The original plan included using the entire corpus of 170 applicants in order to measure precision and recall for top25. However due to time constraints a compromise was made and only the top 25 candidates were used. To compensate, the focus was shifted to the ordering of the list. The order of applicants in the resulting lists are compared using both Spearman correlation and Discounted Cumulative Gain (DCG). Using DCG the order of the most relevant candidates account for more score than the rest.

5   Results and Discussion

5.1   Experiment 1

The combined results from all the evaluations can be seen in Fig. 6. As we expected, the random list did not get favoured in any evaluation, we also expected the HR list and the HR-CRS list to be relatively close. The combined results seem to indicate that human recruiters would not necessarily prefer lists created by other recruiters when stacked up against the CRS and Okapi. We see that our CRS was placed first of the five lists (CRS, HR, BM25, CRS-HR, Random) in 7 out of 14 evaluations for 4 different job cases, each of these cases containing a list created by a unique human recruiter. In second place, we had the HR.

The evaluators were asked in their opinion the number of relevant candidates in each list. The number of relevant candidates over 14 evaluations: CRS with 53, HR with 52, BM25 came out a bit lower with 40 relevant.

5.2   Experiment 2

Since the Spear-man Correlation coefficient does not distinguish the distance between two positions in the ranking, we opted to utilize Discounted Cumulative Gain. This method allows for several shades of relevance. The top 10 candidates of the 25 had been called in for interviews. These were given a descending rank from 10 to 1. The remaining 15 would account for 1 each.

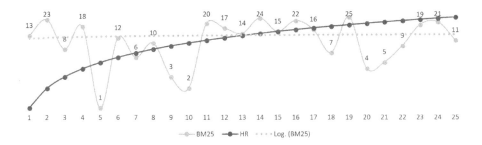

Fig. 7. The correlation between Okapi BM25 and the Recruiter

Figure 7 shows the spearman correlation between the recruiter and Okapi BM25. As we can see from the figure, Okapi shows in several cases to be close for some positions. However, the coefficient turned out to be 0.1284. Comparing the results with the CRS, we see from Fig. 8 that the correlation is stronger. The CRS achieved a score of 0.65.

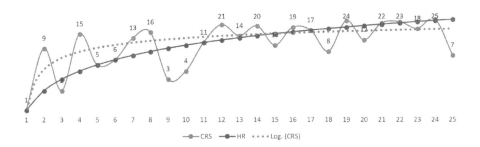

Fig. 8. The correlation between CRS and the Recruiter

Based on the correlation, we see that the CRS using CBR with modest training and naïve case adaptation is able to produce promising results when compared to Okapi BM25. From Fig. 9 we see that the CRS scored a 80% versus the BM25 score of 60%, the CRS was able to rank applicants deemed suitable by the recruiter higher than BM25.

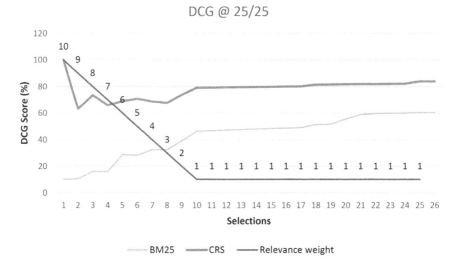

Fig. 9. Discounted Cumulative Gain: Okapi BM25, CRS and Recruiter

6 Conclusion and Future Work

In this research we have developed, deployed, and evaluated our Candidate Ranking System based on CBR. We have evaluated the system by comparing it to Okapi BM25 as a baseline. The results gathered running our experiments show the CRS consistently outperforming Okapi BM25. The amount of data gathered is not ideal for drawing conclusions based on statistical significance. However the results were produced with only a modest amount of training and using a naïve case adaptation implementation. This shows promise for using lazy learning based techniques in candidate ranking applications. Lazy learning techniques also have the advantage of not requiring a large amount of data to train upfront.

Though our CRS system achieved a higher overall evaluation than that of the human experts. The results if anything, indicates that the evaluators did not distinguish the lists produced by HR from the ones produced by CRS.

For future work we would like to look at the case description used in our implementation, which does not consider directly the seniority of the position or the company hiring. Expanding the case description would increase the system's ability to locally optimise. The current implementation requires integration in the recruitment process. As a part of future work an approach using NLP to extract information from CV's could be used. A comparison of different case adaptation techniques as the current adaptation is quite naïve.

References

1. Aamodt, A., Plaza, E.: Case-based reasoning; oundational issues, methodological variations, and system approaches. AI Commun. **7**(1), 39–59 (1994)
2. Arnulf, J.K., Tegner, L., Larssen, Ø.: Impression making by résumé layout: Its impact on the probability of being shortlisted. Eur. J. Work Organ. Psychol. **19**(2), 221–230 (2010)
3. Cole, M.S., Feild, H.S., Giles, W.F.: Interaction of recruiter and applicant gender in resume evaluation: a field study. Sex Roles **51**(9), 597–608 (2004)
4. European-Parliament: European qualifications framework. Official Journal of the European Union (2008)
5. Derous, E., Ryan, A.M., Nguyen, H.H.D.: Multiple categorization in resume screening: examining effects on hiring discrimination against Arab applicants in field and lab settings. J. Organ. Beh. **33**(4), 544–570 (2012)
6. Faliagka, E., et al.: On-line consistent ranking on e-recruitment: seeking the truth behind a well-formed CV. Artif. Intell. Rev. **42**(3), 515–528 (2014)
7. Faliagka, E., Karydis, I., Rigou, M., Sioutas, S., Tsakalidis, A., Tzimas, G.: Taxonomy development and its impact on a self-learning e-recruitment system. In: Iliadis, L., Maglogiannis, I., Papadopoulos, H. (eds.) AIAI 2012. IAICT, vol. 381, pp. 164–174. Springer, Heidelberg (2012). https://doi.org/10.1007/978-3-642-33409-2_18
8. Faliagka, E., Ramantas, K., Tsakalidis, A., Tzimas, G.: Application of machine learning algorithms to an online recruitment system. In: Proceedings of the International Conference on Internet and Web Applications and Services (2012)
9. Faliagka, E., Rigou, M., Sirmakessis, S.: An e-recruitment system exploiting candidates' social presence. Current Trends in Web Engineering. LNCS, vol. 9396, pp. 153–162. Springer, Cham (2015). https://doi.org/10.1007/978-3-319-24800-4_13
10. Faliagka, E., Tsakalidis, A., Tzimas, G.: An integrated e-recruitment system for automated personality mining and applicant ranking. Internet Res. **22**(5), 551–568 (2012)
11. Gil, J.M., Paoletti, A.L., Pichler, M.: A novel approach for learning how to automatically match job offers and candidate profiles. CoRR abs/1611.04931 (2016)
12. Kessler, R., Béchet, N., Roche, M., El-Bèze, M., Torres-Moreno, J.M.: Automatic profiling system for ranking candidates answers in human resources. In: Meersman, R., Tari, Z., Herrero, P. (eds.) OTM 2008. LNCS, vol. 5333, pp. 625–634. Springer, Heidelberg (2008). https://doi.org/10.1007/978-3-540-88875-8_86
13. Kessler, R., Béchet, N., Torres-Moreno, J.-M., Roche, M., El-Bèze, M.: Job offer management: how improve the ranking of candidates. In: Rauch, J., Raś, Z.W., Berka, P., Elomaa, T. (eds.) ISMIS 2009. LNCS (LNAI), vol. 5722, pp. 431–441. Springer, Heidelberg (2009). https://doi.org/10.1007/978-3-642-04125-9_46
14. Kessler, R., Torres-Moreno, J.M., El-Bèze, M.: E-gen: automatic job offer processing system for human resources. In: Gelbukh, A., Kuri Morales, Á.F. (eds.) MICAI 2007. LNCS (LNAI), vol. 4827, pp. 985–995. Springer, Heidelberg (2007). https://doi.org/10.1007/978-3-540-76631-5_94
15. Kesslera, R., Béchet, N., Roched, M., Torres-Morenob, J.M., El-Bèze, M.: A hybrid approach to managing job offers and candidates. Inf. Process. Manag. **48**(6), 1124–1135 (2012)
16. Kmail, A.B., Maree, M., Belkhatir, M., Alhashmi, S.M.: An automatic online recruitment system based on exploiting multiple semantic resources and concept-relatedness measures. In: 2015 IEEE 27th International Conference on Tools with Artificial Intelligence (ICTAI), pp. 620–627 (2015)

17. Martinez-Gil, J., Paoletti, A.L., Pichler, M.: A novel approach for learning how to automatically match job offers and candidate profiles. arXiv:1611.04931 (2017)
18. Menon, V.M., Rahulnath, H.A.: A novel approach to evaluate and rank candidates in a recruitment process by estimating emotional intelligence through social media data. In: 2016 International Conference on Next Generation Intelligent Systems (ICNGIS), pp. 1–6 (2016)
19. Office, I.L. (ed.): International Standard Classification of Occupations (ISCO-08 Standard). International Labour Office (2008)
20. Salazar, O.M., Jaramillo, J.C., Ovalle, D.A., Guzmán, J.A.: A case-based multi-agent and recommendation environment to improve the e-recruitment process. In: Bajo, J., et al. (eds.) PAAMS 2015. CCIS, vol. 524, pp. 389–397. Springer, Cham (2015). https://doi.org/10.1007/978-3-319-19033-4_34
21. Siraj, F., Mustafa, N., Haris, M.F., Yusof, S.R.M., Salahuddin, M.A., Hasan, M.R.: Pre-selection of recruitment candidates using case based reasoning. In: 2011 Third International Conference on Computational Intelligence, Modelling Simulation, pp. 84–90 (2011)
22. Siting, Z., Wenxing, H., Ning, Z., Fan, Y.: Job recommender systems: a survey. In: 2012 7th International Conference on Computer Science Education (ICCSE), pp. 920–924 (2012)
23. Turing, A.M.: Computing machinery and intelligence. In: Epstein, R., Roberts, G., Beber, G. (eds.) Parsing the Turing Test, pp. 23–65. Springer, Dordrecht (2009). https://doi.org/10.1007/978-1-4020-6710-5_3

On the Generalization Capabilities of Sharp Minima in Case-Based Reasoning

Thomas Gabel[(✉)] and Eicke Godehardt

Faculty of Computer Science and Engineering,
Frankfurt University of Applied Sciences,
60318 Frankfurt am Main, Germany
{tgabel,godehardt}@fb2.fra-uas.de

Abstract. In machine learning and numerical optimization, there has been an ongoing debate about properties of local optima and the impact of these properties on generalization. In this paper, we make a first attempt to address this question for case-based reasoning systems, more specifically for instance-based learning as it takes place in the retain phase. In so doing, we cast case learning as an optimization problem, develop a notion of local optima, propose a measure for the flatness or sharpness of these optima and empirically evaluate the relation between sharp minima and the generalization performance of the corresponding learned case base.

1 Introduction

Powered by a number of recent empirical successes, the field of deep learning has become one of the most noticed sub-fields of artificial intelligence during the past few years. Training deep neural networks means solving a complex, non-convex optimization problem. Interestingly, gradient-based optimization of such networks takes place in an over-parameterized setting, where the target function has, in general, a vast number of local and multiple global optima. All of them minimize the train error, but typically many of them generalize poorly. Consequently, just minimizing the train error is merely adequate, since a poorly chosen minimum may bring about bad performance on independent test data. It has been generally accepted that the generalization capabilities do implicitly depend on the algorithm used for minimizing the train error – since that algorithm determines which minimum it gets attracted by – and it is an ongoing debate to which extent properties of the attained minimum can be indicative for generalization.

In this paper, we transfer these thoughts to case-based reasoning. After providing some background and pointers to related work in Sect. 2, we start by searching for a related (optimization) problem in CBR and find one in the retain phase where case base editing and maintenance can naturally be cast as a discrete NP-hard optimization problem. Put simply, the question addressed here is which cases from a given set of train cases shall be retained in the case base

K. Bach and C. Marling (Eds.): ICCBR 2019, LNAI 11680, pp. 79–94, 2019.
https://doi.org/10.1007/978-3-030-29249-2_6

and which not, while optimizing some objective function. In Sect. 3, we model this problem formally and, for measuring an edited case base's competence, we derive an error function around which we center our further analyses made in this paper. As a first contribution, we then develop four variants of two hill-climbing case base editing algorithms (FHC and MHC) which, by design, are attracted by a local optimum of the hitherto derived objective function. The second main contribution of this paper follows in Sect. 4. There, we aim at characterizing an edited case base configuration (and, hence, a possibly attained minimum in the error landscape) as being "flat" or "sharp". In so doing, we derive an appropriate measure of sharpness which, informally speaking, approximates gradient information in the near neighborhood of the edited case base to the extent that this is feasible in the given non-continuous optimization task. The last part of the paper (Sect. 5) is devoted to an empirical evaluation of our approach using a large set of established classification problems. To this end, our hypothesis is that there is some correlation between the sharpness of a case base configuration as defined before and the generalization capabilities of the resulting case-based classifier. A secondary objective also covered in our evaluation concerns the actual empirical power of the hill-climbing case base editing schemes that we proposed in Sect. 3.

2   Background and Related Work

We start by providing some general basics as well as a nearly chronological, brief survey on related work on issues that relate to maintaining case bases. We then adopt an optimization point of view and briefly introduce some foundations on function minimization including the notion of sharp and flat minima.

2.1   Case Base Maintenance and Instance-Based Learning

Case base maintenance (CBM [10]) is known for addressing two goals in case-based reasoning: (1) controlling the number of cases in the case base and, hence, avoiding a performance degradation due to outrageous growth and (2) assuring a high competence of the case base. As a matter of fact, CBM is probably that component of the CBR realm that bears the strongest[1] relation to learning in the classical machine learning sense. Issues of case base maintenance arise early in almost any CBR system which is why this sub-field of CBR has attracted so much attention during the past decades. Starting with the initial proposal of the nearest neighbor rule [3] several authors subsequently aimed at reducing the size of the set of stored instances [5,6,18]. Later, Aha and varying co-authors proposed the family of instance-based learning algorithms (IBL [1]) where new instances are stored subject to different criteria, as well as subtractive counterparts (e.g. [9]). Other approaches more intensively focused on the location of retained instances, such as preferably keeping those near the center of clusters

---

[1] Though learning can take place also in one of the other knowledge containers of a CBR system, e.g. when learning similarity measures or adaption knowledge.

rather than near to decision borders [20]. Another piece of work which – at least from the algorithm's name – suggests to be related to the algorithms we introduce in Sect. 3.4 seems to be random mutation hill-climbing [15]. However, as it turns out our algorithmic approach differs in various ways, most importantly as we do not fix the case base size beforehand, such that the remaining commonality is the hill-climbing nature of the procedure. The family of decremental reduction optimization procedures (DROP [19]) takes the opposite approach and iteratively decides which cases to delete from a case base without deteriorating competence (cf. Sect. 3.4.3 for more details on this). In another line of research, Smyth and varying co-authors approached the problem of case base size limitation using the notion of coverage and reachability of cases [16], concepts that were later exploited by the COV-FP [17] and ICF [2] algorithms. Finally, the idea that case base maintenance represents an optimization problem with multiple goals to be pursued simultaneously appears most pronounced in [14].

2.2 Sharp and Flat Minima of an Error Function

Given some real-valued function f over some domain X, f is said to have a minimum at $m \in X$, if there exists some $\varepsilon > 0$ such that $f(m) \leq f(x) \forall x \in X$ with $d_X(x, m) < \varepsilon$ where d_X measures the distance of x and m. The idea of characterizing minima as sharp or flat has received much attention in the literature especially in the context of the recent rise of deep learning systems [4,8,12], but dating back at least to the seminal work of Schmidhuber [7]. A flat minimum m_f is said to be a point where the function f varies slowly in a relatively large neighborhood of m_f. By contrast, a sharp minimum m_s is a point where the function f increases rather strongly in the near vicinity of m_s. While exact numerical values for measuring the sharpness of a minimum would strongly depend on the scaling of the function f and its inputs as well as on the definition of what a large neighborhood or near vicinity means, it is intuitive that the large sensitivity of a sharp minimum (little changes to m_s yield strong changes to f) may negatively impact the generalization capabilities of the system. Even if an exact numerical score of the sharpness of a minimum may not be as informative in itself, it still may be important when selecting between different minima, e.g. two minima m_1 and m_2 with $f(m_1) = f(m_2)$, but with strongly different levels of sharpness. Figure 1 aims at visualizing this issue conceptually in a one-dimensional space [8]: The function f to be minimized (as it may stem from a set of given training samples) is shown with a solid line, whereas the ground truth g, i.e. the true relation to be learned as it may be represented by a (large) independent set of test data, is depicted with a dashed line. While both, the flat and the sharp minimum have the same training performance, the testing performance, i.e. the generalization capability, of m_s is much worse.

3   Case Base Maintenance as Optimization Problem

In what follows, we aim at an in-depth analysis of what it means for a CBR system to "learn cases" (or to delete some), bridging the gap to issues such as

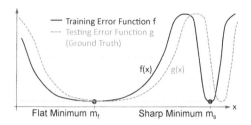

Fig. 1. Visualization of Sharp vs. Flat Minima: The abscissa shows the domain of the search space, the ordinate shows the value of the error function.

optimization, minima found during optimization, generalization, and analyzing the implications for (more or less) established CBM strategies.

3.1   Case Base Editing Problem

Let us denote the set of all cases as $\mathcal{M}$ and, as usual, each case $c = (p, s) \in \mathcal{M}$ be composed of a problem part $p \in P$ and a solution part $s \in S$, i.e. $\mathcal{M} = P \times S$. Now assume, we are given a set $\mathcal{T} \subset \mathcal{M}$ of training cases. Then, the case base editing problem we are focusing on means finding a case base $\mathcal{S}$ as a subset of $\mathcal{T}$ that features as much of the following properties as possible:

- $\mathcal{S} \subset \mathcal{T}$, i.e. $\mathcal{S}$ should be a (desirably small) subset of $\mathcal{T}$ for reasons of retrieval efficiency
- $\mathcal{S}$ should be as competent as possible where competence is typically measured as its problem-solving capability on a disjoint set $\mathcal{U} \subset \mathcal{M}$ of test cases (i.e. $\mathcal{T} \cap \mathcal{U} = \emptyset$)

From a global point of view, this setting gives rise to $2^{|\mathcal{T}|}$ possible configurations for the resulting case base $\mathcal{S}$ since each $c \in \mathcal{T}$ can be either contained or not contained in $\mathcal{S}$, yielding the power set $\mathfrak{P}(\mathcal{T})$ of $\mathcal{T}$ as the search space. Needless to say, that *any* CBM strategy described in the literature performs some kind of search through that space $\mathfrak{P}(\mathcal{T})$, being guided either heuristically or by certain performance criteria.

Definition 1 (Case Base Configuration). *Given training cases $\mathcal{T}$ any non-empty subset $\mathcal{S} \subset \mathcal{T}$ represents a valid case base configuration (CBC) in the context of case base editing, i.e. $\mathcal{S} \in \mathfrak{P}(\mathcal{T})$.*

Accordingly, searching the whole space of all case base configurations is intractable in general, except for toy problems. However, this setting allows us to develop an intuition for local versus non-local changes to a case base.

Definition 2 (Case Editing Operator). *Given some case base configuration $\mathcal{S}$ and a case $c \in \mathcal{T}$, the case editing operator $\mathcal{E} : \mathfrak{P}(\mathcal{T}) \times \mathcal{T} \to \mathfrak{P}(\mathcal{T})$ returns a new case base configuration $\mathcal{S}'$ such that*

$$S' = \begin{cases} S \setminus \{c\} & \text{if } c \in S \\ S \cup \{c\} & \text{else} \end{cases}$$

Clearly, the change that $\mathcal{E}$ introduces to $\mathcal{S}$ is the smallest one possible – we could also say, it is a local change to $\mathcal{S}$ –, since only the membership of a single case c is swapped. By contrast, any case base $\mathcal{S}_1$ can be changed to any $\mathcal{S}_2$ by applying a sequence of such atomic operations where the length of that sequence is, informally, given by the Hamming distance of $\mathcal{S}_1$ and $\mathcal{S}_2$. Note that Definition 2 is not tailored to data sets where a case is contained multiple times in $\mathcal{S}$.

3.2 Introspective Problem-Solving Quality

Different authors have employed different measures for the problem-solving quality of an edited case base. For example, Lupiani et al. [14] define a multi-objective error function combining error, noise, and redundancy which is to be minimized by an evolutionary algorithm, while Smyth and McKenna [16] center competence around the notion of coverage and reachability. A widespread measure for estimating the problem-solving capability of a case base $\mathcal{S}$ is the leave-one-out error (or accuracy) where each case c is used as query once using $\mathcal{S} \setminus \{c\}$ as leave-one-out case base (LOO [11,19]). Throughout the rest of this paper, we stick to a slight modification of this established measure for case base competence due to its simplicity and due to the fact that no further knowledge is needed (e.g. for generating sample solutions) which eases the empirical evaluation.

Definition 3 (Leave-One-Out Train Error of a Case Base Configuration). *For a training set of cases* $\mathcal{T} = \{c_1, \ldots, c_{|\mathcal{T}|}\}$ *with each case* $c_i = (p_i, s_i)$ *consisting of a problem and solution part and for a given case base configuration* $\mathcal{S}$, *the* leave-one-out train error *is defined as*

$$\mathbb{E}_{\mathcal{T}}^{loo}(\mathcal{S}) = \frac{1}{|\mathcal{T}|} \cdot \sum_{i=1}^{|\mathcal{T}|} 1 - Correct(Adapt(Retrieve(\mathcal{S} \setminus \{c_i\}, p_i), p_i), s_i) \quad (1)$$

In that definition, the *Retrieve* function performs case-based retrieval over the leave-one-out case base $\mathcal{S} \setminus \{c_i\}$ using p_i as query. To this end, no restrictions on the retrieval algorithm or the used similarity measures or the value of k in case of a k-nearest neighbor retrieval are made. The *Adapt* function takes the set of nearest neighbors returned by the retrieval and performs adaptation to form a single unique suggested solution or does nothing, if no adaption knowledge is available or necessary. Finally, that returned solution is checked against the solution s_i of case c_i whose problem part p_i was used as query in the first place. If function *Correct* finds that both solutions are identical (or sufficiently identical), it returns 1, otherwise 0. In fact, we will focus on classical classification domains in the remainder of this paper such that indeed no adaptation will be performed and the correctness check is simplified to the matching of class labels. Also note, the small difference to the standard LOO definition for case base competence is that our measure iterates not just over the case base itself, but over all cases in the training set expecting to have a larger sample of the entire problem space $\mathcal{M}$. This is in compliance with the representative assumption for the competence of case bases first proposed in [16].

3.3 Local Optima in Case Base Editing

The search space $\mathfrak{P}(\mathcal{T})$ of case base configurations comprises $|\mathcal{T}|$ dimensions, along each of which only two values are possible (case is "in" or "out"). As a consequence, a case base configuration $\mathcal{S}$ is a *local optimum* (minimum) in the error landscape of $\mathfrak{P}(\mathcal{T})$, if

$$\forall c \in \mathcal{T} : \mathbb{E}_{\mathcal{T}}^{loo}(\mathcal{S}) \leq \mathbb{E}_{\mathcal{T}}^{loo}(\mathcal{E}(\mathcal{S}, c)) \tag{2}$$

since $\mathcal{E}(\mathcal{S}, c)$ on the right-hand side refers to a case base that represents a minimal adaptation to $\mathcal{S}$. Accordingly, a strict minimum is attained, if we replace the less-equal sign by a strict less. Certainly, $\mathfrak{P}(\mathcal{T})$ is full of configurations $\mathcal{S}$ where switching on/off a single case no further reduces the value of the error; in that case $\mathcal{S}$ is a local minimum. Switching on/off a number of cases simultaneously might, however, still bring about improvements. This is exactly what Lupiani et al. [14] are exploiting using evolutionary algorithms where a "more global" search through $\mathfrak{P}(\mathcal{T})$ can be done using crossover. By contrast, the hunt for a global optimum (or a nearly optimal local one) is not our primary concern in this paper. Instead, we are more interested in characterizing the properties of different local minima. Therefore, in the next section, we suggest a number of case base editing schemes, that are based on $\mathbb{E}_{\mathcal{T}}^{loo}$ and that will, by definition, find various local minima easily.

3.4 Hill-Climbing Case Base Editors

For analyzing the properties of local minima in case base editing more thoroughly, it is comfortable to have access to a way for generating such optima easily. Thus, we suggest a set of greedy algorithms for case base editing, which are – because they are hill-climbers – designed to converge to local optima in the error landscape quickly. Besides, we also review a set of well-established case base editing methods from the literature and discuss whether they yield local optima of $\mathbb{E}_{\mathcal{T}}^{loo}$ as well. We developed these algorithms mainly for reasons of analyzing sharp/flat minima of the error function, being aware that they are unlikely to yield a global optimum in the search landscape (unlike e.g. [14]). However, their empirical performance matches up to the performance of a number of established CBM methods that we implemented for the purpose of further analysis.

3.4.1 First Improvement Hill-Climber (FHC)

This algorithm is called $FHC_<$ and, similarly as IB2 or CNN (cf. Sect. 3.4.3), starts out with an empty case base configuration $\mathcal{S} = \emptyset$. It then iterates over all cases c in $\mathcal{T}$ and checks for the first case to fulfill the following condition:

$$\mathbb{E}_{\mathcal{T}}^{loo}(\mathcal{E}(\mathcal{S}, c)) < \mathbb{E}_{\mathcal{T}}^{loo}(\mathcal{S}), \tag{3}$$

i.e. the first case whose addition to or removal from the current case base configuration $\mathcal{S}$ reduces the leave-one-out train error over $\mathcal{T}$. If such a case c is found, the case editing operator $\mathcal{E}$ either adds or removes c to/from $\mathcal{S}$ (depending on whether it was already contained or not). This procedure is repeated until there is no more case in $\mathcal{T}$ for which Eq. 3 becomes true. Algorithm 1 shows pseudo-code for an implementation of $FHC_<$.

Input: training set $\mathcal{T} \subset \mathcal{M}$,   **Output:** case base configuration $\mathcal{S}$ ($\mathcal{S} \subseteq \mathcal{T}$)
Strict variant $FHC_<$ *Modification for non-strict variant $FHC_{\leq}$*
1: $\mathcal{S} \leftarrow \emptyset$, $stop \leftarrow false$
2: **while** $stop = false$ **do**
3:      $stop \leftarrow true$
4:      **for** $c \in \mathcal{T}$ **do**   4: ...
5:          **if** $\mathbb{E}_{\mathcal{T}}^{loo}(\mathcal{E}(\mathcal{S}, c)) < \mathbb{E}_{\mathcal{T}}^{loo}(\mathcal{S})$ 5: **if** $\mathbb{E}_{\mathcal{T}}^{loo}(\mathcal{E}(\mathcal{S}, c)) < \mathbb{E}_{\mathcal{T}}^{loo}(\mathcal{S})$ or
6:          **then** $(\mathbb{E}_{\mathcal{T}}^{loo}(\mathcal{E}(\mathcal{S}, c)) = \mathbb{E}_{\mathcal{T}}^{loo}(\mathcal{S})$ and $c \notin \mathcal{S})$
7:              $\mathcal{S} \leftarrow \mathcal{E}(\mathcal{S}, c)$ 6: **then** ...
8:              $stop \leftarrow false$
9:              **quit** for loop
10: **return** $\mathcal{S}$

Algorithm 1: First Improvement Hill-Climber (variants $FHC_<$ and $FHC_{\leq}$)

A variation of $FHC_<$ is attained, if we replace the strict inequality in Eq. 3 by a less-equal comparison; this variant is named $FHC_{\leq}$ accordingly. Clearly, $FHC_{\leq}$ will in general add more cases to the case base than $FHC_<$. It is also obvious that both variants will, due to their hill-climbing nature, be attracted by a local optimum in the error landscape according to Eq. 3. From an algorithmic point of view, one should take care that $FHC_{\leq}$ does not get trapped in an endless loop due to the addition/removal of some "irrelevant" case whose presence/absence does not alter $\mathbb{E}_{\mathcal{T}}^{loo}$. To this end, we decided to allow for a non-strict comparison for the addition of a case and retain a strict change of the error for removing a case (see right part of Algorithm 1).

3.4.2   Maximum Improvement Hill-Climber (MHC)

The main difference between $FHC_<$ and the maximum improvement hill-climber $MHC_<$ introduced next lies in the selection of the next case that is added to or removed from the current case base configuration $\mathcal{S}$. While $FHC_<$ picks the first case $c \in \mathcal{T}$ whose de/activation brings about an improvement of the train error $\mathbb{E}_{\mathcal{T}}^{loo}$, $MHC_<$ performs an entire sweep over $\mathcal{T}$ and selects that c^* that yields the largest improvement.

So, while the pseudo-code in Algorithm 2 seems quite compact, it hides part of its complexity in the $\arg\min$ operator in line 4. Despite having a larger complexity in its outer *while* loop, $MHC_<$ will on average terminate faster than $FHC_<$ as it requires less iterations of that outer *while* loop because each one yields the maximal possible reduction of the train error. Consequently, it tends to create smaller case bases and terminate faster. Again, a less strict variant of $MHC_<$ is realized, if the inequality is replaced by a less-equal in line 5 (called $MHC_\leq$). Finally, it is trivial to see that both MHC variants do always end up in a local minimum of the error landscape of $\mathbb{E}_\mathcal{T}^{loo}$. As a side note we remark that MHC is nearly[2] insensitive to the "presentation order" of cases which is a criticism to most of the established case base maintenance algorithms [14].

Input: training set $\mathcal{T} \subset \mathcal{M}$, **Output:** case base configuration $\mathcal{S}$ $(\mathcal{S} \subseteq \mathcal{T})$
Strict variant $MHC_<$ *Modification for non-strict variant $MHC_\leq$*
1:   $\mathcal{S} \leftarrow \emptyset$, $stop \leftarrow false$
2:   **while** $stop = false$ **do**
3:       $stop \leftarrow true$
4:       $c^* \leftarrow \arg\min_{c \in \mathcal{T}} \mathbb{E}_\mathcal{T}^{loo}(\mathcal{E}(\mathcal{S}, c))$ 4: ...
5:       **if** $\mathbb{E}_\mathcal{T}^{loo}(\mathcal{E}(\mathcal{S}, c^*)) < \mathbb{E}_\mathcal{T}^{loo}(\mathcal{S})$ 5: if $\mathbb{E}_\mathcal{T}^{loo}(\mathcal{E}(\mathcal{S}, c^*)) < \mathbb{E}_\mathcal{T}^{loo}(\mathcal{S})$ **or**
6:       **then** $(\mathbb{E}_\mathcal{T}^{loo}(\mathcal{E}(\mathcal{S}, c^*)) = \mathbb{E}_\mathcal{T}^{loo}(\mathcal{S})$ and $c \notin \mathcal{S})$
7:           $\mathcal{S} \leftarrow \mathcal{E}(\mathcal{S}, c^*)$ 6: **then** ...
8:           $stop \leftarrow false$
9:   **return** $\mathcal{S}$

Algorithm 2: Maximum Improvement Hill-Climber ($MHC_<$ and $MHC_\leq$)

3.4.3   Related Case Base Editing Schemes

As mentioned, we incorporate a set of well-known case base editing methods into our further analyses. We highlight each of them with some remarks, emphasizing that this list is not complete and could easily be extended by many further algorithms from the realm of case base maintenance.

CNN (condensed nearest neighbor [6]) might be called one of the forefathers of CBM. It starts with an empty case base, makes multiple passes over $\mathcal{T}$ and copies a case c from $\mathcal{T}$ to $\mathcal{S}$, if it finds that c cannot be solved by $\mathcal{S}$. CNN has inspired various alternative and more sophisticated case base editing rules. In its original form it aims at finding a subset $\mathcal{S}$ of $\mathcal{T}$ that is as consistent as $\mathcal{T}$. Formally, CNN minimizes, starting with an empty case base configuration, an error function that is similar to $\mathbb{E}_\mathcal{T}^{loo}$, but not defined in a leave-one-out manner. As a consequence and due to the fact that CNN can just add cases to $\mathcal{S}$ and not remove them (like FHC or MHC), the output of CNN will in general not correspond to a local minimum of $\mathbb{E}_\mathcal{T}^{loo}$ as defined above.

---

[2] There remains some sensitivity to the presentation order since in line 4 multiple cases c may reduce $\mathbb{E}_\mathcal{T}^{loo}$ equally in which case one of those cases must be selected, e.g. randomly or by some convention.

RNN is an extension of the aforementioned CNN which adds a case removal phase during which cases are deleted from S whose removal does not impair the problem-solving competence of S on all cases from T [5]. The output of RNN might in general be expected to be closer to an optimum of $\mathbb{E}_T^{loo}$ than CNN's output. However, RNN is not an optimizer of $\mathbb{E}_T^{loo}$ since case addition and removal are strictly split into two separate phases and because the error is not measured in a leave-one-out manner.

IBL Algorithms denote instance-based learning algorithms [1]. IB2, as one instance of this family of algorithms, iterates over T and adds a case c to S, if c's problem part would not be solved correctly by the cases in S. As a consequence, it is susceptible to noise, but it can also be termed a greedy algorithm in the sense that it tries to add as little cases as possible.

DROP Algorithms denote decremental reduction optimization procedures [19]. Different variants exist; DROP1 starts out with a case base S that is set to the full set of train cases, $S = T$. Then, it iteratively removes individual cases from T whose deletion does not worsen the leave-one-out performance over S (note, not over T). DROP2 is an extension of DROP1 whose leave-one-out performance is measured over the whole set T. Insofar, DROP2 comes close to our FHC and MHC algorithms, except for its inability to re-add cases after having deleted them. Moreover, DROP2 employs also a specialized preference heuristic regarding which cases to remove first, namely those which have the smallest similarity to their "nearest enemy" (which means a case in T whose solution does not match or cannot be adapted). As a consequence, DROP2 is more likely than all other algorithms listed here to yield a local optimum in the sense that we defined in Sect. 3.3.

4   Sharpness of a Case Base Configuration

In the preceding section, we have formalized case base editing as an optimization problem where we (a) defined an error measure $\mathbb{E}$ over the training set that relates to the leave-one-out competence of the system over a training set T and (b) proposed discrete editing operations (case editing operator $\mathcal{E}$) for searching for a minimum of $\mathbb{E}_T^{loo}$. All these steps were necessary to path the way for a further analysis of local optima in the error landscape that we are intending to describe now. Additionally, for performing the actual search, we suggested two hill-climbing algorithms (FHC and MHC, more specifically four variants of them) which are, by definition, designed to find local optima for $\mathbb{E}$.

4.1   Characterizing Flat and Sharp Case Base Editing Optima

As described in Sect. 2.2, a sharp minimum m_s of some function f over domain X is characterized by the observation that f changes rapidly in the near vicinity o m_s. We have also highlighted that in related research fields sharp minima are known to correspond to models with poor generalization capabilities.

The case base editing problem, as defined in Sect. 3.1 is, however, of discrete nature. Instead of the mentioned numeric domain X, for a given set of training cases T, the search space contains the $2^{|T|}$ elements of T's power set $\mathfrak{P}(T)$. One might say, the space to be searched is $|T|$-dimensional with two possible values in each dimension. Acknowledging this and aiming at establishing a notion of the vicinity around some minimum, we thus define:

Definition 4 (Vicinity of a Case Base Configuration). *Given a set of training cases T and a case base configuration S with $S \subseteq T$, the vicinity of S is defined as*

$$\mathcal{V}_T(S) = \{\mathcal{E}(S, c) | c \in T\}$$

Hence, the vicinity of a case base configuration contains all $|T|$ case bases that are formed, if we either leave out exactly one $c \in S$ from S or if we add exactly one case from $T \setminus S$ to S. If S is known to be a local optimum (e.g. as the result of applying FHC or MHC, cf. Sect. 3.4), then by construction it holds that the leave-one-out train error for S is smaller than or equal to the error for any case base configuration within the vicinity set $\mathcal{V}(S)$.

The vicinity definition allows us to derive a numeric estimation of the sharpness of some case base configuration – a notion that of course covers local optima of case base editing as well.

Definition 5 (Sharpness of a Case Base Configuration). *Given a set of training cases T and a case base configuration S with $S \subset T$, the sharpness of S is defined as*

$$\mathbb{S}_T(S) = \sqrt{\frac{1}{|T|} \sum_{V \in \mathcal{V}_T(S)} \left(\mathbb{E}_T^{loo}(V) - \mathbb{E}_T^{loo}(S)\right)^2}$$

So, essentially the sharpness of some case base configuration mirrors how much isolated modifications to the case base (by including/removing a single case) influence the leave-one-out problem solving capabilities of S.

4.2   Discussion of the Sharpness Measure

Using a root mean square definition instead of, for example, a sum over differences in Definition 5 serves two purposes. On the one hand, it allows for adequately assessing the level of sharpness of case base configurations S that are not local minima, i.e. where the inner difference is not guaranteed to be positive. On the other hand, it puts an emphasis on those case base configurations within the vicinity set $\mathcal{V}_T(S)$ where the addition/removal of a single case has an above-average impact on the change in the error. It is also worth mentioning that calculating $\mathbb{S}_T(S)$ is computationally costly with an effort of $O(|T|^3)$ given that a simple linear retrieval is used in Eq. 1.

While we have focused on a set of training cases T so far, we shall now put our attention more to an independent, held-out set $\mathcal{U}$ of test cases, i.e. $\mathcal{U} \subsetneq \mathcal{M}$ and $T \cap \mathcal{U} = \emptyset$. Accordingly, our interest will be on the test error that some case base configuration S yields when its problem-solving capabilities are tested on $\mathcal{U}$. Hence, in analogy to Definition 3 we define:

Definition 6 (Test Error of a Case Base Configuration). *For a test set of cases* $\mathcal{U} = \{c_1, \ldots, c_{|\mathcal{U}|}\}$ *with each case* $c_i = (p_i, s_i)$ *consisting of a problem and solution part and for a given case base configuration* $\mathcal{S}$, *the* test error *is defined as*

$$\mathbb{E}_{\mathcal{U}}(\mathcal{S}) = \frac{1}{|\mathcal{U}|} \cdot \sum_{i=1}^{|\mathcal{U}|} 1 - Correct(Adapt(Retrieve(\mathcal{S}, p_i), p_i), s_i)$$

Our conjecture is that the sharpness of a case base configuration is related to the testing performance of this case base. Thus, besides the actual value of the train error, the sharpness might help us in assessing the generalization capabilities of a case base configuration.

5   Empirical Evaluation

The first goal of our experimental evaluation is to empirically investigate the correlation between sharpness and test error, i.e. answering the question to what extent sharpness is suitable as a predictor of the generalization capability. In the second part of the evaluation, we aim at an empirical analysis of the four hill-climbing CBM variants proposed in Sect. 3.

We selected 21 classification domains from the UCI Machine Learning Repository [13] with varying amounts of case data, classes, and numbers and types of features. In all experiments, we split the available data set into two disjoint sets $\mathcal{T}$ and $\mathcal{U}$ where for the number of cases in the training set we focused on three settings ($|\mathcal{T}|$ being 50, 75, and 100, respectively). For k, as the number of nearest neighbors to be considered during retrieval we focused on $k = 1$ and $k = 3$.

5.1   Correlation Between Sharpness and Generalization

In machine learning, the training error is usually assumed to be strongly correlated to the error on an independent test set, except if overfitting has occurred. This general observation is also true for instance-based learning systems, expressing itself in a positive sample Pearson correlation coefficient $r_{x,y}$, where here x stands for the train error $\mathbb{E}_{\mathcal{T}}^{loo}(\mathcal{S})$ of a specific case base configuration and y for the test error $\mathbb{E}_{\mathcal{U}}(\mathcal{S})$ that $\mathcal{S}$ yields on an independent test set.

To this end, the interesting question is whether the sharpness $\mathbb{S}_{\mathcal{T}}(\mathcal{S})$ of a case base configuration $\mathcal{S}$ (cf. Definition 5) is also correlated with $\mathbb{E}_{\mathcal{U}}(\mathcal{S})$. In order to answer this question, we generated a large number of random case base configurations with random sizes by randomly adding any $c \in \mathcal{T}$ to $\mathcal{S}$ or not. For each domain, we processed 1000 such random case bases and determined $x = \mathbb{E}_{\mathcal{T}}^{loo}(\mathcal{S})$, $y = \mathbb{E}_{\mathcal{U}}(\mathcal{S})$, as well as sharpness values $z = \mathbb{S}_{\mathcal{T}}(\mathcal{S})$ (for brevity, we use x, y, and z as shorthand notation, subsequently). In so doing, we found that case base configurations with high sharpness tend to yield a higher test error, and vice versa. More specifically, the Pearson correlation $r_{z,y}$ is nearly identical to $r_{x,y}$ (see Table 1). In other words, the measure of sharpness introduced above

Table 1. Average Pearson correlations over all classification domains. For all settings examined, the correlation between train and test error can be improved (gain in brackets), when adding sharpness information to the train error.

	$\|\mathcal{T}\| = 50$			$\|\mathcal{T}\| = 75$			$\|\mathcal{T}\| = 100$		
	$r_{x,y}$	$r_{z,y}$	$r_{x+z,y}$	$r_{x,y}$	$r_{z,y}$	$r_{x+z,y}$	$r_{x,y}$	$r_{z,y}$	$r_{x+z,y}$
$k=1$	0.608	0.624	0.675 (+0.066)	0.628	0.644	0.701 (+0.073)	0.647	0.646	0.697 (+0.050)
$k=3$	0.652	0.617	0.687 (+0.034)	0.688	0.646	0.697 (+0.009)	0.706	0.685	0.728 (+0.021)

is approximately as meaningful in assessing the generalization capability of a case base configuration as its leave-one-out train error.

Most interestingly, if we additively combine the train error $\mathbb{E}_{\mathcal{T}}^{loo}(\mathcal{S})$ and the sharpness $\mathbb{S}_{\mathcal{T}}(\mathcal{S})$ and determine the correlation of $\mathbb{E}_{\mathcal{T}} + \mathbb{S}_{\mathcal{T}}$ with $\mathbb{E}_{\mathcal{U}}$, i.e. $r_{x+z,y}$, we find that this is even higher than the correlation of both components alone (cf. Table 1). This is a strong indication that the sharpness can be helpful in estimating the generalization capabilities of the system. Figure 2 visualizes the gain in correlation for all the domains and all variations of k and $\|\mathcal{T}\|$ considered.

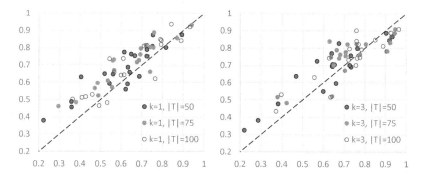

Fig. 2. Each data point represents the average over 1000 random case base configurations for one domain. The value on the abscissa refers to the average train-test correlation $r_{x,y}$, the value on the ordinate to the average sharpness-enhanced correlation $r_{x+z,y}$. Points above the identity function thus refer to runs where the incorporation of sharpness information brought about a better estimate of the generalization capabilities.

5.2 Hill-Climber Variants and Their Optima

Different optimization algorithms yield different minima. Depending on their nature they may tend to end up in rather flat or sharp local minimum of the error landscape. Given our observations reported in the preceding section, we conjecture that the sharpness of attained minima may help in evaluating the generalization capabilities of different algorithms. Besides this, we also want to empirically compare the performance of our proposed FHC and MHC variants with established CBM algorithms.

For this series of experiments, we applied all of the algorithms outlined above for each of the 21 classification domains, repeating this entire procedure 500 times for randomly re-initialized sets $\mathcal{T}$ of training cases (for both, $|\mathcal{T}| = 50$ as well as $|\mathcal{T}| = 100$, keeping $k = 1$ throughout).

First, we report the performance of the greedy case base editing schemes FHC and MHC in order to convey a feeling of their performance. As can be seen from Table 2, all four variants perform well, specifically $MHC_{\leq}$ features the best average test error while at the same time utilizing only less than one third of the cases given as input in $\mathcal{T}$. Domain identifiers are specified in a footnote[3].

Table 2. For each of the case base editing algorithms and for each of the considered domains two performance measures are given (averaged over 500 experiment repetitions, i.e. as many random sets $\mathcal{T}$ of train cases): the test error $\mathbb{E}_{\mathcal{U}}(\mathcal{S})$ of the learned case base configuration $\mathcal{S}$ as well as the achieved case base compactification in percent, i.e. $100\% \cdot \frac{|\mathcal{S}|}{|\mathcal{T}|}$. The table reports results for $|\mathcal{T}| = 50$ and $k = 1$, plus an average μ_{50} over all domains. Additionally, in the last line the average μ_{100} for $|\mathcal{T}| = 100$ is reported. For each domain, the two top-performing algorithms are highlighted. The first column contains domain identifiers (see footnote for plain text names).

Dom.	CNN		RNN		IB2		DROP1		DROP2		$FHC_<$		$FHC_\leq$		$MHC_<$		$MHC_\leq$	
A	.415	53.1	.422	49.6	.420	47.1	.430	**9.4**	.376	17.3	.355	22.2	**.350**	54.8	.360	**13.6**	**.340**	32.9
B	**.067**	22.0	.072	18.7	.090	19.4	.204	**13.2**	.078	21.5	.104	17.4	**.063**	55.0	.115	**11.3**	.085	26.1
C	.354	49.4	.361	43.9	.381	40.9	.355	**7.2**	.312	14.5	**.288**	14.7	.294	53.8	**.282**	**9.0**	.289	32.0
D	.344	47.6	.352	44.0	.356	43.3	.446	**9.5**	.346	18.2	.336	25.9	**.308**	67.0	.344	**14.5**	**.322**	42.7
E	.604	70.6	.606	64.1	.608	62.2	.625	**9.7**	.600	18.5	.597	19.8	**.592**	50.4	.598	**13.9**	**.592**	31.9
F	.258	41.5	.268	34.9	.278	36.1	.315	**15.6**	.238	**17.6**	.222	25.7	**.211**	68.1	.224	21.5	**.215**	37.9
G	.209	38.1	.222	30.7	.236	32.7	.250	**9.0**	.168	12.4	.164	19.1	.159	62.4	**.150**	**10.2**	**.156**	27.5
H	.380	53.3	.386	45.9	.398	41.1	.361	**8.2**	.323	13.7	**.305**	12.6	.328	53.5	**.298**	**6.5**	.319	28.3
I	**.365**	55.3	**.363**	41.4	.368	49.8	.538	17.5	.434	39.1	.466	23.8	.420	54.4	.487	**16.9**	.447	36.6
J	.278	43.6	.289	36.8	.305	33.2	.321	**7.7**	.229	13.4	**.208**	14.1	.209	46.2	.209	**10.2**	**.204**	23.4
K	.090	17.9	.094	14.6	.101	15.1	.147	**11.6**	.073	**12.7**	.073	15.3	**.067**	46.6	.076	13.0	**.068**	23.4
L	.298	41.3	.281	21.7	.359	36.8	.318	**9.5**	.294	33.1	**.236**	13.1	.243	50.4	**.236**	**9.7**	.240	27.4
M	.321	46.7	**.322**	43.1	.331	39.7	.441	**11.1**	.370	21.6	.382	19.1	.351	51.4	.381	**12.5**	.355	33.1
N	.356	51.6	.363	44.0	.373	39.3	.374	**7.7**	.333	14.3	.314	14.4	.317	50.5	**.312**	**10.3**	**.311**	26.8
O	.035	11.7	.036	10.1	.040	10.7	.114	**8.1**	**.035**	12.9	.038	13.4	**.026**	35.4	.043	**9.3**	.039	12.2
P	**.543**	73.2	**.542**	66.2	.563	62.9	.644	**18.2**	.606	27.4	.600	20.6	.583	56.4	.604	**16.3**	.589	36.9
Q	**.317**	48.4	**.323**	44.9	.335	41.5	.441	**12.0**	.372	22.7	.366	18.0	.325	59.1	.366	**9.5**	.344	40.1
R	.300	50.0	.306	43.0	.316	42.7	.465	**15.8**	.332	24.4	.351	27.3	**.309**	66.3	.355	**20.3**	**.327**	44.3
S	.324	49.2	.332	41.6	.345	40.2	.429	**11.2**	.316	16.6	.308	19.2	**.298**	59.4	.311	**14.0**	**.298**	36.4
T	.088	18.4	.094	14.9	.096	16.1	.137	**11.8**	.094	14.9	.074	16.6	**.068**	42.2	.078	**13.4**	**.071**	18.2
U	.584	72.0	.590	65.2	.592	64.5	.658	12.2	.580	**18.2**	.560	30.4	**.553**	62.3	.566	25.6	**.553**	42.3
μ_{50}	.311	45.5	.315	39.0	.328	38.8	.382	**11.3**	.310	19.3	.302	19.2	**.289**	54.5	.305	**13.4**	**.294**	31.5
μ_{100}	.305	43.6	.309	36.1	.326	36.5	.380	**7.9**	.290	16.4	.275	14.5	**.267**	56.7	.276	**8.5**	**.267**	31.0

[3] A-Balance, B-BanknoteAuth, C-Cancer, D-Car, E-Contraceptive, F-Ecoli, G-Glass, H-Haberman, I-Hayes, J-Heart, K-Iris, L-MammogrMass, M-Monks, N-Pima, O-QualBankruptcy, P-TeachAssistEval, Q-TicTacToe, R-UserKnowledge, S-VertebralCol, T-Wine, U-Yeast.

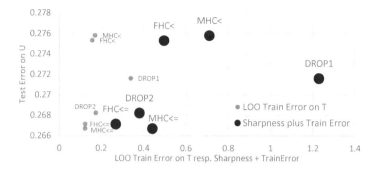

Fig. 3. Each data point stands for an average over the set of domains we considered and visualizes the correlation between the leave-one-out train error and the sharpness-enriched train error information with the generalization error measured on an independent test set $\mathcal{U}$. The plot refers to the setting of $k = 1$ with $|\mathcal{T}| = 100$, where for DROP1/2 only those runs were considered that yielded a case base configuration $\mathcal{S}$ that is a local optimum of $\mathbb{E}_{\mathcal{T}}^{loo}$.

We emphasize, however, that this comparison is of course not comprehensive; implementing other powerful case base editing algorithms from the literature (cf. Sect. 2) and matching our hill climbers' performance with those is an open topic for future work. Nevertheless, the numbers reported allow for concluding that FHC and MHC might be qualified as usable algorithms for case base maintenance.

Second, the hill climbing case base optimizers are designed to attain a local optimum of $\mathbb{E}_{\mathcal{T}}^{loo}$. Incidentally, some of our established case base editing algorithms do so, too, at least in some cases. Specifically, IB2 and CNN never converge to a local optimum of $\mathbb{E}_{\mathcal{T}}^{loo}$, but RNN (0.01%), DROP1 (4.3%), and DROP2 (24.7%) do so occasionally. The numbers in brackets refer to the share of the 21.000 experimental runs during which the respective algorithm attained an optimum. To this end, Fig. 3 indicates that algorithms with a small sum of the train error and the sharpness (i.e. flat minima) correspond to smaller test error and, hence, better generalization than those that yield sharper minima. In particular, the train error alone is only of little use in predicting the test error (gray data points).

6   Conclusion

In this paper, we have made a first attempt to investigate the presence and properties of sharp/flat minima in an error landscape of a case base maintenance scenario. We observed that sharp case base configurations feature poorer generalization properties than those that correspond to flat regions in the domain of the error function. Our analyses came along with two new case base maintenance procedures which, being hill-climbing optimizers, are by design attracted by local optima of the error function. We empirically found that their general performance is competitive in terms of case base competence and compactification and that the sharpness of local minima can be used to better predict the generalization error. An interesting avenue for future work is to design case base editing methods that incorporate sharpness information instantaneously, when adding or removing cases, and hence can be guided to attain flatter minima.

References

1. Aha, D., Kibler, D., Albert, M.: Instance-based learning algorithms. Mach. Learn. **6**, 37–66 (1991)
2. Brighton, H., Mellish, C.: On the consistency of information filters for lazy learning algorithms. In: Żytkow, J.M., Rauch, J. (eds.) PKDD 1999. LNCS (LNAI), vol. 1704, pp. 283–288. Springer, Heidelberg (1999). https://doi.org/10.1007/978-3-540-48247-5_31
3. Cover, T., Hart, P.: Nearest neighbor pattern classification. IEEE Trans. Inf. Theory **13**, 21–27 (1967)
4. Dinh, L., Pascanu, R., Bengio, S., Bengio, Y.: Sharp minima can generalize for deep nets. In: Proceedings of the 34th International Conference on Machine Learning, Sydney, Australia, pp. 1019–1028. JMLR.org (2017)
5. Gates, G.: The reduced nearest neighbor rule. IEEE Trans. Inf. Theory **18**(3), 431–433 (1972)
6. Hart, P.: The condensed nearest neighbor rule. IEEE Trans. Inf. Theory **14**(3), 515–516 (1968)
7. Hochreiter, S., Schmidhuber, J.: Flat minima. Neural Comput. **9**(1), 1–42 (1997)
8. Keskar, N., Mudigere, D., Nocedal, J., Smelyanskiy, M., Tang, P.: On large-batch training for deep learning: generalization gap and sharp minima. In: Proceedings of the 5th International Conference on Learning Representations, Toulon, France (2017)
9. Kibler, D., Aha, D.: Learning representative exemplars of concepts: an initial case study. In: Proceedings of the Fourth International Workshop on Machine Learning, pp. 24–30. Morgan Kaufmann (1987)
10. Leake, D.B., Wilson, D.C.: Categorizing case-base maintenance: dimensions and directions. In: Smyth, B., Cunningham, P. (eds.) EWCBR 1998. LNCS, vol. 1488, pp. 196–207. Springer, Heidelberg (1998). https://doi.org/10.1007/BFb0056333
11. Leake, D., Wilson, M.: How many cases do you need? Assessing and predicting case-base coverage. In: Ram, A., Wiratunga, N. (eds.) ICCBR 2011. LNCS (LNAI), vol. 6880, pp. 92–106. Springer, Heidelberg (2011). https://doi.org/10.1007/978-3-642-23291-6_9
12. Li, H., Xu, Z., Taylor, G., Studer, C., Goldstein, T.: Visualizing the loss landscape of neural nets. In: Annual Conference on Neural Information Processing Systems 2018 (NeurIPS 2018), Montréal, Canada, pp. 6391–6401 (2018)
13. Lichman, M.: UCI Machine Learning Repository (2013). archive.ics.uci.edu/ml
14. Lupiani, E., Craw, S., Massie, S., Juarez, J.M., Palma, J.T.: A multi-objective evolutionary algorithm fitness function for case-base maintenance. In: Delany, S.J., Ontañón, S. (eds.) ICCBR 2013. LNCS (LNAI), vol. 7969, pp. 218–232. Springer, Heidelberg (2013). https://doi.org/10.1007/978-3-642-39056-2_16
15. Skalak, D.: Prototype and feature selection by sampling and random mutation hill climbing algorithms. In: Proceedings of the 11th International Conference on Machine Learning, New Brunswick, NJ, USA, pp. 293–301 (1994)

16. Smyth, B., McKenna, E.: Building compact competent case-bases. In: Althoff, K.-D., Bergmann, R., Branting, L.K. (eds.) ICCBR 1999. LNCS, vol. 1650, pp. 329–342. Springer, Heidelberg (1999). https://doi.org/10.1007/3-540-48508-2_24
17. Smyth, B., McKenna, E.: Competence guided incremental footprint-based retrieval. Knowl.-Based Syst. **14**(3–4), 155–161 (2001)
18. Wilson, D.: Asymptotic properties of nearest neighbor rules using edited data. IEEE Trans. Syst. Man Cybern. **2**(3), 408–421 (1972)
19. Wilson, D., Martinez, T.: Reduction techniques for instance-based learning algorithms. Mach. Learn. **38**(3), 257–286 (2000)
20. Zhang, J.: Selecting typical instances in instance-based learning. In: Proceedings of the 9th International Workshop on Machine Learning, Aberdeen, UK, pp. 470–479 (1992)

CBR Confidence as a Basis for Confidence in Black Box Systems

Lawrence Gates$^{(\boxtimes)}$, Caleb Kisby, and David Leake

School of Informatics, Computing, and Engineering, Indiana University,
Bloomington, IN 47408, USA
{gatesla,cckisby,leake}@indiana.edu

Abstract. Determining when to trust black box systems is a well-known challenge. An important factor affecting users' trust is confidence in system solutions. Previous case-based reasoning (CBR) research has developed criteria for assigning confidence to the solutions of a CBR system. This paper investigates whether such analysis, coupled with factors such as CBR system competence, can be used to predict confidence in the outputs of a black box system, when the black box and CBR systems are provided with the same training data. The paper presents initial strategies for using CBR confidence to predict black box system confidence. An evaluation explores the ability of the strategies to provide useful information and suggests future questions.

Keywords: Case base competence · Case-based reasoning ·
Neural network · Explainable artificial intelligence · Confidence ·
Competence

1 Introduction

Advances in machine learning, and in particular in deep networks, have led to widespread applications of AI systems with powerful performance achieved through methods that are largely opaque to their human users. Such systems, often referred to as *black box* systems, accept an input and propose an output without an account of how the output was generated. This can be especially troubling when the black box systems, despite overall strong performance, sometimes perform unexpectedly. For example, it is well known that deep networks may exhibit unexpected behaviors on *adversarial examples*; two images that a human sees as identical may receive different classifications [20,29]. Such behavior and the inability to explain the performance of black box systems has been widely acknowledged as a concern for confidence in their conclusions, which can limit the domains to which they are applied. This in turn has led to an outpouring of research on explainable AI (e.g., [2]), including in the context of case-based reasoning [1].

Explanation of black box systems has a long history of combining the black box systems with more interpretable methods. For example, one approach is

© Springer Nature Switzerland AG 2019
K. Bach and C. Marling (Eds.): ICCBR 2019, LNAI 11680, pp. 95–109, 2019.
https://doi.org/10.1007/978-3-030-29249-2_7

to use interpretable ML methods, such as decision trees, to build a model of the black box system reasoning that can then be used to explain predictions. However, as rules become more complex they become less interpretable, and it may be difficult to capture the black box system's behavior with sufficient fidelity (e.g., [10]). From the early days of case-based reasoning, the ability to explain CBR system reasoning by reference to prior cases has been seen as an important benefit [15]. This makes it appealing to combine case-based reasoning with black box systems, to increase explainability of black box system behavior. For example, Shin et al. [27] propose a CBR/neural network hybrid in which neural-network-generated features are used to retrieve relevant cases, with the goal of explainability. Nugent and Cunningham propose a general framework for case-based explanation of behavior of black box systems [22]. In their approach an artificial case base, seeded with cases generated by the black box system, is used to determine local feature salience, which is used in turn to guide retrieval of real cases as the basis for explanations to increase user confidence in black box system conclusions. Keane and Kenny provide an extensive survey of research on "twinning" CBR and neural network systems to provide explanations [14].

This paper brings together CBR and black box systems in a different way, for a task complementary to the explanation task per se: to assess confidence in black box solutions. In the presented approach, COBB (Case-based cOnfidence for Black Box), both the CBR system and black box system have access to the same training data (or subsets of each other's data); each functions in parallel. However, the goal of the CBR system processing is not to provide the solution, but instead, to ascribe confidence to the black box system output. That confidence judgment can be directly provided to a user, as a unitary confidence judgment, and the confidence (not the solution itself) can explained in terms of characteristics derived from the CBR system. Thus in contrast with, e.g., Nugent and Cunningham: The role of the CBR system is not to replicate the black box system performance, but to provide an independent view, based on the same data, as a "second opinion" based on a more intelligible process that can be examined to assess its conclusions. The confidence information can then be used, for example, to decide when to expend scarce resources on evaluating solutions (e.g., in a financial system, presenting the problem to a human expert, or presenting the case retrieved by COBB as the basis of independent assessment).

An important question for such paired systems is how much their value depends on the relative performance of the CBR and black box systems. The primary use case for the COBB approach is situations in which in general, black box system solutions have higher confidence. Were that not the case, the CBR system, not the black box, should be the primary reasoning system. We discuss this question in more detail in Sect. 6.1.

Given that the CBR system and black box system are independent, with the CBR system potentially having lower accuracy, a natural question is the extent to which the CBR system can ascribe confidence to the black box system results. The answer is twofold. First, for assessing confidence, independence of the two systems can be a benefit to give a true second opinion. On the other hand, a premise of the approach is that the world basically conforms to the CBR hypothesis that "similar problems have similar solutions," so black box system behavior that conflicts with that premise—as manifested by the CBR system—should be ascribed lower confidence.

This paper proposes and evaluates three potential methods for predicting confidence of a black box system based on a CBR system. The first method proposed is based on a naïve analysis of the relationship between CBR confidence, black box confidence, and the distance between the two solutions. The second method combines several predictors in order to determine the confidence in the black box solution. The third method builds on the extensive work of Cheetham on CBR confidence by applying his confidence indicators approach to the black box system outputs.

Experimental results show that the method with the best overall quality was the second method. It generally had better overall quality than the other two, and for large testing sets had very good quality. The paper closes with directions for extending this work.

2   Previous Work

CBR Confidence Models: In seminal work, Cheetham proposed the development of confidence models for case based reasoning. His goal was twofold: to provide information to help predict whether a solution had low error, and to determine whether the output of the CBR system should be used for a given task. His approach [5,7] explores incorporating a measure of quantitative values for confidence and an error factor into each score. Reilly et al. [26] developed an explicit model of confidence for case-based conversational recommender systems.

Neural Network Confidence Models: Previous work has explored using confidence intervals to determine prediction intervals for Neural Networks (e.g., [4]). We note, however, that use of confidence intervals is different from determining the confidence in a system in the sense pursued by Cheetham. Confidence intervals "are enclosed in prediction intervals and are concerned with the accuracy of our estimates" [4], whereas confidence in Cheetham's (and our) sense is "the degree of belief in the correctness of the result of a CBR system" [7]. Additional approaches for confidence measures of neural networks with confidence intervals have emphasized the use of maximum likelihood error [24] and confidence intervals in classifier models [30].

CBR Integrations: There is a long history of CBR integrations with other types of systems [18], including for black box systems such as neural networks, in which the two systems contribute jointly to problem-solving. For example, in the medical domain to classify skin lesions, a convolutional neural network was used to get features, where those features were passed into the CBR to get retrieved case and output [21]. The proposed integration differs, however, in that the goal is for CBR to contribute to assessment of the other system rather than to the problem-solving process itself. This work is an instance of twinning of CBR and black box systems, such as ANNs [14].

3 Black Box Confidence

3.1 The Notion of Confidence of a Black Box System

Developing an approach to assessing black box system confidence by CBR depends first on understanding what "confidence" should represent. The term *confidence* has been used in CBR to refer to the "degree of belief" that the CBR system's solution is correct [5]. This is a fuzzy notion for which values in the range $[0, 1]$ indicate "percentage belief" in the CBR solution. We distinguish this notion from that of trust; confidence is a technical property of our system, whereas trust is a psychological property of humans using a system [17]. This notion of confidence is well understood for CBR systems [7], and can augment the assessments that could be done by examining the CBR system's internal process of retrievals and adaptation.

Black box systems are widely used, with applications for high-stake scenarios. Unfortunately, it is impossible to examine the internals of a black box system in order for a user to develop a level of trust in it (by definition). Thus it would be useful to be able to evaluate a level of confidence in a black box's solution to a problem.

A simple first approach for black box confidence would be to use a global measure of the black box system's performance, such as its accuracy, to ascribe a level of confidence in the system's solutions. This approach is not satisfactory, however, because the global accuracy provides no per-case information. Based on it, equal confidence would be ascribed to all solutions—which would provide no guidance on which solutions to verify or perhaps reject.

Given the ability to ascribe confidence to CBR solutions, it is appealing to use confidence in a CBR system to assist with determining confidence in the black box. We can twin a CBR system with our black box, training the two on the same set of training cases, for each to provide solutions to each problem. To account for differences in system characteristics, the confidence in the CBR system's solution can then be combined with information from other properties of the CBR and black box systems in order to calculate the confidence in the black box's solution.

In the following sections, we discuss potential predictors for black box confidence. Using these predictors, we present three methods for determining confidence in a black box system's solution.

3.2 Distance from CBR System Solution

A simple indicator for confidence in a black box solution is the distance of that solution to the solution provided by the CBR system itself. We assume that the solution space has some distance metric normalized to the interval $[0, 1]$. Applying this involves complexities discussed in Sect. 3.6.

3.3 CBR Confidence

As mentioned previously, confidence in a CBR system's solution is well understood. We follow Cheetham's approach for calculating the CBR confidence [5–7]. His method involves constructing fuzzy preference functions which map the CBR system's solutions to confidence in those solutions. His approach first sets a confidence scale mapping intervals of error in the CBR solution to confidence intervals. We use his scale, where the Fuzzy Linguistic Term *good* has a confidence interval of 1–0.75 and an error of less than 5%, the term *questionable* has a confidence interval of 0.75–0.5 and an error between 5% and 10%, and the term *poor* has a confidence interval 0.5–0.0 and an error greater than 10%. (Here, "confidence interval" refers to the range of values our "fuzzy" confidence can have. This is distinct from the statistical notion of "confidence interval").

The next step is to pick a few statistical indicators of confidence [6]. We select the following indicators (Cheetham proposes both positive and negative indicators, but we consider only positive):

- Similarity between the given problem case and the most similar retrieved case with the best solution
- Sum of all the similarity scores between the problem case and the k-closest retrieved cases with the best solution
- Number of cases with the best solution out of the k closest retrieved cases
- Percentage of the k closest retrieved cases that have the best solution
- Average similarity score between the problem case and the k closest cases with the best solution

As suggested by Cheetham, we then use the *C4.5* algorithm to construct a decision tree for predicting whether a solution will be correct, based on the values of the indicators. We select the indicators highest in the tree as the most important indicators. The goal is to choose the indicators best at predicting a correct CBR solution, because "the more likely the solution is to be correct the higher our confidence should be" [6].

For each selected indicator, we construct a fuzzy preference function mapping that indicator value to confidence in the CBR system's solution. To do this, we treat each case in the training set as a test case (temporarily removing each in turn from the case base). For each training case, we calculate the value of the indicator and the error in the CBR system's proposed solution. Similarly to Cheetham, we plot the indicator values against the error, and fit a cubic regression to the resulting plot using NumPy [23]. We then construct a piecewise linear function from this regression by obtaining the straight lines that meet at the extrema and inflection points.

We next compose the piecewise function (mapping indicator values to error) with the confidence scale (mapping error to confidence) to construct the fuzzy preference function (mapping indicator values to confidence). The details are spelled out by Cheetham in [5], and involve identifying key indicator values at which we are in a different error interval, and hence in a different confidence interval.

We then can determine the confidence in a CBR solution by calculating the selected indicators for a particular solution, for each fuzzy preference function, taking the confidence value for the value of its respective indicator, and taking the mean of these confidences as the final confidence value.

3.4   CBR Competence

The CBR system's *competence*, the "the range of target problems that a given system can solve" [28], may be another predictor of black box confidence. If the training data is insufficient for CBR coverage of the problem space, it is plausible that it could be insufficient for the black box as well. We follow Smyth and McKenna's model of CBR competence in which the competence of a CBR system depends on the density and distribution of cases in the case base and the strength of the CBR system's retrieval and adaptation.

3.5   Black Box Accuracy

We also expect higher confidence in the black box when the black box itself globally performs well. As a global measure of our black box's performance we use its *accuracy*, i.e. the percent of its own training cases for which the fully-trained black box can successfully provide a solution (within an acceptable threshold). This can be estimated, for example, by leave-one-out testing.

3.6   Proposed Methods for Estimating Black Box Confidence

Given the predictors for black box confidence (black box *accuracy*, CBR *competence*, *confidence* in the CBR solution, and *distance* between the CBR solution and black box solution), we propose three ways to combine them to determine confidence in the black box solution. As emphasized before, each method produces a *fuzzy* confidence value within [0, 1] which represents the degree of belief that the black box's solution is correct.

Naïve Method: Our first approach is based on the insight that if we are very confident in the CBR system's solution and the distance between the two solutions is small, we should also be very confident in the black box system's solution. Similarly, when we are very confident in the CBR system's solution and the distance between the two solutions is large, we should doubt the black box's solution. When we doubt the CBR system's solution and the distance between the two solutions is small, we expect again to doubt the black box's solution. Following this reasoning, we might infer that the confidence of the black box's solution is given by a formula such as:

$$conf_{BB} = |conf_{CBR} - distance| \tag{1}$$

That is, confidence in the black box's solution is the distance between CBR confidence and solution distance, both scaled to [0, 1].

However, there is a problem with this confidence formulation: If we have low confidence in the CBR system and there is a large distance between the two solutions, this method predicts that we will have high confidence in the black box. This is not necessarily the case, because the distant black box solution could still easily be far from the actual solution. In addition, this formula does not make use of the black box accuracy or CBR competence, both of which should affect the confidence in our black box's solution. Because of these issues, we do not expect this method to predict black box confidence well, but we will test it as a simple baseline.

Cheetham's Fuzzy Preference Method: This method provides the most natural extension of CBR confidence to black box confidence. We can simply apply Cheetham's method for determining confidence in a CBR solution to the black box's solution. We use the same confidence scale as for our CBR confidence. For our indicators of black box confidence, we pick the confidence in the CBR system's solution for the same problem and the distance between the black box and CBR solutions. We again construct fuzzy preference functions mapping these indicators to black box confidence (using the training set), and then average the outputs of these fuzzy preference functions for a given black box solution.

Note that we cannot use *accuracy* or *competence* as indicators here, because Cheetham's method requires that indicator values vary per problem case (whereas accuracy and competence are system-global properties). So, like the Naïve Method, this method also does not make use of the black box accuracy or the CBR competence. Rejecting *accuracy* and *competence* as indicators on their own is also justified pragmatically by the fact that they provide no comparative information: They give no indication of which solutions might require further verification.

Weighted Average Method: Unlike the previous two methods, this approach attempts to make use of *all* of our confidence predictors. Each of our predictors is on the same interval $[0, 1]$, and we can propose a weighted average:

$$conf_{BB} = \frac{w_1 \times (1 - distance) + w_2 \times conf_{CBR} + w_3 \times comp_{CBR} + w_4 \times acc_{BB}}{w_1 + w_2 + w_3 + w_4} \quad (2)$$

where *distance* is the distance between the two solutions, $conf_{CBR}$ is our confidence in the CBR solution, $comp_{CBR}$ is our CBR competence, and acc_{BB} is our black box accuracy. Note that we use $1 - distance$ because the relationship between the CBR and black box confidences strengthens as the distance *decreases*.

For any domain, weights may be set by hill climbing (see Sect. 4). For concreteness, in our evaluation, hill climbing resulted in the following weights, which define what we henceforth refer to as the Weighted Average method:

$w_1 = 3.0$ (for $1 - distance$)
$w_2 = 0.25$ (for $conf_{CBR}$)

$$w_3 = 1.5 \qquad\qquad (\text{for } comp_{CBR})$$
$$w_4 = 3.0 \qquad\qquad (\text{for } acc_{BB})$$

Although the w_2 value is comparatively small, its inclusion as a nonzero value improves overall performance, showing that $conf_{CBR}$ provides useful information.

4 Evaluating Methods for Black Box Confidence

4.1 Assessing Quality of Confidence Predictions

In order to evaluate how well the proposed methods predict black box confidence, we propose a measure of confidence function *quality*. This measure is founded on the principle that ideally, confidence should be high if and only if error in the black box solution is low. Hence, a method for determining black box confidence is "good" if it assigns high confidence whenever there is low error in the solution, and low confidence whenever there is high error in the solution. We propose that the quality of a black box confidence predictor *is the degree to which the black box confidence prediction decreases monotonically with black box system error*. We consider the ability of the measure to properly rank cases by confidence as more important than the particular score it assigns, which could be normalized or scaled to fit domain expectations. The primary goal is to be able to assess which solutions should be ascribed more confidence than others, to identify those which might deserve more scrutiny. Spearman's rank correlation coefficient provides a method to assess the ability of the measure to properly order solutions, i.e., to determine the correlation of the measure's assessment with the actual ordering by accuracy [9].

We evaluate the quality of a confidence method as follows. First, we use it to compute the confidence in the black box solution for each problem in the testing set, and also determine the error in the black box system's solution for each problem in the testing set. We then rank the test problems from lowest confidence to highest confidence and rank the problems again from highest error to lowest error. We then computes the Spearman correlation coefficient ρ for these rankings.

A ρ value of 1 implies that, for that confidence method, black box confidence increases monotonically with reverse-ranked error. That is, black box confidence decreases monotonically with error. Similarly, a ρ value of -1 implies that black box confidence increases monotonically with error. We interpret the *strength* of this correlation using the table suggested by Akoglu [3] for Spearman coefficients. Using this table, one may say that the confidence method is "good" whenever we have a ρ value that corresponds to a strong correlation.

4.2 Experimental Questions

Given our measure of black box confidence quality, we can begin to experimentally evaluate the methods for black box confidence prediction. Three key questions are:

1. How effectively do the methods predict confidence in the black box system's solution? How do they compare with baselines of using CBR confidence alone or random confidence assignment?
2. Are the methods able to ascribe confidence successfully even when the black box's accuracy is very low?
3. When the black box outperforms the CBR (as is likely to be the case if the black box system is used instead of relying on CBR alone), is CBR confidence a better or worse predictor of black box confidence?

We perform an evaluation addressing questions 1 and 2 in this paper. We do not answer question 3, but we include a discussion of how this could be done in Sect. 6.1.

The first two questions directly deal with the quality of our black box confidence methods. In order to evaluate Question 1, we first establish baseline methods against which to compare. The baselines are:

- *CBR Confidence:* This baseline simply returns the confidence in the CBR system's solution in lieu of confidence in the black box.
- *Random Confidence:* This baseline returns a randomly generated confidence value on the interval $[0, 1]$.

5   Testing Confidence Methods with COBB

5.1   Overview of COBB System Design

Our testbed system, COBB (Case-based cOnfidence for Black Box), pairs a CBR system and a black box system to predict confidence in the black box system using the methods outlined in Sect. 3.6.

Our particular CBR system is a simple domain-independent retrieval system; it returns the closest case as a solution, with no adaptation. Feature weights were determined by hand, with no attempt to fine-tune weight values. A case is considered to solve a problem c if its solution is within a certain threshold of the solution for c (this threshold is used in competence calculations).

Our black box system is a multi-layer perceptron regressor provided by the SciPy 'scikit-learn' package [13,25]. In order to handle non-numerical attribute values during training, any non-numerical value from a training case is converted using one-hot encoding. If a non-numerical value is encountered during testing that was not seen in the training set, our one-hot encoding codes it as a sequence of zeroes. The COBB system does not rely on any particular properties of this regressor, so is fully general for other black box systems.

5.2   Test Domains

We test COBB with four regression datasets from the UCI Machine Learning Repository [11]: Computer Hardware (7 numerical attributes, 2 text attributes, 209 total cases), Student Portuguese Performance (SPP) [8] (16 numerical attributes, 17 text attributes, 649 total cases), Airfoil (5 numerical attributes, 0 text attributes, 1503 total cases), and SML (23 numerical attributes, 2 text attributes, 4137 total cases). For each domain, we perform 10-fold cross-validation on the domain dataset. For each fold, we train both the CBR and the black box on that fold.

To these domains we apply black box systems of varying accuracies, ranging from very high to very low. The average accuracy (across 10-fold cross-validation) of each black box's domain is as follows: Computer Hardware at 23.7%, SPP at 97.6%, Airfoil as 9.4%, and SML at 70.2%. We treat the Computer Hardware and Airfoil domains as examples for which black box accuracy is low, and similarly the SPP and SML domains as examples for which accuracy is high.

5.3   Results for Quality of Confidence Methods

For each domain and each confidence method, we compute the Spearman correlation function[1] to obtain the ρ value per fold. We then take the mean of the ρ values across the folds, and calculate a 95% confidence interval for the mean of the ρ values.

Figure 1 shows the mean Spearman ρ values (across 10-fold cross-validation) along with their confidence intervals for each of our domains.[2] We also include the mean Spearman ρ values and confidence intervals for our two baseline methods, CBR confidence in lieu of black box confidence and random confidence.

To assess the results, first, we compare our confidence methods to the baselines. As shown, the Naïve Method has positive correlation but low quality on the Computer Hardware and SPP domains. The Fuzzy preference method, on the other hand, has higher quality than the base methods on the Airfoil and SPP domains. The Weighted Average method performed consistently well across domains, maintaining a higher quality than both CBR Confidence and Random Confidence. Interestingly, in the SML domain all three methods have higher quality than Random Confidence, but match the quality of just using CBR confidence.

Next, we compare the quality of the confidence methods with each other. For the Computer Hardware and SPP domains, the weighted average method outperforms the Naïve and Fuzzy Preference confidence methods. Within the Airfoil domain, on the other hand, the Weighted Average and Fuzzy Preference confidence methods have roughly the same quality, and this quality is far higher than that for the Naïve method. Surprisingly, in the SML domain all three methods have roughly the same quality.

For the SML domain, the mean ρ values for all three methods are in the "very strong" range (using [3]). In addition, this domain is the only one in which CBR Confidence (on its own) has very strong quality. In the Airfoil domain, only the Weighted Average and Fuzzy Preference methods have mean ρ values in the "very strong" range, whereas the Naïve method has weak negative correlation.

---

[1] Using SciPy [13].
[2] Plotted using the Matplotlib package [12]).

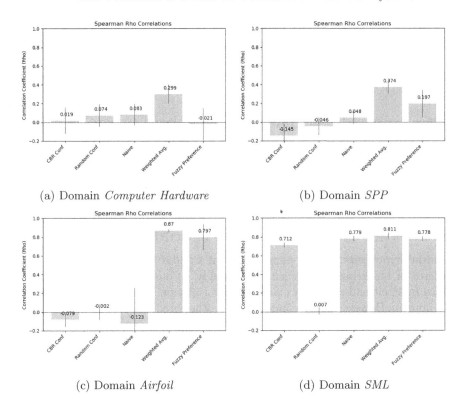

Fig. 1. Spearman ρ values for each confidence method, for each domain.

For the other two domains, we obtain mean ρ values in the weak and moderate range for all three confidence methods. We suspect that this discrepancy is due to how well the black box and CBR systems are paired for each domain, but future work is needed to evaluate this.

6   Discussion

6.1   Answering the Experimental Questions

The previous results suggest preliminary answers to questions 1 and 2 from Sect. 4.2.

Question 1 asks how successfully the three methods predict confidence in the black box solution. The experiments suggest that the Weighted Average method and Fuzzy preference method can give high quality predictions. Compared to the baselines of CBR confidence or random assignment, the Weighted Average method has consistently higher quality, whereas the Fuzzy Preference method only has higher quality in certain domains. The Naïve method has poor quality across our domains and fails to compete with the Weighted Average method in any domain except SML.

Question 2 asks whether the confidence methods can still have high quality even when the black box system has low accuracy. As mentioned in Sect. 5.2, the Airfoil domain black box system has very poor accuracy. The Weighted Average and Fuzzy Preference methods give good results in the Airfoil domain despite low black box accuracy.

Question 3 asks whether the CBR confidence has better or worse quality (as a black box confidence method) when the black box outperforms the CBR system. Answering this question requires considering instances in which CBR accuracy is low and black box accuracy is high. Tests on the current datasets did not produce any such situations. We intend to address this and further analyze results for the prior questions using additional datasets in future work.

6.2   Reflecting on Assumptions Made in COBB

An initial hypothesis for this paper was that CBR confidence could be a good predictor of black box confidence. Surprisingly, in our experiments, CBR confidence by itself had almost no monotonic correlation with error, except in the SML domain. However, when combined with other predictors (as in the Weighted Average method), CBR confidence is a useful predictor. So we must revise the initial hypothesis: The individual indicators *combined* provide a good prediction of black box confidence.

We previously mentioned the potential problem of the Naïve method that it does not apply when our confidence in the CBR is low and there is a large distance between the two solutions. Because the quality of our Naïve method is consistently low across all domains except SML, we conclude that the Naïve method is not a useful approach.

7   Explaining Confidence with COBB

The confidence judgments of COBB can be treated as standalone confidence judgments to aid a user determining trust in black box system conclusions, in the tradition of the confidence literature. However, the information developed by COBB can also be used to provide users with useful explanations of the confidence judgment, in two ways:

– Direct explanation from cases: When COBB retrieves a case for a similar problem, and low confidence suggests that additional scrutiny is needed, that case may be presented to the user either as substantiation (if its solution is in agreement) or as a conflict for the user to examine. Depending on the domain, presentation of the case could be paired with traditional information sources in explainable CBR to help the user assess the proposed conflicting solution (e.g., visualizations of attributes [19]). *Bracketing cases*, the most similar cases with and without the same solution [16], could be presented as well. The key added benefit from COBB compared to normal presentation of a retrieved case is that the user's attention need only be drawn to problems likely to be worthy of scrutiny.

– Explanations based on confidence indicators: The values for the specific confidence indicators from Sect. 3.3 can be presented to the user as additional data for assessing the overall confidence judgment.

8   Conclusion

We have proposed three methods for determining the confidence of a black box system using a paired CBR system. These methods make use of various predictors of black box confidence (i.e. distance between systems' solutions, CBR confidence, CBR competence, and black box accuracy).

We have also provided a test for quality of a black box confidence method. Applying this test to COBB, the black box confidence method with the best quality in general was the Weighted Average method. In certain domains, the Fuzzy Preference method has nearly as high quality as the Weighted Average method. As expected, the Naïve method has low quality in almost all domains (although in one domain it performs just as well as the former methods). We also noted that there is one domain in which both the Weighted Average and Fuzzy Preference methods are high-quality confidence methods, despite poor black box accuracy.

We see multiple future steps. In addition to performing evaluations on additional domains, we intend to incorporate both negative and positive indicators into the CBR confidence calculation. Some neural network systems output a value characterizing strength of a prediction; the quality of this self-assessment could be compared with that of the methods here, and could also be used as an additional input to the calculations of the weighted method. A more substantial extension would involve systematic study of the methods applied to different CBR and black box systems with varying competences and accuracies. This would enable experimentally answering questions such as Question 4.2. A fundamental question is how closely paired the CBR and black box systems must be for the approach to be useful. For practical application, we intend to explore the feasibility of using an initial calibration phase to determine domain suitability for the COBB approach.

COBB can explain its confidence assessment in terms of confidence indicators, as well as presenting cases for user examination when the CBR confidence assessor detects potential problems. A future topic is analyzing the value of explanations aimed directly at confidence.

The COBB approach was envisioned for situations in which a black box system is more accurate, motivating its use as the primary system but also raising the need for confidence assessment and explanation. An interesting question is whether, when the CBR system is more accurate, the black box system could help in assessing the CBR system confidence. The COBB approach could also be applied to two CBR systems that are independent (e.g., due to different similarity metrics) in parallel, with each assessing confidence in the other, explaining its confidence assessment, and presenting both conclusions to the user or to be combined in an overarching system. This could provide the basis for an approach to system- or user-mediated ensemble reasoning in CBR.

Acknowledgment. Testing was performed on a machine supported in part by Lilly Endowment, Inc., through its support for the Indiana University Pervasive Technology Institute.

References

1. Aha, D., Agudo, B.D., Garcia, J.R. (eds.): Proceedings of XCBR-2018: First Workshop on Case-Based Reasoning for the Explanation of Intelligent Systems (2018)
2. Aha, D., Darrell, T., Pazzani, M., Reid, D., Sammut, C., Stone, P. (eds.): Proceedings of the IJCAI-17: Workshop on Explainable AI (XAI) (2017)
3. Akoglu, H.: User's guide to correlation coefficients. Turkish J. Emerg. Med. **18**(3), 91–93 (2018)
4. Carney, J.G., Cunningham, P., Bhagwan, U.: Confidence and prediction intervals for neural network ensembles. In: International Joint Conference on Neural Networks. Proceedings (Cat. No. 99CH36339), IJCNN 1999, vol. 2, pp. 1215–1218, July 1999
5. Cheetham, W.: Case-based reasoning with confidence. In: Blanzieri, E., Portinale, L. (eds.) EWCBR 2000. LNCS, vol. 1898, pp. 15–25. Springer, Heidelberg (2000). https://doi.org/10.1007/3-540-44527-7_3
6. Cheetham, W., Price, J.: Measures of solution accuracy in case-based reasoning systems. In: Funk, P., González Calero, P.A. (eds.) ECCBR 2004. LNCS (LNAI), vol. 3155, pp. 106–118. Springer, Heidelberg (2004). https://doi.org/10.1007/978-3-540-28631-8_9
7. Cheetham, W.E.: Case-based reasoning with confidence. Ph.D. thesis, Rensselaer Polytechnic Institute (1996)
8. Cortez, P., Silva, A.: Using data mining to predict secondary school student performance. In: Brito, A., Teixeira, J. (eds.) Proceedings of 5th FUture BUsiness TEChnology Conference (FUBUTEC 2008), pp. 5–12. EUROSIS (2008)
9. Dodge, Y.: The Concise Encyclopedia of Statistics, chap. 379, pp. 502–505. Springer, New York (2008). https://doi.org/10.1007/978-0-387-32833-1_379
10. Domingos, P.: Knowledge discovery via multiple models. Intell. Data Anal. **2**(3), 187–202 (1998)
11. Dua, D., Graff, C.: UCI machine learning repository (2017). http://archive.ics.uci.edu/ml
12. Hunter, J.D.: Matplotlib: a 2D graphics environment. Comput. Sci. Eng. **9**(3), 90–95 (2007)
13. Jones, E., Oliphant, T., Peterson, P., et al.: SciPy: Open source scientific tools for Python (2001). http://www.scipy.org/
14. Keane, M., Kenny, E.: How case based reasoning explained neural networks: An XAI survey of post-hoc explanation-by-example in ANN-CBR twins. arXiv preprint arXiv:1905.07186 (2019)
15. Leake, D.: CBR in context: the present and future. In: Leake, D. (ed.) Case-Based Reasoning: Experiences, Lessons, and Future Directions, pp. 3–30. AAAI Press, Menlo Park (1996). http://www.cs.indiana.edu/~leake/papers/a-96-01.html
16. Leake, D., Birnbaum, L., Hammond, K., Marlow, C., Yang, H.: An integrated interface for proactive, experience-based design support. In: Proceedings of the 2001 International Conference on Intelligent User Interfaces, pp. 101–108 (2001)
17. Madsen, M., Gregor, S.: Measuring human-computer trust. In: Proceedings of the Eleventh Australasian Conference on Information Systems, pp. 6–8 (2000)

18. Marling, C., Sqalli, M., Rissland, E., Munoz-Avila, H., Aha, D.: Case-based reasoning integrations. AI Mag. **23**(1), 69–86 (2002)
19. Massie, S., Craw, S., Wiratunga, N.: A visualisation tool to explain case-base reasoning solutions for tablet formulation. In: Macintosh, A., Ellis, R., Allen, T. (eds.) SGAI 2004. Springer, Berlin (2004). https://doi.org/10.1007/1-84628-103-2_16
20. Moosavi-Dezfooli, S.M., Fawzi, A., Frossard, P.: Deepfool: a simple and accurate method to fool deep neural networks. In: Proceedings of the IEEE Conference on Computer Vision and Pattern Recognition, pp. 2574–2582 (2016)
21. Nasiri, S., Helsper, J., Jung, M., Fathi, M.: Enriching a CBR recommender system by classification of skin lesions using deep neural networks. In: ICCBR 2018, p. 86, July 2018
22. Nugent, C., Cunningham, P.: A case-based recommender for black-box systems. Artif. Intell. Rev. **24**(2), 163–178 (2005)
23. Oliphant, T.: NumPy: A Guide to NumPy. Trelgol Publishing, USA (2006). http://www.numpy.org/
24. Papadopoulos, G., Edwards, P.J., Murray, A.F.: Confidence estimation methods for neural networks: a practical comparison. IEEE Trans. Neural Netw. **12**(6), 1278–1287 (2001)
25. Pedregosa, F., et al.: Scikit-learn: machine learning in Python. J. Mach. Learn. Res. **12**, 2825–2830 (2011)
26. Reilly, J., Smyth, B., McGinty, L., McCarthy, K.: Critiquing with confidence. In: ICCBR, pp. 436–450 (2005)
27. Shin, C., Yun, U.T., Kim, H.K., Park, S.: A hybrid approach of neural network and memory-based learning to data mining. IEEE Trans. Neural Netw. Learning Syst. **11**(3), 637–646 (2000)
28. Smyth, B., McKenna, E.: Modelling the competence of case-bases. In: Smyth, B., Cunningham, P. (eds.) EWCBR 1998. LNCS, vol. 1488, pp. 208–220. Springer, Heidelberg (1998). https://doi.org/10.1007/BFb0056334
29. Szegedy, C., et al.: Intriguing properties of neural networks. arXiv preprint arXiv:1312.6199 (2013)
30. Zaragoza, H., Buc, D.: Confidence measures for neural network classifiers. In: Proceedings of the Seventh Int. Conf. Information Processing and Management of Uncertainty in Knowlegde-Based Systems (IPMU), vol. 1, pp. 886–893. Editions EDK (Paris), January 1998

Probabilistic Selection of Case-Based Explanations in an Underwater Mine Clearance Domain

Venkatsampath Raja Gogineni[✉], Sravya Kondrakunta,
Danielle Brown, Matthew Molineaux, and Michael T. Cox

Wright State University, Dayton, OH 45435, USA
gogineni.14@wright.edu

Abstract. Autonomous agents should formulate and achieve goals with minimum support from humans. Although this might be feasible in a perfectly static world, it is not as easy in the real world where uncertainty is bound to occur. One approach to solving such a problem is to formulate goals based on cases that explain discrepancies observed in the environment. However, in an uncertain world, multiple such cases often apply (i.e., as alternative explanations). Moreover, agents in the real world often have limited resources to achieve their missions. So, it is risky to generate and achieve goals for every applicable explanatory case. Our solution to these problems is to down-select the retrieved cases based on probabilities derived using Bayesian inference, then to monitor the selected cases' validity based on observed evidence. We evaluate the performance of an agent in an underwater mine clearance domain and compare it to another agent that selects a random case from the candidate set.

Keywords: Case selection · Case-base explanation · Explanation patterns

1 Introduction

Agents in a mine clearance domain must perform critical surveillance tasks with high accuracy in waters where communication and observability are limited. Due to unpredictable and dangerous events such as explosions in unexplored areas, intelligent behavior is required to understand and respond to the environment. In this domain, explanation is useful for both monitoring the environment and engendering trust in human operators who have only intermittent contact with the agent. Trust is not investigated in this paper. However, we consider the problem of selecting an explanatory case from a candidate set of applicable cases for a deliberative mine hunting agent that must respond to the discrepancies.

Our agent for this domain is called *GATAR (Goal-driven Autonomy for Trusted Autonomous Reasoning)* [1]. GATAR plans to achieve its goals, then executes each step in this plan after checking to confirm that its preconditions are met. The actions and the postconditions constitute GATAR's *expectations* about the world. When they do not match the current observations of the world, GATAR tries to recognize the cause of the discrepancy and predict its effects on the agent's goals. However, when

K. Bach and C. Marling (Eds.): ICCBR 2019, LNAI 11680, pp. 110–124, 2019.
https://doi.org/10.1007/978-3-030-29249-2_8

there are multiple cases that might explain a discrepancy, GATAR intelligently selects a case based on the observations it possesses. Furthermore, it also forms *expectations* based on the selected explanatory case to monitor its validity. These cognitive capabilities provide three benefits: first, they help GATAR to intelligently respond to such events and prevent their recurrence; second, they help GATAR adapt its reasoning behind explanation selection to observed evidence; third, they help GATAR communicate the rationale behind its behavior to third parties. This third benefit is critical to building trust when working with humans [2]. While the GATAR agent is the primary focus of this paper, we expect lessons learned and results to be generalizable; we expect intelligent explanation-based behavior with deliberately selected goals to be useful in other critical domains like surveillance, medicine and autonomous driving.

A different approach to responding to discrepancies would be to generate contingent plans in advance that cover all possibilities. Unfortunately, this is computationally intractable in most domains and handling all contingencies for an unexpected event can be overwhelming. Our approach requires additional domain knowledge, but any specific domain of interest, an abstract case-base of explanations defined by domain experts can be retrieved and adapted to explain a discrepancy.

We present a Bayesian approach to select an explanation among the candidate explanatory cases retrieved from the case base. Moreover, expectations are extracted from the selected explanation to monitor its viability. Here explanation provides a causal basis for goal generation and enables creation of communicative rationales for goal changes to third parties. This approach follows *Goal Directed Autonomy (GDA)* principles [3–7], in which agents respond to *discrepancies* (i.e., agent expectation failures) by generating explanations and generating goals based on those explanations.

In Sect. 2, we describe the representation of the explanatory cases, their retrieval, selection, goal formulation and GATAR's algorithmic approach to a discrepancy. A description of the domain, possible explanatory cases in the domain, and their retrieval are followed in Sect. 3. Section 4 presents the working example of the GATAR agent in a sample scenario. Evaluation and empirical results are presented in Sects. 5 and 6. Related work is illustrated in Sect. 7. Finally, the conclusion completes the paper in Sect. 8.

2   Case-Based Explanation Patterns

In our work, we use case-based explanations [8–10]. Each case in the case-base is an abstract *explanation pattern (XP)* [11, 12] engineered for a specific domain (see Fig. 1). An XP is a data structure that represents a causal relationship between two states and/or actions; each action/state is abstractly defined with variables to be adapted during or after case retrieval. In GATAR, an action or state is referred to as a *node* and different types of nodes are described based on their role in an XP.

- *Explains node*: A discrepancy/unknown state that is observed;
- *Pre-XP node*: Action/state that is observed along with the explains node;
- *XP-asserted node*: Action, state or XP contributing to the explanation's cause.

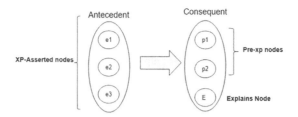

Fig. 1. The explanation pattern (XP) causal structure in which XP-asserted nodes (e1, e2, e3) form an antecedent, and a consequent is made up of pre-XP nodes (p1, p2, p3) and an explains node (E); XP-asserted nodes thus cause the associated explains and pre-XP nodes.

2.1   GATAR's Algorithm to Respond to the Discrepancies

Algorithm 1 represents GATAR's approach toward identifying and responding to a discrepancy. Given a plan $\pi = <a_1...a_n>$ to achieve goals from agenda $\hat{G} = \{g_1...g_m\}$, whenever its current observations (s_c) do not match the expectations (s_e), a discrepancy is detected. These observations (s_c) are obtained (line 1) from the successor function (γ) that takes in the current state (s_c) and action (a_1) [13]. Since, GATAR is currently executing a_1, its plan is updated to the set of remaining actions (line 2). GATAR then adds to its expectations (s_e) the preconditions (a_1^+) and effects (a_1^-) of the current action (line 3). When expectations differ from observations (line 4), the algorithm tries to explain this discrepancy from the case base (lines 5–8). First, a candidate set of explanations $c = \{\chi_1...\chi_k\}$ is retrieved from the case-base $c = \{\chi_1...\chi_k...\chi_l\}$ of explanations. An explanatory-case (χ_s) is selected from these candidates by applying Bayesian inference (line 5). Next, additional expectations are extracted from the XP-asserted nodes of the selected explanation and added to the current set of GATAR's expectations (line 6). This facilitates monitoring the validity of the explanatory case. The interpretation function (beta) sets a new current goal set to respond to the discrepancy (line 7) (see [32]). Finally, the new goals (g_c) are added to the goal agenda $(\hat{G})$ (line 8).

> **Algorithm 1:**
> $DiscrepancyCheck\ (\pi, s_c, s_e, \hat{G}, c, g_c)$
> 1. $s_c \leftarrow \gamma\ (s_c, a_1)$
> 2. $\pi \leftarrow < a_2 ... a_n >$
> 3. $s_e \leftarrow s_e \cup pre(a_1) \cup a_1^+ - a_1^-$
> 4. $if\ s_c \not\models s_e$
> 5. $\chi_s \leftarrow Select(Retrieve\ (c, s_c))$
> 6. $s_e \leftarrow s_e \cup XP\text{-}asserted\ (\chi_s)$
> 7. $g_c \leftarrow \beta\ (s_c, g_c)$
> 8. $\hat{G} \leftarrow \hat{G} \cup g_c$

2.2 Retrieving, Reusing and Revising Explanation Patterns from a Case Base

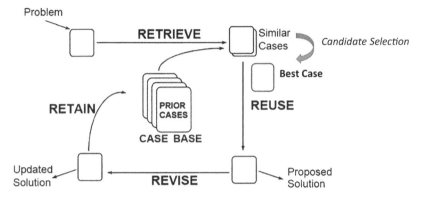

Fig. 2. The CBR process with candidate selection (adapted from [14])

Case-based reasoning follows a four-step process to retrieve, reuse, revise and retain cases [14–16] (see Fig. 2). In the current work, we assume that all cases are defined by domain experts, so we do not consider retention. The following describes how XPs are retrieved, reused and revised.

A set of abstract XPs is retrieved when an unpredicted state or action observed by the agent unifies with each explains node of an XP in the case-base. If the unification turns out to be successful then the pre-XP nodes of the corresponding case are unified with the observations of the corresponding states or actions, if they turn out to be successful then the specific XP is retrieved. The retrieved abstract XP is reused by binding variables in the antecedent to values found during unification of the consequent. However, if the XP-asserted nodes in the reused XP contain hypothetical information they can be revised when the new knowledge is obtained from further observations. We now describe our approach to selecting an explanation from a retrieved candidate set.

2.3 Probabilistic Selection of Explanation Patterns from a Candidate Set

In our work, we use Bayesian inference to select an explanation from the candidate set of retrieved explanations. Bayesian inference takes prior knowledge about the parameter and uses newly collected data or information to update the prior beliefs [17]. The agent uses its observations of states/actions as new information in updating its prior beliefs. Bayes equation adapted to the explanation patterns is given as follows:

$$P(XP|evidence) = \frac{P(evidence|XP) * P(XP)}{P(evidence)}$$

$$P(evidence|XP) = \frac{no.\ of\ times\ the\ evidence\ obtained\ given\ explanation}{total\ no.\ of\ times\ evidences\ obtained\ given\ explanation}$$

$$P(XP) = \frac{no.\ of\ times\ the\ explanation\ is\ selected}{total\ no.\ of\ times\ the\ explanation\ is\ in\ a\ candidate\ set}$$

$$P(evidence) = \frac{no.\ of\ times\ the\ evidence\ is\ obtained}{total\ no.\ of\ times\ evidence\ is\ in\ a\ candidate\ set}$$

Each XP-asserted node is considered as an evidence, so Bayes equation is applied for every XP-asserted node. The probability of the explanation given all evidence is

$$P(XP|evidence_{1..n}) = P(XP|(evidence_1 \cap evidence_2 \ldots evidence_n))$$

For all the explanations, the one with the highest $P(XP|evidence_{1..n})$ is the one to be selected. However, since our explanations are designed manually by domain experts, we assume the domain experts provide the expected values for different prior probabilities i.e. $P(XP), P(evidence_{1..n})$, before the mission starts. Moreover, if there is no evidence obtained then an explanation having the highest $P(XP)$ among the candidate cases is selected. Finally, goals are formulated and the XP monitored to check validity.

2.4   Goal Formulation and Monitoring Explanation Patterns

Goal formulation is essential for an intelligent agent to respond to discrepancies [18, 19]; in GATAR, we perform formulation by preventing the recurrence of one or more explanation antecedent nodes. Antecedent nodes may include actions and/or states; therefore, when GATAR wishes to prevent an undesired consequent from recurring, it considers the elimination of antecedent actors or objects that participate in antecedent states as potential goals. The potential goals are generated using the removal mapping function that takes in the actors or objects and outputs the goals that eliminate them.

Monitoring a selected explanation is essential for an intelligent agent to adapt its beliefs and misclassifications. In our work, each node in the XP-asserted nodes of the selected XP is added to the agent's expectations. These nodes constitute the evidence. Whenever the evidence matches the agent's observations of the world, the agent changes its beliefs. However, when the evidence contradicts the agent's observations, the selected XP is switched with the next probable explanation.

3   Underwater Mine Clearance Domain

Our approach is implemented in a limited mine clearance domain [20], which is simulated using MOOS-IVP [21], software that provides complete autonomy for marine vehicles. Figure 3 shows the simulation of the mine clearance domain with a GATAR agent directing a Remus unmanned underwater vehicle. The Q-route is a safe passage for ships to enter and leave the port and is represented as a rectangular area in Fig. 3. GA1 and GA2 are the two octagonal areas where mines are expected to exist, while the triangular objects are the mines. The goals of the agent are to survey and clear mines in GA1 and GA2. These goals are given to the agent after a reconnaissance mission performed by a different agent across the whole sea route. The Remus has a sonar sensor with a specific width of ten units and a length of five units to detect mines.

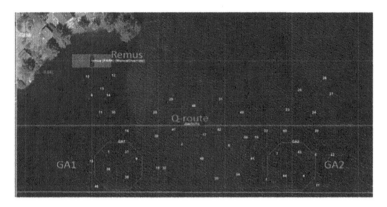

Fig. 3. Underwater Mine Clearance domain with two clearance areas in the Q-route. (labelled items in this figure match later figures)

3.1 Discrepancies in Underwater Mine Clearance Domain

In the underwater mine clearance domain, several events often co-occur simultaneously, and many events cannot be predicted based on knowledge available to an agent. These events might affect the agent itself or the mission of the agent. Explanations help the agent to recognize these events and respond to them. We will look at several uncertain events that might happen.

Events in this domain include minelaying, sensor failure, and reconnaissance failure. Minelaying events occur when an enemy ship, aerial vehicle, or fishing vessel lays traps to hurt friendly ships. Removing such mines within areas GA1 and GA2 is an explicit goal for GATAR in the above scenario. Sensor failure event indicates that a faulty GATAR's sensor is responsible for a misclassification of mine, and the failure of proper reconnaissance mission indicates that an agent prior to GATAR did not identify mines which in turn failed its mission. We will look at some of the explanations in the next section that sheds some knowledge on the discrepancies that happen in this domain.

3.2 Plausible Explanations in Underwater Mine Clearance Domain

In GATAR, explanations are retrieved by a version of the *Meta-AQUA* component [22], a story understanding system that tries to explain discrepancies in a story through use of case-based explanations. We have integrated this system with the *MIDCA (Metacognitive Integrated Dual Cycle Architecture)* component [23], a cognitive architecture that perceives and acts directly on the world, to examine the interaction between explanation generation (by Meta-AQUA) and goal formulation (provided by MIDCA); for the purposes of this paper, we refer to the combined system as the GATAR agent.

As mentioned earlier, each explanation in our case-base is abstractly designed by the domain experts to cover the possible discrepancies in the underwater mine clearance domain. We will look at one of the detailed candidate explanatory case that can occur for a discrepancy of detecting multiple mines at a location.

Figure 4 represents an abstract XP structure, which explains that an enemy ship laid the mines in the clear-area and hence the UUV (unmanned underwater vehicle) detected multiple mines. UUV is an abstraction for GATAR, while Clear-area is an abstraction for areas that are not expected to have mines. As described earlier the XP structure is in the form of an antecedent causing consequent. The consequent contains the explains node "hazard-detection(uuv, mine)" to represent the discrepancy of detecting a mine by the UUV, while the Pre-XP nodes "at-location(mine, clear-area)" and "hazard-checked(mine, clear-area)" are the observational support to the discrepancy. These Pre-XP nodes thus convey that a mine is already checked prior to the currently detected mine, and implicitly conveys that multiple mines exist.

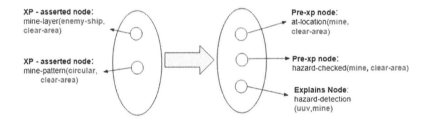

Fig. 4. XP Structure that describes enemy ship caused multiple mines

The antecedent contains the XP-asserted nodes of mine laying activity by the enemy-ship in the clear-area represented by the action "mine-layer(enemy-ship, clear-area)" and the mine pattern being circular is represented by state "mine-pattern(circular, clear-area)". Whenever this *XP* is retrieved the abstracted variables will be replaced by the observations. For example, UUV will be replaced by GATAR, clear-area will be replaced by all the areas in the domain except GA1 and GA2.

In a similar structure there are multiple explanations for multiple discrepancies that might happen in the domain. Below are the high-level descriptions of those explanations along with their responsive behaviors in Table 1.

Table 1. Explanations for the discrepancies along with the behaviors

Explanation	Discrepancy	Behavior
Fisher-XP: Fisher-vessel laid a single mine	Single mine detected at the clear-area	Remove the single mine (Q-route) and report existence of fishing vessel
Sensor-XP: GATAR's sensor failure	Single mine detected at the clear-area	Recalibrate the sensor and continue the mission
Tide-XP: Mines drifted from expected regions GA1 or GA2 with tides	Single or multiple mines detected at clear-area	Clear the mines and continue the mission
Reconnaissance-XP: Reconnaissance failed	Multiple mines detected in the clear area	Survey and clear the mines in the whole Q-route
Enemy-Ship-XP: Enemy ship laid the circular pattern of mines	Multiple mines detected in the clear area	Survey region covering circular area with calculated radius & clear mines
Enemy-Aerial-XP: Aerial vehicle laid the line pattern of mines	Multiple mines detected in the clear area	Survey region across line region with statistically calculated slope & clear mines

4 Example of Selecting a Case-Based Explanation

An example scenario from the underwater mine clearance domain can help us understand the application of selecting explanations from the candidate set by GATAR. Moreover, this will also give an idea of how such an application can improve the performance of GATAR.

Figure 5 represents the example scenario in the underwater mine clearance domain, where there are four minefields: first, at GATAR's transit to the Q-route; second, at the GA1; third, at GATAR's area of transit from GA1 to GA2; and fourth, at the GA2. However, GATAR has a mission to clear the second and fourth minefields, so any mines encountered in the first and the third minefields are discrepancies. When GATAR encounters the first mine in the first minefield, it retrieves a candidate set of explanatory cases. Table 2 shows this set of explanatory cases retrieved as well as the respective selection probabilities.

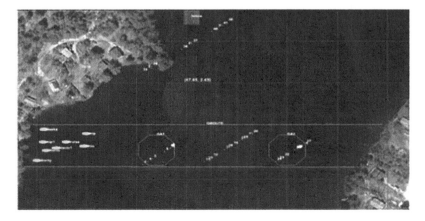

Fig. 5. A scenario where mines are laid by an aerial vehicle

These probability values were chosen to match typical values in the domain. Since Fisher-XP is the only explanation with evidence, the probability to select an explanation is 0.5 and the probabilities of the Sensor-XP and Tide-XP are 0.14 and 0.08 respectively. Thus Fisher-XP is selected due to its higher probability and the table is updated.

Table 3 shows the updated probabilities calculated using the Bayesian inference as described in the Sect. 2.3. Furthermore, a goal is generated to report the existence of a fisher vessel laying mines after the mission. Finally, the evidence is added to GATAR's expectations i.e. there is only one mine in the transit area to the Q-route.

Table 2. Probabilities of the retrieved explanations

Explanation (*XP*)	P (*XP*)	P (*XP* | evidence)
Fisher-XP: Fisher vessel laid a single mine	0.20	P (XP | single-mine) = 0.5
Sensor-XP: GATAR's sensor failure	0.14	No evidence obtained regarding damage of sensor
Tide-XP: Mines drifted by the tides from the mine expected regions (GA1 or GA2)	0.08	No evidence obtained regarding tides and the mines not at expected minefield locations

Table 3. Probabilities after selecting the explanation

Explanation (*XP*)	P (*XP*)	P (*XP* | evidence)
Fisher-XP: Fisher vessel laid a single mine	0.207	P (XP | single-mine) = 0.509
Sensor-XP: GATAR's sensor failure	0.138	No evidence obtained regarding damage of sensor
Tide-XP: Mines drifted by the tides from the mine expected regions (GA1 or GA2)	0.079	No evidence obtained regarding tides and the mines not at expected minefield locations

When GATAR encounters another mine in the same area, its expectation of a single mine is violated, and it retrieves another set of candidate explanatory cases to explain the discrepancy. Since the previously selected explanation is not valid, it updates its probabilities and drops the goal to report about the fisher vessel. Similarly, as described above, GATAR selects the explanation that an aerial vehicle laid the mines, updates its probabilities, formulates goals to report about the vehicle and finally generates an expectation that the mines exist in a straight line.

After clearing all the mines from GA1, GATAR encounters mines in the third minefield, selects the explanation that the mines exist in a straight line, clears all the mines in the Q-route, and updates its probabilities. Finally, it returns to the GA2 and clears all the mines in the GA2. Later after the mission GATAR reports the existence of the aerial vehicle and its behavior to the base which is outside the scope of the paper. However, in this scenario, GATAR intelligently made the Q-route safe for the ships to traverse.

5   Experimental Setup

As previously mentioned, GATAR retrieves explanatory cases that explains any discrepancies. A case-base of ten abstract explanatory cases are used to cover all the behaviors of the GATAR agent in this domain. From the set of the retrieved cases it selects an applicable explanation by applying Bayesian inference. Furthermore, the antecedents of the candidate explanations are monitored to update the beliefs of the agent. GATAR's ability to the above intelligent behavior is evaluated by number of ships that traverse the Q-route without hitting mines. Moreover, the performance of GATAR is compared to a random agent. The random agent is like the GATAR agent in retrieving a candidate set of explanatory cases. However, it differs from GATAR in the selection process. It selects a random explanation from the candidate set to respond to a discrepancy. The experiment is laid out in terms of two scenarios that are differentiated by the placement of mines. Each scenario has three groups of three ships that start at a specific location and ends across the other side of the Q-route at another specific location. The first group will start with incremental delays while the second and third groups will start with a constant delay of 0.25 and 0.50 min respectively following the first group. The agent has the goals to clear mines in the areas GA1 and GA2.

Figure 6 represents the experimental setup of both scenarios 1 and 2. In the first scenario, an aerial vehicle laid mines in a line pattern in the areas of transit, GA1, GA2 and the transit area between GA1 and GA2. The setup gives the agent a proper evidence that an aerial vehicle laid the mines while pursuing its goals. In the second scenario, an enemy ship laid the mines in the area between GA1 and GA2, while the aerial vehicle laid the mines at transit, GA1 and GA2. The first scenario tests GATAR's ability to select the correct explanation by obtaining evidence while the second scenario tests GATAR's ability to switch explanations to clear mines between GA1 and GA2 when the evidence implies an explanation that an aerial vehicle is laying mines.

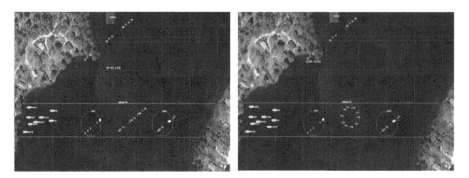

Fig. 6.   (Left) Scenario 1 – mines laid by an aerial vehicle. (Right) Scenario 2 – mines laid by both an aerial vehicle and an enemy ship.

6 Empirical Results

Figure 7 shows the comparison of the performance with GATAR and the random agent in the scenario 1. The X-axis shows the starting time, in increments of 0.25 min, for the first group of ships after the agent's mission starts. The Y-axis is the average number of ships that successfully traversed through the Q-route. The experiment runs five times for every time delay and for every agent and then averaged for each experiment. GATAR outperforms the random agent at every time interval greater than 0.25.

When we look closely into results with a delay of 0.25 min, when the first group of ships start, both the agents are still in their transit to the GA1. By the time the third batch of ships start, both the agents cleared mines in GA1 and are in transit to Q-route. This implies that a delay of 0.25 min is not long enough for both agents to clear enough mines in the Q-route for any ship to survive. However, at an interval of 0.5 min of time, both the agents could clear mines in GA1 as well as some on their transit to GA2 before the third group of ships started, which signifies the steep increase in the curve. The random agent underperforms because of the failure to select the correct explanation leading to a different behavior.

At 0.75 interval of time, the GATAR agent could successfully clear GA1 and some mines on its transit to the GA2 by the time the second group of ships start. The GATAR agent completely clears all the mines in the Q-route before the third batch of ships start. The random agent underperformance is due to its wrong choice in behavior. Finally, at 1.25 interval of time, GATAR could completely clear all the mines in the Q-route, so all the ships survived. Even at 2.5 intervals of time, the random agent could not clear all the mines because of its selection of behaviors that ignore the mines on their transit to the Q-route.

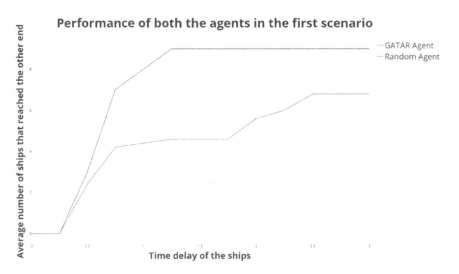

Fig. 7. Performance of the GATAR and the random agent in scenario 1

Figure 8 compares the performance of GATAR and the random agent in scenario 2. The experiment pattern follows the same as described above. GATAR outperforms the random agent at every interval of time greater than the 0.75. In this scenario, GATAR clears the mines in GA1 and when it happens to identify a mine between GA1 and GA2 in Q-route, it generates a behavior to traverse in a line based on the prior evidence it obtained. After observing that the pattern is not a line, the GATAR agent changes its behavior to a deep search pattern. Later it continues to clear the mines in GA2.

At 0.25 min of time, the random agent happens to select a behavior to clear only the mines that it came across during two of the five random runs before the third group of ships started. This resulted in clearing a mine at the upper part of the circle of mines between the areas GA1 and GA2 in the Q-route, which allowed one of the ships to survive. However, this is not the case with GATAR where all ships sank at time .25. At 1.25 min, GATAR clears all mines in the Q-route which allowed all ships to traverse the route. At a delay of 2.5 min of time, the random agent seems to complete all its goals before the ships start. However, because of its wrong choice of explanations, the formulated goals could not allow for safe passage for all the ships.

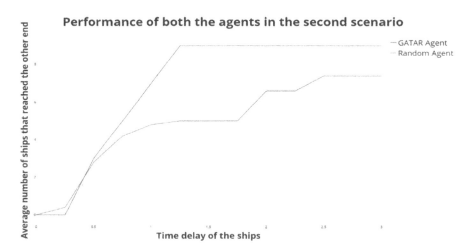

Fig. 8. Performance of the GATAR and the random agent in scenario 2

7   Related Work

Explanation patterns were introduced by Schank in 1982 [8] and were later used in the story understanding systems SWALE [24, 25] and AQUA [26]. SWALE is a case-based approach to explanation of discrepancies in a story that retrieves, adapts and stores explanation patterns. SWALE demonstrated an early technique for ruling out competing explanations using memory knowledge. AQUA (Asking Questions and Understanding Answers) operates by first questioning missing knowledge in a story, then using explanation patterns to understand the answers.

Gentner and Forbus in [27], present the MAC/FAC approach, which closely aligns to our approach. It is a two-phase retrieval process to improve the performance of the retrieval process. MAC (Many are chosen) refers to using flat structures to eliminate irrelevant cases outright while FAC (Few are chosen) refers to applying computationally complex algorithms to rank the cases from MAC. Our approach aligns closely with the two-phase retrieval process, where we select a candidate set of *XP's* using explains and Pre-XP nodes and apply probabilities to select a single *XP*. Moreover, Morwick and Leake in [28], show a performance increase in having such two-phase retrieval approach.

Roth-Berghofer et al's [29] work on classifying explanations and their use-cases according to the user's intentions is one of the theoretical research directions towards explanations in case-based reasoning (see also [30]). This paper introduces the concept of "explanation goals" that are used to decide when and what the system should explain to users based on their expectations. In future research, we will investigate application of these techniques to prevent the system from repeatedly explaining the same type of unexpected events to a user who is already familiar with them. This paper also talks about different kinds of explanations and classifies them into four different knowledge containers, all of which are used to generate explanations based on the user's goals or intentions.

In [2], a robot adapts its behavior to gain trust in human machine teaming using the approach of case-based reasoning. In addition, Floyd and Aha [31] presented an approach to explain such adaptations based on an operator's feedback, and evaluated their system based on how closely the explanations aligned with the operator's feedback. Our interest in generating explanations of the intelligent behavior of an agent aligns closely with the interests of this paper.

8   Conclusion

In this paper we discussed a probabilistic approach of selecting an explanatory case from the candidate set of retrieved cases for the discrepancy. Moreover, we have also presented an approach to monitor the selected case, which helps GATAR to adapt its beliefs and switch cases if a selection error occurs. Finally, the results show that the performance of GATAR is better than the random agent. The causal structure of explanations helps GATAR communicate and justify its behavior to human users.

In some cases, there can be more than one explanation relevant to a discrepancy. For example, if both an enemy ship and an aerial vehicle laid mines in the same area then it would be incorrect to choose one explanation. In future research, we would like to reason about the probability of co-occurrence of causal events leading to discrepancies.

We also acknowledge that our current explanatory cases will provide only abstract mine patterns which are evenly spaced or uniformly distributed. However, the real world contains noise leading to misclassification of the evidence. In the future, we would like to use statistical learning algorithms to predict mine patterns as well as their distributions. Furthermore, GATAR should reason about the tradeoff between the time required to clear the mines in GA1 and GA2 and the time to pursue its additional goals from the selected explanatory case. We expect that such reasoning will improve GATAR's performance.

Moreover, we also want GATAR to explain the rationale behind its intelligent behavior to the human operator and obtain some feedback related to the hypothetical evidence. This will improve the quality of explanatory cases selected from the case-base.

Finally, we want GATAR to reason about the tradeoff between immediately formulating goals from a selected explanatory case and formulating goals after obtaining evidence. We expect this functionality to help GATAR to adapt after selection failures.

Acknowledgements. This research was supported by AFOSR under grant FA2386-17-1-4063 and by ONR under grant number N00014-18-1-2009. We thank the anonymous reviews for the comments and suggestions.

References

1. Gogineni, V., Kondrakunta, S., Molineaux, M., Cox, M.T.: Application of case-based explanations to formulate goals in an unpredictable mine clearance domain. In: Proceedings of the ICCBR-2018 Workshop on Case-Based Reasoning for the Explanation of Intelligent Systems, Stockholm, Sweden, pp. 42–51 (2018)
2. Floyd, M.W., Drinkwater, M., Aha, D.W.: Trust-Guided Behavior Adaptation Using Case-Based Reasoning. Naval Research Laboratory, Washington, United States (2015)
3. Paisner, M., Cox, M., Maynord, M., Perlis, D.: Goal-driven autonomy for cognitive systems. In: Proceedings of the Annual Meeting of the Cognitive Science Society, vol. 36, no. 36 (2014)
4. Molineaux, M., Klenk, M., Aha, D.W.: Goal-driven autonomy in a navy strategy simulation. In: AAAI, pp. 1548–1554 (2010)
5. Dannenhauer, D., Munoz, A.H.: Raising expectations in GDA agents acting in dynamic environments. In: IJCAI, pp. 2241–2247 (2015)
6. Cox, M.T.: Goal-driven autonomy and question-based problem recognition. In: Proceedings of the 2nd Annual Conference on Advances in Cognitive Systems, Maryland, USA, pp. 29–45 (2013)
7. Munoz-Avila, H., Aha, D.W., Jaidee, U., Klenk, M., Molineaux, M.: Applying goal driven autonomy to a team shooter game. In: FLAIRS Conference (2010)
8. Schank, R.C., Kass, A., Riesbeck, C.K.: Inside Case-Based Explanation. Psychology Press, London (2014)
9. Cox, M.T., Burstein, M.H.: Case-based explanations and the integrated learning of demonstrations. Künstliche Intelligenz (Artif. Intell.) **22**(2), 35–38 (2008)
10. Ram, A.: Indexing, elaboration and refinement: incremental learning of explanatory cases. Mach. Learn. **10**, 201–248 (1993)
11. Schank, R.C.: Explanation Patterns: Understanding Mechanically and Creatively. Psychology Press, London (2013)
12. Ram, A.: A theory of questions and question asking. J. Learn. Sci. **1**(3 and 4), 273–318 (1991)
13. Ghallab, M., Nau, D., Traverso, P.: Automated Planning: Theory and Practice. Elsevier, Amsterdam (2004)
14. de Mántaras, R.L., et al.: Re-trieval, reuse and retention in case-based reasoning. Knowl. Eng. Rev. **20**(3), 215–240 (2006)

15. Aamodt, A., Plaza, E.: Case-based reasoning: foundational issues, methodological variations, and system approaches. AI Commun. **7**(1), 39–52 (1994)
16. Kolodner, J.: Case-Based Reasoning. Morgan Kaufmann, San Francisco (1993)
17. Box, G.E., Tiao, G.C.: Bayesian Inference in Statistical Analysis. Wiley, Hoboken (2011)
18. Maynord, M., Cox, M.T., Paisner, M., Perlis, D.: Data-driven goal generation for integrated cognitive systems. In: 2013 AAAI Fall Symposium Series (2013)
19. Hanheide, M., et al.: A framework for goal generation and management. In: Proceedings of the AAAI Workshop on Goal-Directed Autonomy (2010)
20. Kondrakunta, S., Gogineni, V., Molineaux, M., Munoz-Avila, H., Oxenham, M., Cox, M.T.: Toward problem recognition, explanation and goal formulation. In: Proceedings of the 6th Goal Reasoning Workshop at IJCAI/FAIM-2018, Stockholm, Sweden (2018)
21. Benjamin, M.R., Schmidt, H., Newman, P.M., Leonard, J.J.: Nested autonomy for unmanned marine vehicles with MOOS-IvP. J. Field Robot. **27**(6), 834–875 (2010)
22. Cox, M.T., Ram, A.: Introspective multistrategy learning: on the construction of learning strategies. Artif. Intell. **112**(1–2), 1–55 (1999)
23. Cox, M.T., Alavi, Z., Dannenhauer, D., Eyorokon, V., Munoz-Avila, H., Perlis, D.: MIDCA: a metacognitive, integrated dual-cycle architecture for self-regulated autonomy. In: AAAI (2016)
24. Schank, R.C., Leake, D.B.: Creativity and learning in a case-based explainer. Artif. Intell. **40**(1–3), 353–385 (1989)
25. Leake, D.B.: Evaluating Explanations: A Content Theory. Psychology Press, London (2014)
26. Ram, A.: AQUA: Questions that drive the explanation process. Georgia Institute of Technology (1993)
27. Gentner, D., Forbus, K.: MAC/FAC: A model of similarity-based retrieval. In: Proceedings of the Thirteenth Annual Conference of the Cognitive Science Society, Chicago, IL, pp. 504–550 (1991)
28. Kendall-Morwick, J., Leake, D.: A study of two-phase retrieval for process-oriented case-based reasoning. In: Montani, S., Jain, L.C. (eds.) Successful Case-Based Reasoning Applications-2, vol. 494, pp. 7–27. Springer, Heidelberg (2014). https://doi.org/10.1007/978-3-642-38736-4_2
29. Roth-Berghofer, T.R., Cassens, J.: Mapping goals and kinds of explanations to the knowledge containers of case-based reasoning systems. In: Muñoz-Ávila, H., Ricci, F. (eds.) ICCBR 2005. LNCS (LNAI), vol. 3620, pp. 451–464. Springer, Heidelberg (2005). https://doi.org/10.1007/11536406_35
30. Aamodt, A.: Explanation-driven case-based reasoning. In: Wess, S., Althoff, K., Richter, M. (eds.) Topics in Case-Based Reasoning, vol. 837, pp. 274–288. Springer, Berlin (1994). https://doi.org/10.1007/3-540-58330-0_93
31. Floyd, M.W., Aha, D.W.: Incorporating transparency during trust-guided behavior adaptation. In: Goel, A., Díaz-Agudo, M.Belén, Roth-Berghofer, T. (eds.) ICCBR 2016. LNCS (LNAI), vol. 9969, pp. 124–138. Springer, Cham (2016). https://doi.org/10.1007/978-3-319-47096-2_9
32. Cox, M.T.: A model of planning, action, and interpretation with goal reasoning. Adv. Cogn. Syst. **5**, 57–76 (2017)

A Data-Driven Approach for Determining Weights in Global Similarity Functions

Amar Jaiswal$^{(\boxtimes)}$ and Kerstin Bach

Department of Computer Science,
Norwegian University of Science and Technology, Trondheim, Norway
`amar.jaiswal@ntnu.no`
`https://www.ntnu.edu/idi`

Abstract. This paper presents a method to discover initial global similarity weights while developing a case-based reasoning (CBR) system. The approach is based on multiple feature relevance scoring methods and the relevance of features within each scoring method. The objective of this work is to utilize the characteristics of a dataset when creating similarity measures. The primary advantage of this method lies in its data-driven approach in the absence of domain knowledge in the early phase of a CBR system development. The results obtained based on the experiments on multiple public datasets show that the method improves the performance of similarity measures for a CBR system in discriminating relevant similar cases. Evaluation of the results is based on the method suitable for unbalanced datasets.

Keywords: Global similarity weights · Feature weights · CBR · Case-based reasoning

1 Introduction

Case-based reasoning [1] (CBR) is a problem solving methodology based on past experiences. It is based on the assumption that similar problems have similar solutions. With this assumption, a CBR system is designed to retrieve similar cases for a new problem. The solution of the retrieved cases are used to solve the new problem. Hence, it becomes a key to retrieve the correct and relevant cases. This paper proposes a data-driven approach to address this issue in the early phases of a CBR system development, where the domain knowledge might not be initially available.

When developing CBR systems today, we often have access to datasets containing experiences. Those experiences are often structured for various purposes and not necessarily all information are relevant to represent a case. The relevant attributes can often be determined in collaboration with experts or using data driven approaches, while the definition of initial similarity measures are more challenging. This task has been addressed by researchers before, and learning or deriving similarity measures is an active field in CBR research [11,27].

© Springer Nature Switzerland AG 2019
K. Bach and C. Marling (Eds.): ICCBR 2019, LNAI 11680, pp. 125–139, 2019.
https://doi.org/10.1007/978-3-030-29249-2_9

In this paper we will investigate whether we can derive global similarity measures from a given dataset. In paper [22] similarity measures have been learned using feedback and similarity teacher. Local similarity measures as well as the learning of comprehend similarity measures have been obtained using Artificial Neural Networks is presented in [2,12].

However, deriving the weights for global similarity measures from a given dataset is a novel approach and has the potential to improve building initial CBR systems. In this paper, we will address how those similarity measures can be automatically defined and we show how the proposed methods works on open datasets.

The hypothesis of the paper is using an ensemble of feature relevance scoring methods to discover initial feature weights for a CBR system. This can be used in early phases of a CBR system development, where the researcher has little or no guidance for the domain knowledge.

The paper is organised as follows: Sect. 2 discusses the related work about finding feature weights, Sect. 3 presents the core of the paper, how to discover feature weights using a data-driven approach. Section 4 provides the details of the experiments' setup, datasets used, and evaluation process. Section 5 presents the experimental results. Section 6 is dedicated to the interpretation of the results and its relevance to our hypothesis. The last section concludes the paper and projects the future work.

2   Related Work

Extracting feature weights is a well known research problem area since multiple decades [4]. Multiple methods and references are mentioned in the paper that are used in feature weight extraction. In another paper [5] Aha and Goldstone demonstrate that the feature weights in the similarity setting are context dependent, with the help of 40 human subjects in their experiment. Thus, a universal algorithm for feature weight extraction might not be possible in this context.

The work in [8] describes the challenges involved with the symbolic features to be used for k-NN, and claims that the weighted k-NN is advantageous in simplicity, training speed, and perspicuity. The paper [21] is focused on learning a non-symmetric local similarity metrics, which is based on the learning approach.

Stahl and Gabel [23] discusses the challenges involved in developing a CBR system and points out that the required knowledge, many a times, is unavailable during the developmental phase. He also describes optimising the similarity measures with the help of a *similarity teacher*, which might not be available in the initial phases of the development.

Cost and Salzberg [8] discusses the importance of k-NN for classification tasks, where features have symbolic values. It also presents the experimental results based on three techniques PEBLS, back propagation, and ID3 for the comparison. Novakovic et al. [17] compare six feature ranking methods and their experimental results, which shows that different ranking methods assigns different ranks to the features. This supports our hypothesis that using ensemble of multiple methods could provide improvement in discovering the correct feature weights.

Prati's paper [19] investigates and proposes a general framework for ensemble feature ranking based on different ranking aggregation methods. It suggests that the ensemble feature ranking improves the quality of feature ranks. It also elucidates on the merits and demerits of using score aggregation versus rank aggregation for discovering the feature weights.

Multiple papers have presented the evaluation of feature weights based on mean absolute error, mean squared error, or accuracy [21, 23, 25]. However, when datasets are unbalanced, these evaluation methods might not be suitable, due to the well known issues of class-imbalance and accuracy paradox [26]. Thus, for the evaluation of the results, we have used the F1-scores and 10-fold cross-validation, based on the confusion matrix.

3   Relevance-Based Feature Weights

This section presents our approach for discovering global similarity weights for a given classification dataset. It is primarily based on the scores from multiple feature relevance scoring methods.

The global similarity function is the weighted sum of all the local similarity scores. The global similarity function used in this paper is shown in Eq. 1, where w_i is the weight of the feature i. The $sim(Q, C)$ describes the global similarity function between a query Q and a case C. Further, for each attribute i a local similarity function is defined as $sim_i(q, c)$, where q is the attribute value from the query and c is the respective attribute value from the case. The result of this global similarity function is a similarity score in the range $[0,1]$. The paper is focused on data-driven approach to discover the value of w_i for the feature i.

$$\mathbf{sim}(\mathbf{Q}, \mathbf{C}) = \frac{1}{\sum w_i} \cdot \sum_{i=1}^{n} w_i . sim_i(\mathbf{q}, \mathbf{c}) \tag{1}$$

3.1   Proposed Method

We will refer to "feature relevance scoring methods" as "**scoring methods**", and "feature relevance scores" as "**scores**" going forward.

The entire method is described as a flowchart shown in Fig. 1. The process of discovering feature weights starts by selecting a classification dataset, a set of scoring methods, and percentage of features to be used. The percentage of features, *percent*, defines the proportion of features that are considered in the feature weight computation. Thus, a *percent* = 100 refers to all the features, while a *percent* = 25 refers to 25% of features with highest ranks, by each scoring method.

The scores are computed for all the features over each scoring method. However, only the *percent* of features with highest scores in each scoring method

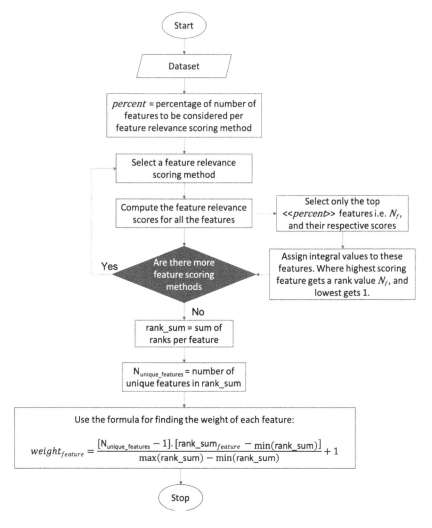

Fig. 1. Flow chart of the proposed method for discovering the subset features and their weights for a given dataset and featuring scoring methods.

are considered. It is represented as $max\_number\_of\_features$ shown in Eq. 2. The top $max\_number\_of\_features$ are sorted in descending order with respect to their scores.

The following procedure is executed for each scoring method. A rank, as per Eq. 2, is assigned to each feature. A feature with a highest score receives the highest rank which is equal to the value of $max\_number\_of\_features$. The rank of the lowest scoring feature is assigned to 1. Additionally, in case of a collision with equal scores, the rank of the previous feature (in descending order) is assigned to the colliding feature. While the succeeding non-colliding feature receives a rank with respect to its position in descending order.

$$rank = \{x \in \mathbb{N} \mid 1 \leq x \leq max\_number\_of\_features\} \qquad (2)$$

Once the ranks are computed for all the scoring methods, they are summed up with respect to each feature and stored in a $rank_{sum}$ vector. The size of $rank_{sum}$ vector is assigned to the variable N, in Eq. 3.

Finally, the computation of the feature weight is performed as shown in Eq. 3. The $Weight(\mathbf{f})$ represents the weight of the feature f. These feature weights are used as global similarity weights for the respective CBR system.

$$Weight(\mathbf{f}) = \left[\left(N - 1 \right) \cdot \left(\frac{rank_{sum}(\mathbf{f}) - min(\mathbf{rank_{sum}})}{max(\mathbf{rank_{sum}}) - min(\mathbf{rank_{sum}})} \right) \right] + 1 \qquad (3)$$

3.2 Relevance-Based Feature Weighting Algorithm

In this section we present the relevance-based feature weighting algorithm. The Algorithm 1 requires three parameters: the target classification dataset, a list of scoring methods, and the percentage of features to be considered. These parameters are defined as arguments for the function $computeFeatureWeights$. This function returns a map of feature weights where the feature names are the keys.

The ranks for the features are computed for every scoring method, and are stored in the variable $feature_{rank}$. The ranks are assigned in the descending order per scoring method, thus the most relevant feature gets the highest rank. If multiple features possess the same score then all of them are assigned with the same rank whereas the subsequent feature gets a rank with respect to its position in the descending order. The algorithm computes the sum of all the ranks with respect to each feature and stores it in the variable $rank_{sum}$.

In the last step of the algorithm, the value for N is the number of unique features in the $rank_{sum}$. And, the $min()$ and $max()$ functions provide the minimum and maximum values. Once the feature weights are successfully computed, they are used as feature weights in modeling the global similarity function.

The size of the final feature list and hence the attributes that receive a global weight depends on multiple factors, such as:

- the value of the *percent* variable,
- the relevance of a feature for the classification,
- the number of scoring methods used,
- the scores from various scoring methods

Algorithm 1. Relevance-based feature weighting algorithm

Input: Dataset
Input: $methods \leftarrow feature\ relevance\ scoring\ methods$
Input: $percent \leftarrow percentage\ of\ features\ to\ be\ considered\ per\ method$
Output: $Weights$

1 $Weights \leftarrow computeFeatureWeights(Dataset, methods, percent)$
2 **Function** computeFeatureWeights($Dataset,\ methods,\ percent$):
3     $feature_{rank} \leftarrow getRanks(Dataset, methods, percent)$
4     $rank_{sum} \leftarrow sum(feature_{rank})$ `// per feature`
5     $Weights \leftarrow 0$
6     $N \leftarrow size(rank_{sum})$ `// N scaling factor`
7   **for** $feature$ **do**
8        $Weights(feature) \leftarrow [N-1] \cdot \left[\frac{rank_{sum}(feature) - \mathbf{min}(rank_{sum})}{\mathbf{max}(rank_{sum}) - \mathbf{min}(rank_{sum})}\right] + 1$
9   **end**
10 **return** $Weights$

One of the inherent properties of this algorithm is that it also performs feature selection, which could be influenced in multiple ways. Two of the primary ways are by changing the value of *percent* or by varying the number of scoring methods to be used in the algorithm.

4 Experiments

In this section we present a set of experiments where we used openly available datasets to evaluate our method. The criteria for the considered datasets were that they fit a classification task, have different numbers of features as well as a variation of cases vs. the number of features.

The experimental setup uses myCBR tool [24] including its workbench and REST API module. The myCBR tool is used for modeling similarity, generating ephemeral case bases, and performing retrievals. The evaluation of the experimental results are based on 10-fold cross-validation.

The following subsections briefly describe datasets, feature relevance scoring methods, and confusion matrices used in our experiments.

4.1 Datasets

Table 1 lists four public datasets used in our experiments. They are available on "UCI Machine Learning Repository"[1]. The chosen datasets are for multivariate classification tasks and consist of features of type categorical, numerical, or a combination of both. One can see that the number of cases and target classes vary, and therewith pose different challenges for a CBR classifier.

---

[1] https://archive.ics.uci.edu/ml/index.php.

Table 1. Description of the datasets used in the experiment

Sln.	Dataset	Task type	Data types	Samples	Features	Missing values	Target classes
1	Car evaluation [6]	Multivariate classification	Categorical	1728	6	0	4
2	Pima Indians Diabetes [28]	Multivariate classification	Float, Integer	768	8	0	2
3	Tic-Tac-Toe Endgame [3]	Multivariate classification	Categorical	958	9	0	2
4	Zoo [10]	Multivariate classification	Categorical, Integer	101	17	0	7

4.2 Feature Relevance Scoring Methods

To get the feature relevance scores we used Orange [9], an open source tool. With the help of Rank widget, Orange version 3.20.1, the scores from the six scoring methods are obtained at the default settings for each dataset. As described in Sect. 3, one can use multiple scoring methods, for the experiments we used the default six scoring methods from the tool. A brief description of these scoring methods are as follows:

- *Information Gain* [16]: measures the gain in information entropy by using a feature with respect to the class.
- *Gain Ratio* [20]: a ratio of the information gain and the attribute's intrinsic information, which reduces the bias towards multi-valued features that occurs in the information gain.
- *Gini* [7]: is a measure commonly used in decision trees to decide what is the best attribute to split the current node for an efficient decision tree construction. It is a measure of statistical dispersion and can be interpreted as a measure of impurity for a feature or the inequality among values of a frequency distribution.
- *Chi2* [18]: this method evaluates each feature individually by measuring the chi-squared statistic with respect to the class.
- *ReliefF* [15]: this method uses the ability of an attribute to distinguish between classes on similar data instances.
- *FCBF* [29]: (Fast Correlation Based Filter) entropy-based measure, which also identifies redundancy due to pairwise correlations between features.

4.3 Confusion Matrix

The results of the retrievals are represented as a confusion matrix (CM). For instance, a retrieval result for 26 classes of a dataset can be represented using a CM as shown in Eq. 4, where $A, B, ..., and\ Z$ are the class labels. An element of this matrix, Φ (a positive integer value ($\mathbb{Z}^{\geq 0}$)), is the number of times a query resulted in a class pair. A class pair represents the location of an element in a CM, and is represented by lower subscripts of Φ as Φ_{tp}. Where, t represents the true class and p represents the predicted class. Additionally, the t and p represents the rows and columns of the CM, respectively.

$$CM = \begin{bmatrix} \Phi_{AA} & \Phi_{AB} & \cdots & \Phi_{AZ} \\ \Phi_{BA} & \Phi_{BB} & \cdots & \Phi_{BZ} \\ \cdot & \cdot & \cdot & \cdot \\ \Phi_{ZA} & \Phi_{ZB} & \cdots & \Phi_{ZZ} \end{bmatrix} \qquad (4)$$

4.4 Confusion Matrix for k-Fold Cross-Validation

A CM is constructed with respect to each dataset before the evaluation process begins and is initialized to 0 (element-wise).

When a query is executed, the CM_{init} is updated as shown in Eq. 5 with respect to the true and predicted class labels.

$$CM_{query} = CM_{init}[predicted\_class][true\_class] + 1 \qquad (5)$$

Thereafter, according to the Eq. 6, the confusion matrix CM_k for k^{th} iteration of the k-fold cross-validation is computed. In this equation, the variable M represents the total number of query cases in the kth iteration.

$$CM_k = \sum_{m=1}^{M} CM_{query_m} \qquad (6)$$

Finally, the confusion matrix for the entire k-fold cross-validation is computed as shown in Eq. 7. Thus, at the end of all k iterations the CM_{k-fold} contains all predictions with respect to the entire case base.

$$CM_{k-fold} = \sum_{i=1}^{k} CM_k \qquad (7)$$

The experiments performed in this paper are with *percent* values equal to 50, 75, and 100.

4.5 Evaluation

For the process of evaluation, we create a case base for each dataset where all the features are included. The local similarity measures are modeled using the interquartile ranges for a numerical feature (see [27] for details), and pair-wise similarity for a categorical feature. As a baseline system each dataset has been provided to a CBR engineer to model the global and local similarities manually. Additionally, a equal weighted global similarity function is implemented for each these datasets. The basis for the evaluation of the experimental results is a confusion matrix generated from 10-fold cross-validation as per Eq. 7 and the F1-scores.

The datasets selected for the experiments are unbalanced, thus we use F1-score as an evaluation measure. The F1-scores are computed for 10 runs over the 10-fold cross-validation confusion matrices. The computation of F1-scores are based on the Eq. 8.

$$F1\_score = 2 \cdot \frac{precision \cdot recall}{precision + recall} \tag{8}$$

5   Results

This section presents the results obtained from the experiments described in the previous sections.

The naming convention used for representing a global similarity function is <**name**>_<**percentage**>. Where the <**name**> describes the type of similarity function, explained as below:

- **manual_***: global similarity function with manually modeled feature weights, based on domain knowledge.
- **eq_***: global similarity function with equal feature weights.
- **rank_***: global similarity function with discovered feature weights, which uses sum of ranks for weight computation.
- **score_***: global similarity function with discovered feature weights, which uses sum of scores for weight computation.
- **info_gain_*, gain_ratio_*, gini_*, chi_sq_*, relief_f_***,*and***fcbf_***: global similarity functions for the individual scoring methods, described in Sect. 4.2.

The <**percentage**> or * is a place holder for percentage of features selected, per scoring method, that was used for weight computation. The percentages considered for this paper are **50%**, **75%**, and **100%** (**all**). All the features with respect to various global similarity functions are same for a given percentage value.

Figure 2 presents the confusion matrices for of the Zoo dataset. Each confusion-matrix is obtained based on 10-fold cross-validation for a global similarity function. The title of each matrix describes the name of the dataset and the global similarity function used for the retrieval. Likewise, the y-axis represents the true class labels (label of the query case), and the x-axis represents the predicted class labels (label of the retrieved case). The Zoo dataset poses the most challenging classification task since the classifier needs to distinguish 7 different classes while only having 101 cases available, which leads to very low support cases during the evaluation. However, the general trend shows that the more features are included, the better the classifier is performing. This can be seen in Fig. 2 where there are less misclassifications in the first row (all features

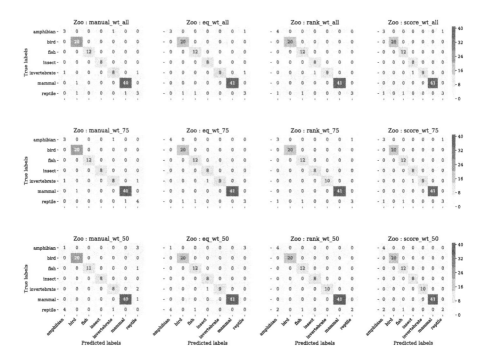

Fig. 2. Zoo dataset confusion matrices after one run of 10-fold cross validation with respect to the global similarity functions.

included in the global similarity function), compared to the second row (75% of features included) and third row (50% of features included).

Figure 3 presents the F1-score distributions for the 10-fold cross-validation, where none of the individual scoring methods perform consistently across all the datasets. Thus, in absence of domain knowledge our approach for predicting global similarity weights (*rank_all*) performs reasonably well across all the datasets.

Figure 4 presents the F1-score distributions with respect to the reduced feature percentages for all the four datasets. In this figure each row contains three sub-plots with respect to the *percent* values. The F1-scores are obtained based on 10-fold cross-validation for each global similarity function. The title of each plot describes the name of the dataset.

6   Discussion

The results are in accordance with the hypothesis of this paper: we can use distributions and statistical relationships within a dataset to define an initial global similarity measure. Our method can help to identify whether all or only a subset of features is necessary to carry out the desired classification task.

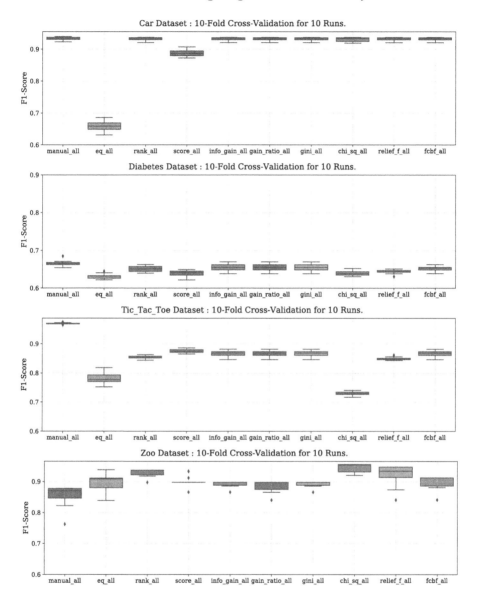

Fig. 3. Box-plots of evaluation metrics for 10-fold cross-validation over 10 runs. The plots are for datasets: Car, Diabetes, Tic Tac Toe, and Zoo respectively. All the plots are plotted with respect to the aforementioned global similarity functions, based on: manual, equal-weighted, rank, score, and the individual scoring methods.

As the paper describes a method for discovering feature weights based on the data-driven approach with a possibility of feature reduction. The F1-score distributions in Fig. 4 shows whether a reduction of features has an effect on the retrieval. Since the entire approach of generating the similarity measures

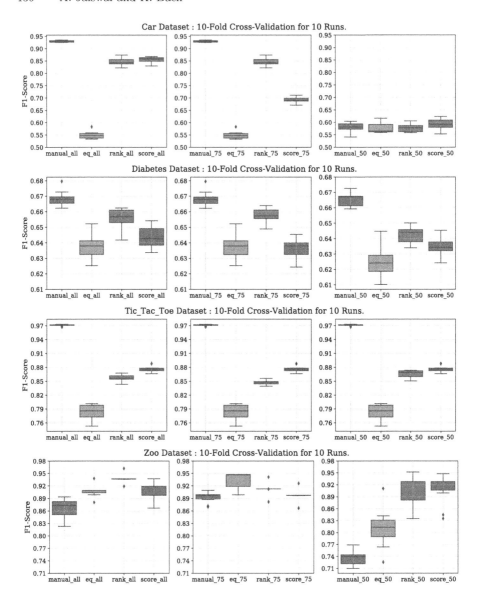

Fig. 4. Box-plots of F1-scores for 10-fold cross-validation with varying *percent* values. The plots are for datasets: Car, Diabetes, Tic Tac Toe dataset, and Zoo respectively. All the plots are plotted with respect to the aforementioned global similarity functions, on the x-axis.

is automatic, we can now gradually reduce the number of features (selecting smaller steps than presented here) and therewith find the best possible system, before discussing the details with domain experts. This will save effort and has the advantage of collaborating and incrementally improving a CBR system.

The proposed method of the paper does not challenge the similarity modeling based on the domain knowledge. With one exception of the Zoo dataset, the manual knowledge engineering performed better than the automatic one - obviously, because a knowledge engineer can encode domain characteristics and is not dependent on the distribution within the dataset. However, in the absence of the correct domain knowledge, the presented approach holds better than equally weighted features. A similar pattern occurs when the number of features are reduced. As the features of the datasets are highly representative of the class, the significant difference might not be noticeable for the selected datasets.

In general, we can see that the global similarity function based on the rank and feature relevance scores are higher than the global similarity functions based on equal weights. However, the global similarity function based on manual modeling of the local and global similarities outperforms the automatically created CBR systems, except in the case of Zoo dataset. This is an expected outcome as the manual modeling is based on the domain knowledge.

In the absence of publicly available reference CBR systems we could not perform any comparison or bench-marking of our results. In order to allow other researchers comparisons with their work, we share our projects used in this paper[2].

7   Conclusion and Future Work

We have presented our method for discovering feature weights for modeling global similarity function for a CBR system using a data-driven approach. This method is well suited in the initial phases of a CBR system development. This method also provides an opportunity for the developer of the CBR system to discuss the setup with domain experts and present comparisons of the results to them in various configurations. Moreover, the feature selection is also supported and it's results can be compared against multiple choices of the *percent* values, as proposed in the paper. We comment that the method brings reduction in time of the development and prototyping phase of a CBR system. The approach does not involve iterations of learning, thus reduces the chances of over-fitting, which is also supported by the use of ensemble of multiple feature relevance scoring methods. With the publishing of the developed case bases and its similarity functions, the experiments become fully reproducible and can serve as reference implementations in the future.

Inspired by the results of the present method, we would use the approach over multiple other public datasets, and publish them to be used by researchers of CBR community. The future work in continuation with paper is to research on discovering the local similarities for symbolic features. Additionally, we currently apply this approach to the dataset of our ongoing research described in the paper [13,14], where a more complex dataset has been presented and different application scenarios are discussed.

---

[2] https://github.com/ntnu-ai-lab/cbr-benchmark-projects.

References

1. Aamodt, A., Plaza, E.: Case-based reasoning: foundational issues, methodological variations, and system approaches. Artif. Intell. Commun. **7**(1), 39–59 (1994)
2. Abdel-Aziz, A., Strickert, M., Hüllermeier, E.: Learning solution similarity in preference-based CBR. In: Lamontagne, L., Plaza, E. (eds.) ICCBR 2014. LNCS (LNAI), vol. 8765, pp. 17–31. Springer, Cham (2014). https://doi.org/10.1007/978-3-319-11209-1_3
3. Aha, D.W.: Tic-tac-toe endgame database (1991). https://archive.ics.uci.edu/ml/datasets/Tic-Tac-Toe+Endgame
4. Aha, D.W.: Feature weighting for lazy learning algorithms. In: Liu, H., Motoda, H. (eds.) Feature Extraction, Construction and Selection. The Springer International Series in Engineering and Computer Science, vol. 453, pp. 13–32. Springer, Heidelberg (1998). https://doi.org/10.1007/978-1-4615-5725-8_2
5. Aha, D.W., Goldstone, R.L.: Concept learning and flexible weighting. In: Proceedings of the Fourteenth Annual Conference of the Cognitive Science Society, pp. 534–539. Erlbaum (1992)
6. Bohanec, M.: Car evaluation database (1997). https://archive.ics.uci.edu/ml/datasets/Car+Evaluation
7. Ceriani, L., Verme, P.: The origins of the gini index: extracts from variabilità e mutabilità (1912) by Corrado Gini. J. Econ. Inequality **10**(3), 421–443 (2012)
8. Cost, S., Salzberg, S.: A weighted nearest neighbor algorithm for learning with symbolic features. Mach. Learn. **10**(1), 57–78 (1993)
9. Demšar, J., et al.: Orange: data mining toolbox in python. J. Mach. Learn. Res. **14**, 2349–2353 (2013). http://jmlr.org/papers/v14/demsar13a.html
10. Forsyth, R.: Zoo database (1990). https://archive.ics.uci.edu/ml/datasets/Zoo
11. Gabel, T.: On the use of vocabulary knowledge for learning similarity measures. In: Althoff, K.-D., Dengel, A., Bergmann, R., Nick, M., Roth-Berghofer, T. (eds.) WM 2005. LNCS (LNAI), vol. 3782, pp. 272–283. Springer, Heidelberg (2005). https://doi.org/10.1007/11590019_32
12. Hüllermeier, E.: Exploiting similarity for supporting data analysis and problem solving. In: Hand, D.J., Kok, J.N., Berthold, M.R. (eds.) IDA 1999. LNCS, vol. 1642, pp. 257–268. Springer, Heidelberg (1999). https://doi.org/10.1007/3-540-48412-4_22
13. Jaiswal, A.: Personalized treatment recommendation for non-specific musculoskeletal disorders in primary care using case-based reasoning. In: Minor, M. (ed.) Workshop Proceedings of ICCBR 2018, pp. 214–218 (2018)
14. Jaiswal, A., Bach, K., Meisingset, I., Vasseljen, O.: Case representation and similarity modeling for non-specific musculoskeletal disorders - a case-based reasoning approach. In: The Thirty-Second International Florida Artificial Intelligence Research Society Conference (FLAIRS-32), pp. 359–362. AAAI Press (2019)
15. Kononenko, I., Šimec, E., Robnik-Šikonja, M.: Overcoming the myopia of inductive learning algorithms with relieff. Appl. Intell. **7**(1), 39–55 (1997)
16. Kullback, S.: Information Theory and Statistics. Wiley, New York (1959)
17. Novaković, J., Strbac, P., Bulatović, D.: Toward optimal feature selection using ranking methods and classification algorithms. Yugoslav J. Oper. Res. **21**(1), 119–135 (2011)
18. Pearson, K.: On the criterion that a given system of deviations from the probable in the case of a correlated system of variables is such that it can be reasonably supposed to have arisen from random sampling. London Edinburgh Dublin Philos. Mag. J. Sci. **50**(302), 157–175 (1900)

19. Prati, R.C.: Combining feature ranking algorithms through rank aggregation. In: Proceedings of the International Joint Conference on Neural Networks, pp. 1–8 (2012)
20. Quinlan, J.R.: Induction of decision trees. Mach. Learn. **1**(1), 81–106 (1986)
21. Ricci, F., Avesani, P.: Learning a local similarity metric for case-based reasoning. In: Veloso, M., Aamodt, A. (eds.) ICCBR 1995. LNCS, vol. 1010, pp. 301–312. Springer, Heidelberg (1995). https://doi.org/10.1007/3-540-60598-3_27
22. Stahl, A.: Learning similarity measures: a formal view based on a generalized CBR model. In: Muñoz-Ávila, H., Ricci, F. (eds.) ICCBR 2005. LNCS (LNAI), vol. 3620, pp. 507–521. Springer, Heidelberg (2005). https://doi.org/10.1007/11536406_39
23. Stahl, A., Gabel, T.: Optimizing similarity assessment in case-based reasoning. In: 21st National Conference on Artificial Intelligence, AAAI 2006, vol. 21, pp. 1667–1670 (2006)
24. Stahl, A., Roth-Berghofer, T.R.: Rapid prototyping of CBR applications with the open source tool myCBR. In: Althoff, K.-D., Bergmann, R., Minor, M., Hanft, A. (eds.) ECCBR 2008. LNCS (LNAI), vol. 5239, pp. 615–629. Springer, Heidelberg (2008). https://doi.org/10.1007/978-3-540-85502-6_42
25. Tamoor, M., Gul, H., Qaiser, H., Ali, A.: An optimal formulation of feature weight allocation for CBR using machine learning techniques. In: SAI Intelligent Systems Conference, IntelliSys 2015, pp. 61–67. IEEE (2015)
26. Valverde-Albacete, F.J., Peláez-Moreno, C.: 100% classification accuracy considered harmful: the normalized information transfer factor explains the accuracy paradox. PloS One **9**(1), e84217 (2014)
27. Verma, D., Bach, K., Mork, P.J.: Modelling similarity for comparing physical activity profiles - a data-driven approach. In: Cox, M.T., Funk, P., Begum, S. (eds.) ICCBR 2018. LNCS (LNAI), vol. 11156, pp. 415–430. Springer, Cham (2018). https://doi.org/10.1007/978-3-030-01081-2_28
28. VincentSigillito: Pima indians diabetes database (1997). https://archive.ics.uci.edu/ml/datasets/pima+indians+diabetes
29. Yu, L., Liu, H.: Feature selection for high-dimensional data: a fast correlation-based filter solution. In: Proceedings of the 20th International Conference on Machine Learning (ICML 2003), pp. 856–863 (2003)

Personalized Case-Based Explanation of Matrix Factorization Recommendations

Jose Jorro-Aragoneses$^{(\boxtimes)}$, Marta Caro-Martinez$^{(\boxtimes)}$,
Juan Antonio Recio-Garcia$^{(\boxtimes)}$, Belen Diaz-Agudo$^{(\boxtimes)}$,
and Guillermo Jimenez-Diaz$^{(\boxtimes)}$

Department of Software Engineering and Artificial Intelligence,
Universidad Complutense de Madrid, Madrid, Spain
{jljorro,martcaro,jareciog,belend,guille}@ucm.es
http://gaia.fdi.ucm.es

Abstract. Matrix factorization is an advanced recommendation strategy based on characterizing both items and users on a vector of latent factors inferred from rating patterns. These vectors represent, somehow, a characterization of the user preferences in a lower dimensionality space. Although matrix factorization is more accurate that other recommendation strategies, the main problem associated with this approach is that the discovered factors are opaque and difficult to explain to the final user. In this paper we propose a personalized case-based explanation strategy that uses the latent factors to find similar explanatory cases already rated by the user.

Keywords: Case-based explanation · Personalised explanation · Matrix factorization

1 Introduction

Recommender systems are typically based on one of two strategies. The *content filtering* approach creates a profile for each user or product to characterize its contents and recommends a similar product that matches the user profile. For example, a movie profile could include attributes regarding its genre, year, director, actors, and so forth. An alternative approach is *collaborative filtering* that is more flexible and generally more accurate than content-based techniques. Collaborative filtering relies only on user ratings and analyzes relationships between users and items, or between items to identify new user-item associations [1]. Recommendations resulting from content-based strategies are more comprehensible for users, as they are based on the explicit user preferences.

Since its success during the Netflix prize challenge the *matrix factorization* algorithm [2] has became one of the most successful algorithms to generate personalized recommendations. Matrix factorization is an advanced strategy that

Supported by the UCM (Research Group 921330), the Spanish Committee of Economy and Competitiveness (TIN2017-87330-R) and the fundings provided by Banco Santander in UCM (CT17/17-CT17/18) and (CT42/18-CT43/18).

K. Bach and C. Marling (Eds.): ICCBR 2019, LNAI 11680, pp. 140–154, 2019.
https://doi.org/10.1007/978-3-030-29249-2_10

Table 1. Recommended movie using matrix factorization

Movie Id.	Movie title	Year	Director	Stars	Predicted
223	Clerks	1994	Kevin Smith	Jason Mewes Jeff Anderson	4.042

attempts to merge the content and collaborative information in a single model based on characterizing both items and users on a vector of factors inferred from the ratings patterns. Although these vectors represent, somehow, a characterization of the user preferences, they are opaque collections of numeric values computed by the algorithm. In this paper we propose using these vectors to define a personalized similarity metric between items for every user. Case-based explanations focus primarily on finding explanatory cases that are *similar* to the recommended item [20]. Then, we use these cases to interpret the opaque output of the matrix factorization recommendation algorithm.

From the point of view of recommender systems, we propose an item-based explanation, since it uses items to justify a recommendation [16]. The main advantage of this approach is that it allows users to assess the quality of the recommendation by comparing items, that ideally should be *similar* according to the user's criteria. The main challenge of these case-based explanation strategies is to find a similarity metric that matches the user's criteria. Current content-based approaches [13] are based on the comparison of item's features, leaving aside the user's interpretation of these features. Therefore, in this paper we use the vectors of factors that characterize the user preferences to compute a similarity metric that finds related items in order to explain the recommendation.

Let's motivate our approach with an example. Given a user that has rated several movies in a dataset, the matrix factorization algorithm recommends "Clerks". Table 1 shows its features and Table 2 shows the most similar rated movies using as similarity metric the cosine of the vectors of factors extracted from the matrix factorization.

Here, "The usual suspects" is the most similar but there is not a clear intuition about the reasons for this similarity from the point of view of the canonical content-based distance. According to that distance, "The usual suspect" won't be chosen as an item for comparison as there are no common features between both movies (leaving aside the year). However, our hypothesis is that the vector of factors resulting from the matrix factorization is able to capture relations that make sense from the user's point of view. For example, the user may like politically incorrect movies and the matrix factorization has captured that factor, and therefore making both movies similar.

Section 2 reviews the related work in explanations in recommender systems. Section 3 explains the matrix factorization method. Section 4 describes how to define a personalized similarity metric between items for every user that is used to retrieve the explanatory cases. Section 5 evaluates the similarity metric associated to our case-based explanation model demonstrating how to get relevant explanatory cases without additional knowledge on the item features. Section 6 concludes the paper and describes the ongoing lines of work.

Table 2. Most similar movies according to the vectors of factors resulting from the matrix factorization

Movie Id.	Movie title	Year	Director	Stars	Rating	Similarity
50	The Usual Suspects	1995	Bryan Singer	Kevin Space Kevin Pollak	5.0	0.796
163	Desperado	1995	Robert Rodriguez	Quentin Tarantino Salma Hayek	5.0	0.750
596	Pinocchio	1940	Norman Ferguson	Mel Blanc Cliff Edwards	5.0	0.665
151	Rob Roy	1995	Michael Caton-Jones	Liam Nesson Eric Stoltz	5.0	0.646
101	Bottle Rocket	1996	Wes Anderson	Andrew Wilson Lumi Cavazos	5.0	0.465

2   Related Work

Using explanations in recommendation systems is an important area of research in this type of systems. One of the main problems with recommendation systems is that users do not know why a product has been recommended to them. Recommender systems that use explanations improve user confidence in those recommendations [20]. In addition, users consume more products resulting from a explainable recommendation process [7].

Nowadays there are many works that apply explanations in recommender systems. In a previous work [3], we carried out an in-depth study of the explanation systems applied to recommendation systems. As a result of this study, we developed a theoretical model to classify the explanation systems according to their characteristics. According to this model, explanation systems employ different methods to obtain the knowledge needed to generate explanations. The model we present in this paper is knowledge-light and the only knowledge container employed is the algorithm, and more precisely, the similarity between items and the user's experiences.

In [12] we find explanation system for movie recommendation systems based on the similarity between plots. Movie similarity is based on the characteristics that are in common between the characters and the interactions of the characters in the plot. The IMVEX system [5] is a rule-based system that personalizes the explanations for different types of users. The knowledge base used is the user profile. The system developed by [11] shows an explanation system for a recommendation system for groups, based on the similarity of preferences among the members of the group. In [17], we found a system that displays the recommendations along with the characteristics that have been involved in the selection of the best candidates for the recommendation. Another example of a system that takes into account similarities between user preferences and item characteristics is the framework presented in [23].

We are particularly interested in experience-based explanations, which use the past actions of the user and her history of interactions as a source of knowledge to generate explanations. CBR-based explanations are an example of experience-based explanations. There are different works based on CBR. The work in [4] reviews classic systems that use CBR as a way to find similar cases that are used as an explanation of recommendations. In [19], the attribute with the highest weight in the similarity metric is selected in order to find the similar cases that may be of interest to the user as an explanation of the recommendation. In [8] we found a case-based system to explain the detection of healthcare-associated infections. The work in [15] describes a case-based recommender system for hotels, where cases are obtained from users' reviews. The explanations of the recommendations are based on features obtained from this information. The PSIE (Personalized Social Individual Explanation) approach [18] includes explanations to group recommender systems and social explanations with the aim of inducing a positive reaction to users in order to improve their perception of the recommendations. In [14] we found a CBR system that uses the difference between the query and the case descriptions to explain all recommendations.

Finally, there are some works to explain recommendations provided by systems based on latent factors. This is due to the fact that these systems work very well, but they are difficult to explain. In [10] the authors describe the TriRank system, which extracts information from the reviews to improve the transparency of the recommender system. Another work that tries to explain the recommendations obtained from matrix factorization is [24]. The explanation model consists of determining which movies have influenced the rating predicted by the matrix factorization algorithm. In [21], authors propose a method called Tree-enhanced Embedding Method (TEM) that uses embedding-based and tree-based models to extract explanations of recommenders systems based on collaborative filtering and latens factors.

In the following section we explain how a recommendation system based on matrix factorization works. In addition, we explain what information we will be able to use from this algorithm to generate the explanations.

3   Recommendation Using Matrix Factorization

Matrix factorization is one of the most commonly used methods for creating a latent factor model applied to recommendation systems. To create the model, the algorithm uses a $R \in \mathbb{R}^{U \times I}$ matrix that contains the ratings that users (U) have made on a set of items (I). The main problem with the R matrix is that it is very sparse, that is, it only contains a small part of the ratings. The goal of matrix factorization is to complete the R matrix by relating users to items through latent factors of N dimensionality.

To do this, we apply the Simons Funk's model [6]. We define $P \in \mathbb{R}^{U \times N}$ matrix, which relates each user from U to the factor dimensions (N), and $Q \in \mathbb{R}^{I \times N}$ matrix, which relates the set of items I to each factor dimension (N). This way, a user $u \in U$ is associated with a vector $p_u \in P$ that measures the preferences of the user on items according to the corresponding latent factors. On the other hand, an item $i \in I$ is associated with a vector $q_i \in Q$ that measures how the item is reflected according to the latent factors. The dot product of both vectors will give us the user's u rating prediction (r'_{ui}) of item i, as illustrated in Fig. 1:

$$r'_{ui} = p_u q_i^T \tag{1}$$

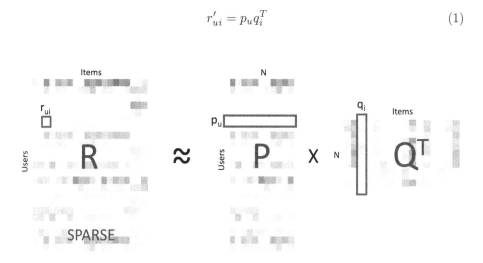

Fig. 1. Matrix factorization general schema.

A recommender system uses a $R'^{U x I}$ matrix, which contains the estimations for each user and each item. This matrix is the result of multiplying P and Q^T matrices.

$$R' = PQ^T \tag{2}$$

From this matrix we will obtain the items that will be recommended to a specific user. To learn the values of P and Q the system minimizes the error between the rating prediction and the known ratings. In our learning process we use the stochastic gradient descent method. In this process, the algorithm runs through the known rating set ($r_{ui} \in R$). For each rating, the system computes the error between this rating and its prediction.

$$e_{ui} = r_{ui} - p_u q_i^T \tag{3}$$

Once the error is known, the values of p_i and q_u are modified by a magnitude proportional to γ in the opposite direction of the gradient. The new values will be:

$$q_i \leftarrow q_i + \gamma \cdot (e_{ui} \cdot p_u - \lambda \cdot q_i) \tag{4}$$

$$p_u \leftarrow p_u + \gamma \cdot (e_{ui} \cdot q_i - \lambda \cdot p_u) \tag{5}$$

Once we have described the general schema of the matrix factorization recommendation technique, following sections will depict our proposal for using the Q matrix to find explanatory cases, because this matrix captures user preferences through the factor vectors.

4   Retrieval of Explanatory Cases Using the Q Space

Case-based explanation requires a set of similar items that will be presented as explanatory examples. These items must be similar to the item recommended by the system according to the user preferences. As we described in the previous section, P matrix describes users as factor vectors, meanwhile, Q matrix contains factor vector representations for every item, both of them using a N dimensional space. The dot product of user and item vectors, $p_u q_i^T$ computes the estimated rating for a user u and item i. This way, p_u contains the description of the user, and q_i a general description of the item according to the preferences of all the users in the dataset. As the goal of the explanation process is to obtain explanation items in a personalized way for each user, we need initially to transform the Q matrix to represent the items according to the concrete user u. To do so, we transform the Q matrix into a collection of vectors where each N-dimensional vector represents the description of an item q_i multiplied by the user preferences p_u:

$$Q^u = \{q_1^u, \ldots, q_M^u\} \tag{6}$$
$$\text{where } q_i^u = p_u q_i$$

Here, $q_i \in \mathbb{R}^N$ and $M = |I|$ is the number of items in the dataset. This collection of vectors summarizes the user u preferences, where several factor vectors are more discriminant that others. The example in Fig. 2 shows that the vectors represented in columns 1, 12 and 14 are the most discriminant in order to compute the predicted rating of an item. It is important to note that these vectors are personalised for every user as it is the result of multiplying Q by p_u. Therefore, the Q^u matrix is completely different for every user.

This fact is illustrated by Fig. 3, which shows the factor value distribution of the Q^u vectors for two different users given the same set of movies. We can clearly observe that the characterization of both users is different, allowing us to use Q^u as a description of the user's profile. However, the characteristics of the matrix factorization algorithm does not provide a symbolic description of

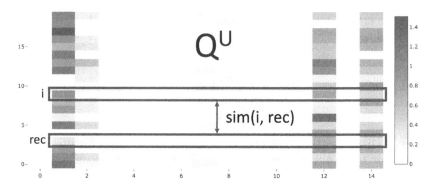

Fig. 2. Visualization of the Q^u matrix capturing the user preferences in a collection of M vectors (number of items) with dimension N (in this example $M = 20 \times N = 14$).

these factors that could be used to explain the results. We cannot even have an intuition of what these vectors exactly mean for each user, as they are numeric values computed by the algorithm. But we can exploit this Q^u matrix to define a personalized similarity metric between items for every user.

Matrix Q^u describes the items according to the user rating patterns. But, to generate the explanations using a case-based approach, the system will only use the items that the user has previously rated. That is, the system filters the items that the user has not rated yet from the Q^u matrix. The result is a new matrix $Q^{u'}$:

$$Q^{u'} = \{q_i^u \in Q^u : r_{ui} \neq \emptyset\} \tag{7}$$

Now, we can define a similarity metric over this space to calculate the similarity between two items according to the user's perception. We propose using the cosine similarity function to compare q_i^u vectors of each item. The benefit of using this similarity function is that it does not take into account vector magnitudes, which allows item comparison without having to obtain a prior knowledge about the latent factors for each user:

$$sim^{Q^u}(i, rec) = cos(q_i^u, q_{rec}^u) = \frac{q_i^u \cdot q_{rec}^u}{|q_i^u| \cdot |q_{rec}^u|} \tag{8}$$

Once the similarity metric is defined over the $\mathbb{R}^N$ vector factors space, the set of explanatory cases is obtained by selecting the most similar rated items. The explanatory case set (Exp) includes the k items of $Q^{u'}$ that are more similar to the recommended item (rec) as described in Algorithm 1.

5   Evaluation

We have described a case-based explanation model where the explanatory examples are retrieved from the Q^u matrix that captures the user preferences in a

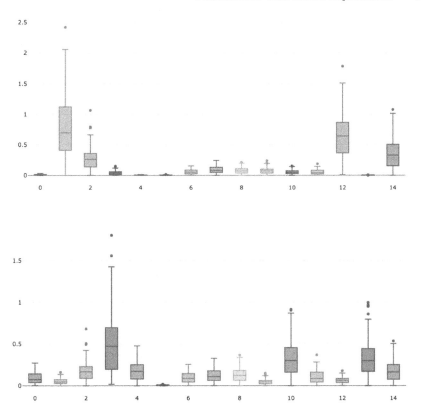

Fig. 3. Factor value distribution of the Q^u vectors that summarizes two users' profiles (preferences) with dimension $N = 14$. Top figure corresponds to the matrix shown in Fig. 2.

N-dimensional space. To evaluate our model we will prove that the explanatory examples retrieved using sim^{Q^u} are more relevant to the user than the items that we would retrieve using sim^I, that is, a content-based approach that can compute the similarity for every pair of items. A benefit of our approach is that it is knowledge-light, in opposition to the classical content-based approach using I.

Our experiments demonstrate that our model provides personalized results without requiring any knowledge about the items' description. It overcomes one of the main problems associated with content-based approaches, namely that they require gathering external information that might not be available. Our model does not need the I description matrix, but only the R matrix that includes the users' ratings.

Algorithm 1. Computation of the explanatory cases set

Input: R, P, Q, u, rec, k

Output: *Exp*

1 $p_u \leftarrow P[u]$

2 $r_{ui} \leftarrow R[u, i]$

3 $Q^u \leftarrow \{p_u q_1, \ldots, p_u q_m : q_i \in Q\}$

4 $Q^{u\prime} \leftarrow \{q_i^u \in Q^u : r_{ui} \neq \emptyset\}$

5 $Exp \leftarrow \{\}$

6 **while** $k > 0$ **do**

7 $\quad Exp \leftarrow Exp \cup \{i : \underset{q_i^u \in Q^{u\prime}}{\operatorname{argmax}}\, sim(q_i^u, q_{rec}^u)\}$

8 $\quad Q^{u\prime} \leftarrow Q^{u\prime} \setminus q_i^u$

9 $\quad k \leftarrow k - 1$

10 **end**

11 **return** *Exp*

5.1 Datasets

To test our hypothesis we have used the popular movie domain. In this evaluation we used two public datasets. The first one is the 100k MovieLens dataset [9], which contains 100,000 ratings made by users in the MovieLens recommendation system. This dataset will be used by the matrix factorization algorithm. The second dataset contains the features of 5,000 movies [22]. These descriptions have been extracted from IMDB[1]. More concretely, the movie features that we used in the evaluation are: genres, directors, actors, screenwriters and the decade in which the movies were released. This second dataset let us to compare the quality of our examples compared to a classical content-based approach.

In the evaluation we selected the movies that both datasets have in common. The final dataset used for the evaluation contains 11,477 ratings made by 587 users on 164 movies. 90% of the dataset has been used to train the P and Q matrices of the recommender system. Regarding the sparsity of the training matrix, it represents the 11% of the complete matrix. The remaining 10% of the dataset has been used to perform the evaluation. Moreover, in order to perform a stratified evaluation according to the rating values, we have created another dataset where each fold has the same rating value. To create the stratified dataset, we have selected 34 items for each rating value[2]. We have made this selection randomly, and we have repeated it 100 times. Then, we have got 3400 items for each rating value. This second evaluation set will verify that the system works better by eliminating the bias of the most popular ratings.

[1] https://www.imdb.com/.

[2] This is the highest possible value as the dataset only contains 34 items rated with 2.5 as shown in Table 5.

5.2   Methodology

As we have explained before, in this evaluation we try to demonstrate that the items we recover using the sim^{Q^u} metric are more relevant as a personalized explanation of a recommendation. To prove it, we have to compare the retrieved examples to those cases we would retrieve using a classical content-based approach.

In order to define a content-based similarity metric using the I matrix we need a binary representation of the item description, where each vector position represents if the item has that description feature or not. To build these descriptions we have converted the multivalued features of the film descriptions (genres, directors, actors, ...) into binary values. This way we avoid the bias of knowledge-rich approaches that use more elaborated metrics to compute these multivalued features. Another advantage is that we could use the same cosine metric to compare both item descriptions in Q^u and I:

$$sim^{Q^u}(i, rec) = cos(q_i^u, q_{rec}^u) \tag{9}$$
$$sim^I(i, rec) = cos(I[i], I[rec]) \tag{10}$$

To estimate the quality of the recovered explanatory cases, we are going to compare them with the recommended item. Our evaluation metric will compute the Root Mean Square Error (RMSE) between the estimated rating r'_{ui} for the recommended item and the average of the actual user ratings for the k explanatory cases, either retrieved using sim^{Q^u} or sim^I. In the evaluation we used different k values, with $k \in \{1, 2, 3, 5, 10\}$.

The intuition behind this evaluation is that, given a recommended item to be explained, the explanatory cases should have a real rating given by the user similar to the estimated rating provided by the recommender system for the recommended item. As we are using a stratified evaluation, this approach let us validate if the proposed method could be useful to explain both positive (high estimated rating) and negative recommendations (low estimated rating).

5.3   Results

Table 3 shows the RMSE values that we have obtained using both similarity metrics. We observe that the use of the sim^{Q^u} metric to retrieve the explanatory cases decreases the RMSE value. In other words, the rating given by the user to the cases that are recovered with the descriptions of q_i^u are more similar to the rating estimated for the recommended item than using binary descriptions in a content-based style. The third column shows the improvement percentage using the methodology proposed in this paper. In the table we see that the best result is with the value of $k = 1$ where the improvement is 5.5%.

The corresponding results of the stratified evaluation are shown in Table 4. We observe again the best results with low k values. The explanation for this behaviour, both in the complete and the stratified dataset, is the highest performance of the sim^{Q^u} metric when presenting few explanatory cases. On the other

Table 3. RMSE using the complete evaluation set.

k	sim^I	sim^{Q^u}	Improvement (%)
1	1.211	1.144	5.53
2	1.194	1.144	4.16
3	1.189	1.136	4.47
5	1.190	1.143	3.94
10	1.179	1.148	2.67

Table 4. RMSE using the stratified dataset.

k	sim^I	sim^{Q^u}	Improvement (%)
1	1.229	1.125	8.47
2	1.208	1.139	5.71
3	1.211	1.134	6.34
5	1.199	1.146	4.42
10	1.180	1.148	2.71

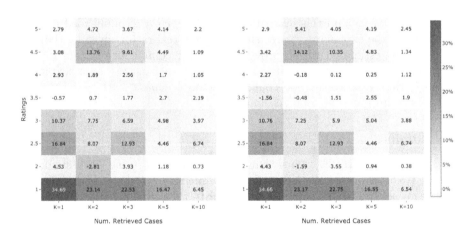

Fig. 4. Graphical representation of the improvement percentage of the sim^{Q^u} metric with respect to the content-based approach sim^I. Left heatmap corresponds to the results of the complete dataset shown in Table 5, whereas heatmap on the right corresponds to the stratified dataset detailed in Table 6. Red cells represent negative improvement (content-based approach is better than latent factors metric). (Color figure online)

hand, as the number of explanatory cases increases, the content-based approach is able to leverage its worse performance.

Next, Tables 5 and 6 show the results segmented by the rating value[3]. As a general result, we can conclude that the latent factors obtain better results than the content-based approach. The corresponding improvements (in percentage) are illustrated by Fig. 4. This figure let us observe that the similarity metric based on the vector of factors is not only able to explain a movie that the user may like (high predicted ratings), but also to explain why the user won't like a move. It is specially remarkable for those movies with a very low rating, where our approach achieves the highest performance. This figure also illustrates the behaviour of the proposed similarity metric when presenting to the user few explanatory cases, that was summarized by Tables 3 and 4.

Table 5. Detailed RMSE using the complete dataset segmented by the rating value.

Rating	Size	k = 1		k = 2		k = 3		k = 5		k = 10	
		sim^I	sim^{Q^u}	sim^I	sim^{Q^u}	sim^I	sim^{Q^u}	sim^I	sim^{Q^u}	sim^I	sim^{Q^u}
1	37	1.18	0.77	1.08	0.83	1.10	0.85	1.07	0.89	0.99	0.93
2	85	1.04	1.09	0.99	1.02	1.02	0.98	0.98	0.97	0.98	0.97
2.5	34	1.09	0.91	1.03	0.95	1.06	0.93	0.98	0.94	0.99	0.92
3	295	0.98	0.89	0.95	0.87	0.93	0.87	0.94	0.89	0.93	0.89
3.5	95	0.84	0.84	0.81	0.81	0.82	0.81	0.86	0.84	0.86	0.84
4	304	0.90	0.88	0.93	0.91	0.91	0.89	0.92	0.90	0.91	0.90
4.5	66	0.97	0.94	1.05	0.91	1.02	0.92	1.00	0.96	0.94	0.93
5	219	1.06	1.03	1.03	0.98	1.02	0.99	1.01	0.97	1.00	0.98

Table 6. Detailed RMSE using the stratified dataset segmented by the rating value.

Rating	Size	k = 1		k = 2		k = 3		k = 5		k = 10	
		sim^I	sim^{Q^u}	sim^I	sim^{Q^u}	sim^I	sim^{Q^u}	sim^I	sim^{Q^u}	sim^I	sim^{Q^u}
1	3400	1.18	0.77	1.08	0.83	1.10	0.85	1.07	0.89	0.99	0.93
2	3400	1.04	0.99	1.00	1.01	1.02	0.98	0.97	0.97	0.98	0.97
2.5	3400	1.09	0.91	1.03	0.95	1.06	0.93	0.98	0.94	0.99	0.92
3	3400	0.98	0.88	0.94	0.87	0.93	0.87	0.93	0.89	0.93	0.89
3.5	3400	0.83	0.85	0.81	0.81	0.82	0.80	0.86	0.84	0.86	0.84
4	3400	0.92	0.90	0.93	0.93	0.91	0.91	0.92	0.92	0.92	0.91
4.5	3400	0.99	0.95	1.06	0.91	1.02	0.92	1.01	0.96	0.94	0.93
5	3400	1.05	1.02	1.02	0.97	1.02	0.98	1.01	0.97	1.00	0.97

---

[3] Ratings 0.5 and 1.5 have been removed due to the low number of items.

6 Conclusions and Future Work

Case-based explanation approaches in recommender systems typically use previous items rated by the user in order to explain a recommendation. The novelty of the approach described in this paper is that it infers a similarity measure from the vector of factors obtained by the matrix factorization algorithm and uses this similarity measure to capture the preferences of the user from previous rating patterns (cases). We have proposed a case-based explanation model where the explanatory cases are retrieved from the Q^u matrix, computed by the matrix factorization algorithm, instead of using an item description space, as the Q^u captures the user preferences in a N-dimensional space.

Matrix factorization decomposes the user-item interaction matrix into the dot product of two lower dimensionality matrices, P and Q, using N latent features. This way, each row in P represents the strength of the associations between a user and the latent features. Similarly, each row in Q represents the strength of the associations between an item and the latent features. In this paper we propose combining both P and Q matrices to get a personalized representation of the items in a lower dimensional space according to the user preferences. Although these vectors of latent features are not easy to understand and they cannot be directly exploited to explain the recommendation, we propose to use them in a case-based explanation style. We use the personalized Q^u matrix in order to find those past items rated by the user that are related to the current recommendation in that latent factor space.

The empirical evaluation that we have conducted compares the quality of the explanatory cases obtained by our proposal with a canonical content-based approach. Results reveal a clear improvement specially remarkable for explanatory cases with low ratings. This way, we can provide explanations with a positive or negative perspective, showing examples of why an item is interesting to the user or not. The similarity metric based on the latent factors also achieves good results when proposing explanation based on very few items.

As future work we would like to evaluate this approach with real users and to compare our similarity metric to retrieve the explanatory cases with other similarity distances. We also plan to evaluate this approach using an external recommender system acting as a black box. This way, we do not need to know the underlying recommender algorithm, and compute the Q matrix to obtain explanatory examples instead of using this matrix to provide recommendations.

We would also explore the possibility of associating a semantic description to the latent feature vectors that capture the user preferences. If we could correlate these vectors to a semantic description of the items or the user's profile we could provide a more detailed explanation about the recommended item. However, as explained in this paper, this correlation is not intuitive and very difficult to obtain.

References

1. Bell, R., Koren, Y., Volinsky, C.: Matrix factorization techniques for recommender systems. Computer **42**(08), 30–37 (2009). https://doi.org/10.1109/MC.2009.263
2. Bennett, J., Lanning, S., et al.: The netflix prize. In: Proceedings of KDD Cup and Workshop, New York, NY, USA, vol. 2007, p. 35 (2007)
3. Caro-Martinez, M., Jimenez-Diaz, G., Recio-Garcia, J.A.: A theoretical model of explanations in recommender systems. In: ICCBR 2018, p. 52 (2018)
4. Doyle, D., Tsymbal, A., Cunningham, P.: A review of explanation and explanation in case-based reasoning. D2003, TCD CS report (2003)
5. Finch, I.: Knowledge-based systems, viewpoints and the world wide web. In: IEE Colloquium on Web-Based Knowledge Servers (Digest No. 1998/307), pp. 8/1–8/4, June 1998
6. Funk, S.: Netflix update: Try this at home (2006). https://sifter.org/simon/journal/20061211.html
7. Gedikli, F., Jannach, D., Ge, M.: How should I explain? A comparison of different explanation types for recommender systems. Int. J. Hum.-Comput. Stud. **72**(4), 367–382 (2014)
8. Gómez, H., et al.: A case-based reasoning system for aiding detection and classification of nosocomial infections. Decis. Support Syst. **84** (2016). https://doi.org/10.1016/j.dss.2016.02.005
9. Harper, F.M., Konstan, J.A.: The movielens datasets: history and context. ACM Trans. Interact. Intell. Syst. **5**(4), 19:1–19:19 (2015). https://doi.org/10.1145/2827872
10. He, X., Chen, T., Kan, M.Y., Chen, X.: Trirank: review-aware explainable recommendation by modeling aspects. In: Proceedings of the 24th ACM International on Conference on Information and Knowledge Management, CIKM 2015, pp. 1661–1670. ACM, New York (2015). https://doi.org/10.1145/2806416.2806504
11. Jameson, A.: More than the sum of its members: challenges for group recommender systems. In: Proceedings of the Working Conference on Advanced Visual Interfaces, AVI 2004, 25–28 May 2004, Gallipoli, Italy, pp. 48–54 (2004). https://doi.org/10.1145/989863.989869
12. Lee, O., Jung, J.J.: Explainable movie recommendation systems by using story-based similarity. In: Joint Proceedings of the ACM IUI 2018 Workshops Co-located with the 23rd ACM Conference on Intelligent User Interfaces (ACM IUI 2018), 11 March 2018, Tokyo, Japan (2018). http://ceur-ws.org/Vol-2068/exss5.pdf
13. Lops, P., de Gemmis, M., Semeraro, G.: Content-based recommender systems: state of the art and trends. In: Ricci, F., Rokach, L., Shapira, B., Kantor, P.B. (eds.) Recommender Systems Handbook, pp. 73–105. Springer, Boston, MA (2011). https://doi.org/10.1007/978-0-387-85820-3_3
14. Mcsherry, D.: Explanation in recommender systems. Artif. Intell. Rev. **24**(2), 179–197 (2005). https://doi.org/10.1007/s10462-005-4612-x
15. Muhammad, K., Lawlor, A., Rafter, R., Smyth, B.: Great explanations: opinionated explanations for recommendations. In: Hüllermeier, E., Minor, M. (eds.) ICCBR 2015. LNCS (LNAI), vol. 9343, pp. 244–258. Springer, Cham (2015). https://doi.org/10.1007/978-3-319-24586-7_17
16. Papadimitriou, A., Symeonidis, P., Manolopoulos, Y.: A generalized taxonomy of explanations styles for traditional and social recommender systems. Data Mining Knowl. Discov. **24**(3), 555–583 (2012). https://doi.org/10.1007/s10618-011-0215-0

17. Pu, P., Chen, L.: Trust-inspiring explanation interfaces for recommender systems. Knowl.-Based Syst. **20**(6), 542–556 (2007). https://doi.org/10.1016/j.knosys.2007.04.004
18. Quijano-Sanchez, L., Sauer, C., Recio-Garcia, J.A., Diaz-Agudo, B.: Make it personal: a social explanation system applied to group recommendations. Expert Syst. Appl. **76**, 36–48 (2017)
19. Ray, S., Sharma, R.: Explanations in recommender systems: an overview. Int. J. Bus. Inf. Syst. **23**, 248 (2016). https://doi.org/10.1504/IJBIS.2016.10000276
20. Tintarev, N., Masthoff, J.: Explaining recommendations: design and evaluation. In: Ricci, F., Rokach, L., Shapira, B. (eds.) Recommender Systems Handbook, pp. 353–382. Springer, Boston, MA (2015). https://doi.org/10.1007/978-1-4899-7637-6_10
21. Wang, X., He, X., Feng, F., Nie, L., Chua, T.S.: TEM: tree-enhanced embedding model for explainable recommendation. In: Proceedings of the 2018 World Wide Web Conference, WWW 2018, pp. 1543–1552. International World Wide Web Conferences Steering Committee, Republic and Canton of Geneva, Switzerland (2018). https://doi.org/10.1145/3178876.3186066
22. Yueming: Imdb 5000 movie dataset (2018). https://www.kaggle.com/carolzhangdc/imdb-5000-movie-dataset
23. Zanker, M., Ninaus, D.: Knowledgeable explanations for recommender systems. In: 2010 IEEE/WIC/ACM International Conference on Web Intelligence, WI 2010, 31 August–3 September 2010, Toronto, Canada, Main Conference Proceedings, pp. 657–660 (2010). https://doi.org/10.1109/WI-IAT.2010.131
24. Zhang, Y., Lai, G., Zhang, M., Zhang, Y., Liu, Y., Ma, S.: Explicit factor models for explainable recommendation based on phrase-level sentiment analysis. In: The 37th International ACM SIGIR Conference on Research and Development in Information Retrieval, SIGIR 2014, 06–11 July 2014, Gold Coast, QLD, Australia, pp. 83–92 (2014). https://doi.org/10.1145/2600428.2609579

How Case-Based Reasoning Explains Neural Networks: A Theoretical Analysis of XAI Using *Post-Hoc* Explanation-by-Example from a Survey of ANN-CBR Twin-Systems

Mark T. Keane[1,2,3(✉)] and Eoin M. Kenny[1,2]

[1] School of Computer Science, University College Dublin, Dublin, Ireland
{mark.keane, eoin.kenny}@insight-centre.org
[2] Insight Centre for Data Analytics, University College Dublin, Dublin, Ireland
[3] VistaMilk SFI Research Centre, University College Dublin, Dublin, Ireland

Abstract. This paper proposes a theoretical analysis of one approach to the eXplainable AI (XAI) problem, using *post-hoc* explanation-by-example, that relies on the twinning of artificial neural networks (ANNs) with case-based reasoning (CBR) systems; so-called ANN-CBR twins. It surveys these systems to advance a new theoretical interpretation of previous work and define a road map for CBR's further role in XAI. A systematic survey of 1,102 papers was conducted to identify a fragmented literature on this topic and trace its influence to more recent work involving deep neural networks (DNNs). The *twin-systems approach* is advanced as one possible coherent, generic solution to the XAI problem. The paper concludes by road-mapping future directions for this XAI solution, considering (i) further tests of feature-weighting techniques, (ii) how explanatory cases might be deployed (e.g., in counterfactuals, *a fortori* cases), and (iii) the unwelcome, much-ignored issue of user evaluation.

Keywords: CBR · Explanation · Artificial neural networks · XAI · Deep learning

1 Introduction

As AI systems impact our everyday lives, jobs, and leisure time, the issue of explaining how these systems actually work has become more acute, the so-called eXplainable AI (XAI) problem. In the last few years, almost every major AI/ML conference has targeted this problem either as a major theme or as a focus for thematic workshops (e.g., NIPS-16, IJCAI-17, IJCAI/ECAI-18, IJCNN-17, ICCBR-18, ICCBR-19) along with the emergence of meetings dedicated solely to it (FAT-ML, FAT*19; see [1]), as well as being a regulatory focus for government (e.g., GDPR in the EU [2, 3]). This paper surveys one particular solution to the XAI problem, where an opaque, black-box AI system is explained by a more interpretable, white-box AI system; the so-called *twin-systems* approach [4]. This survey is used to advance a new theoretical interpretation of previous work and define a road map for CBR's further role in XAI. Specifically, we

© Springer Nature Switzerland AG 2019
K. Bach and C. Marling (Eds.): ICCBR 2019, LNAI 11680, pp. 155–171, 2019.
https://doi.org/10.1007/978-3-030-29249-2_11

review the pairing of artificial neural networks[1] (ANNs) with case-based reasoning (CBR) systems where the explanatory cases of the latter are used to interpret the opaque outputs of the former; so-called ANN-CBR twins (e.g., see Fig. 1). We have discovered a fragmented literature on this topic that deserves to be brought together, if only to avoid unnecessary re-invention. In the next sub-section, we lay out our orientation to "explanation" and the motivation for the present systematic survey.

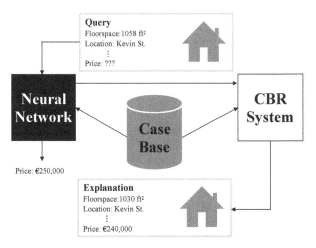

Fig. 1. A simple ANN-CBR twin-system (adapted from Kenny and Keane [4]); a query-case posed to an ANN gives an accurate, but unexplained, prediction for a house price. The ANN is twinned with the CBR system (both use the same dataset), and its feature-weights are analyzed and used in the CBR system to retrieve nearest-neighbor cases to explain the ANN's prediction.

1.1 "Explanation" Needs Explanation

As an area, XAI has many issues; foremost amongst these, perhaps, is some clarity on what "explanation" actually means. Several recent XAI reviews have pointedly noted the lack of clear definitions for the notions of explanation, interpretability and transparency [5–10], echoing long-standing discussions in CBR [11, 12], recommender systems [13], Philosophy [14–16] and Psychology [17]. While the exact meaning of these terms remains a matter of debate, these reviews make useful taxonomic distinctions. For example, Sørmo *et al.*'s [11] review reports the distinction between explaining how the system reached some answer (which they call *transparency*) and explaining why the answer is good (*justification*). More recently, this distinction is echoed by dividing *interpretability* into (i) *transparency* (or *simulatability*) which tries to reflect *how* the AI system produced its outputs, and (ii) *post-hoc interpretability* which is more about *why* the AI reached its outputs, providing some after-the-fact

---

[1] Here, ANN is used to label all neural network techniques; older neural networks will be labelled as multi-layered perceptrons (MLP) and newer deep learning techniques will be called "deep neural networks" (DNNs; following [5]).

rationale/evidence for system outputs [7]. CBR systems are notable in this respect, through their use of examples/cases/precedents to explain system outputs [7, 10–12, 18, 19]. So, here, when a CBR system's cases are used to explain an ANN's opaque predictions, it is classed as "*post-hoc* explanation-by-example". As such, twin-systems are just one possible solution to interpretability in the XAI problem, but one that deserves more attention.

1.2   Motivation for a Systematic Review

There are several reasons why a systematic review of ANN-CBR twins for XAI is both timely and necessary. First, if citation patterns are any indication, there is clear evidence that the literature on twin-systems is fragmented. For instance, many recent reviews of XAI make little or no reference to key twin-systems papers in the CBR literature [1, 6, 7], while referencing papers outside CBR canon [20, 21]. Second, if we do not know the literature on this CBR-solution to XAI then, arguably, we are doomed to re-invent its findings and mistakes. Third, the XAI area requires a systematic, general framework to bring the literature together and focus future efforts. As Pedreschi *et al.* [22] point out "the state of the art to date still exhibits *ad-hoc*, scattered results, mostly hardwired to specific models…[and]… a widely applicable, systematic approach has not emerged yet". The twin-systems idea represents one possible general solution to a broad class of systems. Fourth, a systematic survey should allow us to know where we currently stand, and then to strategically road map future directions for this XAI solution.

2   Defining ANN-CBR Twins

ANN-CBR twin-systems can be found at the intersection between research on ANNs [23, 24], CBR [25–27], and hybrid systems [28–30] when explanation is important.

Artificial Neural Networks (ANNs). Biologically inspired, these AI systems typically consist of layers of nodes with non-linear activation functions and a bias term, which are connected by weights [23, 24]. Here, we distinguish between traditional neural networks of the multilayer perceptron (MLP) or backpropagation (BP) variety, and deep neural networks (DNNs) which include a wide variety of techniques; such as, recurrent neural networks (RNNs), convolutional neural networks (CNNs) and generative adversarial networks (GANs) [31–34]. MLPs typically consist of three layers, an input feature layer, a hidden layer (aka, its latent features), and an output layer. At their simplest, these ANNs learn an input-output mapping over a training set, so that when a test-case is presented, its features are used to accurately predict/classify at the MLP output layer. Significantly, the model's learning of an input-output mapping depends on modifying the weights connecting the nodes in these layers and the bias terms within the nodes. DNNs are a menagerie of many different techniques; notably, they advance beyond MLPs by being able to learn features in unstructured data (such as images or video). However, the non-linear nature of all of these ANNs make them difficult to interpret and poor at explaining their outputs [35–37]. Attempts to make ANNs more interpretable use many different "explanation methods" that are often specific to a given architecture (see reviews [5, 9, 37]). Arguably, DNNs are even less interpretable than MLPs, because of their complexity and difficulties in surfacing their extracted features. Currently, major efforts at "explaining" DNNs hinge on visualizing what specific neurons have learned or indicating "where the DNN is looking" in an image using saliency maps [36–40]. However, these methods are often quite specific to particular DNN-techniques and do not reflect the model's "reasoning process" [83]. So, a key question for the field is whether any approach can explain all ANNs – both MLPs and DNNs – in a general, unified way [5]; arguably, ANN-CBR twins are one possible solution [4].

Case-Based Reasoning (CBR). These systems perform a type of reasoning from examples or cases using a *retrieval, reuse, revise,* and *re-train* cycle [25–27, 41]. At its simplest, in CBR, when a query-case is presented the most similar cases to it are retrieved before being adapted (or used directly) to make a prediction/classification. Typically, the *retrieval* step finds cases by matching the features of the query-case and cases in the case base using k-nearest neighbor (k-NN). Retrieval accuracy (and, hence, the success of the system) can depend heavily on the weights given to these features, weights that reflect their importance in the domain. Notably, CBR is claimed to have a "natural" transparency as its reasoning-from-precedent or reasoning-from-example parallels what human experts sometimes do [18, 25]; though these claims have not always been extensively user-tested [11]. Accordingly, CBR has a substantial literature on explanation [11, 12, 19], as does its sister area of recommender systems [13, 42].

ANN-CBR Twins. These systems are a special-case of a hybrid system, that combines ANN and CBR modules, when both accuracy and interpretability are primary requirements of the overall system. Though ANNs and CBR were coupled as early hybrid systems [43–45], it is not really until the late-1990s that "true" *twins* emerge [20, 46–54]. Figure 1 shows one simple example of an ANN-CBR twinning. The task, here, is the prediction of house prices, where one has some dataset of training examples (i.e., a case base of prior cases) describing houses and their prices from previous years. The ANN accurately learns to predict the price of unseen houses (i.e., query-cases) having computed the input-output mapping from house-features to their price using the training set. To explain the ANN's prediction, its feature-weights are (in some way) extracted and used in the CBR's k-NN retrieval-step, to identify a nearest-neighbor case (or cases) to "explain" the ANN's prediction. In essence, the explanation step is asserting: price-x is predicted, because these other houses, that have very similar features, have these prices (that are close to the predicted one). Of course, the success of this whole enterprise depends on a number of factors: (i) the ANN has to be reasonably accurate in its predictions, (ii) the feature-weights extracted from the ANN have a high fidelity to the ANN's function (iii) the nearest neighbors found do not bear an overly complex relationship to the query-case (iv) and the user has sufficient expertise to easily relate these explanatory nearest neighbors to the query case (e.g., as in Figs. 1 and 2; see also [11]) and so on.

Definition of ANN-CBR Twin. Accordingly, we can define an ANN-CBR twin-system, precisely, as a system with:

- *Two Techniques.* A hybrid system in which (at least) two techniques[2] – ANN (MLP or DNN) and CBR techniques (notably, k-NN) – are combined to meet system requirements of accuracy and interpretability.
- *Separate Modules.* Where these techniques are run as separate modules, independently but, as it were, side-by-side.
- *Common Dataset.* The two techniques are run on the same dataset (i.e., they are twinned by this common usage).
- *Feature-Weight Mapping.* Some description of the ANN's functionality, typically described as its *feature-weights*, that "reflect" what the ANN has learned, is mapped to the k-NN retrieval step of the CBR-system.
- *Bipartite Division of Labor.* In the ANN and CBR modules, the former delivers predictions and the latter provides interpretability, explaining the ANN's outputs (for classification or regression).

As we shall see in our subsequent survey, though this is quite a simple definition, it excludes many hybrid systems that combine ANNs and CBR, as well as many CBR systems that do explanation *without* any ANN-aspect. For example, there are many systems that combine ANNs and CBR in a *pipelined* way where the ANN is used to extract features or feature weights that then improve the CBR's performance, using both MLPs [55–57] and DNNs [58]; these are *not* twin-systems because the CBR module is making the predictions (though the ANN improves these predictions), and the CBR system is not explaining the ANN's predictions. Similarly, there are some systems that use ANN and CBR modules in a single system, where *both* make predictions [59–61]. For instance, several agent-based systems for predicting oceanographic events (e.g., sea temperature, oils spills, red tides) alternate between ANN and CBR sub-systems, where the predictions from both are monitored to ensure continuing accuracy over time [59–61]; here, both systems are tasked with accuracy and the CBR sub-system is not specifically tasked with explanation (i.e., there is no mapping of feature weights). In the next section, we survey the somewhat abandoned regions of the hybrid-systems literature, relating to *true* ANN-CBR twins that specifically address explanation.

At this point some brief caveats are perhaps required. First, it should be said that there are domains in which k-NNs are more accurate than ANNs, where twin-systems will clearly not apply. More subtly it could be argued that if the ANN is accurate and its weights are used in the CBR system to make the same prediction and retrieve explanatory cases, then surely the ANN is redundant (or we should just pipeline it).

---

[2] Note, there are many other systems that combine CBR with other techniques, that are not considered here (e.g., with Genetic Algorithms, Rule-Based systems, Bayesian techniques).

This scenario could arise but we believe that the CBRs will, typically, lack the accuracy of the ANNs, and their predictions will be proximate (depending on choice of k).

Table 1. Five distinct searches performed in GoogleScholar (Jan–Mar, 2019) showing the keywords used, the number of results checked and the total number of unique papers that were identified for closer reading.

Search terms	#	Paper-results relevance checked	Unique papers selected for reading
"hybrid systems for explanation" "survey" "review"	1	200	12
"hybrid" "CBR" "explain" "explanation"	2	250	211
"ANN" "CBR" "explanation"	3	100	79
"NN" "Neural Network" "CBR" "explanation"	4	200	57
None (Manual check of citations to 8 key papers)	5	352	20
Totals		1,102	379

3   A Systematic Review: Methodology

A systematic search of the literature on ANN-CBR twins for explanation was done with a number of top-down searches using relevant keywords, supplemented by bottom-up, citation-based searches from key papers (see Table 1). In total 1,102 papers were checked (title and abstract) and filtered down to 379 papers; from this latter set a close reading of 90 papers was carried out to identify all the ANN-CBR twins in the literature.

3.1   Method: Search Procedure

Five systematic searches were carried out on https://scholar.google.com between January 6th, 2019 and March 24th, 2019: four top-down searches using keywords and one bottom-up search through papers that cited key articles (as it seemed to have the best coverage over Scopus/WoS). Table 1 shows the string searches used in each of the top-down searches with (i) the number of results considered in a given search (ii) the unique papers selected across these searches that were considered further (N = 379). From the latter set, a final set of papers (N = 90) were selected to be read in full to determine if the paper described a twin-system, as defined (see Fig. 1 for PRISMA flow chart). In all searches, review papers were excluded as we sought original system papers.

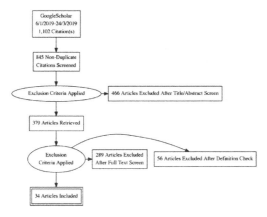

Fig. 2. PRISMA flow diagram for our systematic review of twin-systems literature.

3.2 Results Summary

In total 1,102 distinct GoogleScholar results were initially checked and filtered down to 379 potentially relevant papers, that were downloaded and examined for evidence of being twin-systems. From this set, only 90 were read in full to see if they match the twin-system definition. Of these 90 papers, only 34 were identified as *true* ANN-CBR twin-systems (n.b., only 21 of these papers report unique systems). Many systems combine ANN and CBR techniques, but fail to meet some key property of the twin-system definition (e.g., they were a pipeline, they did not work over the same dataset, or explanation was not a major concern). There was some indication that the top-down searches identified a fairly complete set of relevant papers because (i) many of the same papers recurred across searches, (ii) several, apparently, plagiarized papers were found, where the same paper was published with non-overlapping author names [62, 63], and (iii) many of the papers found in top-down searches cited the key papers on the twinning topic in the bottom-up search. The character and profile of the identified papers is discussed as a history in the next section.

4 A History of ANN-CBR Twins

Our survey of the landscape of ANN-CBR twins reveals a fragmentary and subdued development of the twinning idea. A close reading of 90 articles on hybrid ANN-CBR systems that made some mention of explanation, found only 34 papers (21 unique systems) that were *true* twin-systems. The remaining papers tend to be ANN-CBR pipelines where the ANN is used to compute features and/or feature-weights that are then used in the CBR's k-NN for predictive purposes. Even though the idea of combining ANN and CBR is first referenced in 1989 [43, 44], it is not until 1999 that the first true twin-systems emerge [20, 46]. From this beginning, there is a very modest development of the idea over the intervening 20 years. Indeed, citation patterns are quite inconsistent and lookbacks from more recent papers are patchy. The history divides into three distinct periods: (i) a major piece of work by a Korean Group in the late-1990s (with a parallel proposal in the USA), (ii) a significant addition by an Irish Group through the mid-2000s and, then (iii) more recent work in deep learning that revisits related approaches, often with no or poor reference to the prior literature.

4.1 Korean Developments (1999–2007): Feature-Weighting Tests of Twins

Around 1999, a South Korean group working at the Korean Advanced Institute of Science and Technology (KAIST), explored a range of feature-weighting techniques in comparative tests of the twin-system idea, in a framework they called "Memory Based Neural Reasoning" [46–54]. Shin and colleagues [46, 47, 51] paired MLPs with CBR operating over the same dataset, proposing that this hybrid system "can give example-based explanations together with prediction values of the neural network" [47, p. 637]. Initially, they tested these twin-systems on a semiconductor-yield dataset before moving on to tests on many benchmark datasets (e.g., Iris and Wisconsin Diagnostic Breast Cancer datasets) for classification and regression tasks. In this work, they perform competitive tests of four different feature-weighting schemes for capturing the MLPs activation patterns (i.e., *sensitivity*, *activity*, *relevance* and *saliency*). For example, in *sensitivity* a feature's weight is calculated by taking the absolute difference between the normal prediction of the MLP and its prediction with that input feature set to zero; this is repeated and summed for the entire training set, and the final figure is then divided by the number of instances in the training data to normalize it for the final feature-weight value [46]. For each weighting scheme, the feature-weights were used in the k-NN to retrieve cases, noting which matched the prediction for the query-case.

There are three significant contributions in this Korean work: (i) the researchers explicitly talk about a division of labor, where the ANN provides accuracy, using its feature-weights, and the CBR system provides explanations using nearest-neighbor cases, (ii) they recognize that there are many different ways to describe the ANN (i.e., different feature-weighting methods) that need to be tested[3], and (iii) they understand that there are two classes of feature-weighting methods (global and local).

Shin and colleagues [46, 47, 51] appear to be the first in the literature, to explicitly pair ANNs and CBR systems in a twinned way for purposes of explanation and to perform comparative tests of different feature-weighting schemes. Shin and colleagues [46, 47] proposed that the *sensitivity* and *activity* measures seemed to perform best, (though conclusions are different for different datasets) arguing that their fidelity to the function being computed by the MLP was better. Park *et al.* [51] extend the earlier tests with a new feature-weighting scheme based on the important distinction between global and local weighting schemes. *Global feature-weighting* assumes the input space is isotropic, deriving a single ubiquitous feature-weight vector for the entire domain (i.e., weights do not change for different query cases), whereas *local feature-weighting* weights each specific query-case (and sometimes each training case) differently to help case retrieval. Park *et al.* [51] find that local-weighting schemes perform markedly

---

[3] A fact overlooked in most papers, even very recent ones.

better than global-weighting methods, presumably because the former captures information about a specific area of the input space for a given query in a more fine-grained way. However, their local-weighting technique is not applicable to *post-hoc* explanations in MLPs, as it is specifically trained to generate query-specific feature-weight-sets rather than giving predictions. Hence, it cannot be considered to be a twin-system; however, it does show the potential for local-weighting methods. Later work extends global weighting schemes to datasets having symbolic feature-values [53, 54]. Overall, the global methods tend to produce very similar results for different values of k but the results from local methods are found to be demonstrably better.

These nine Korean papers did not attract huge levels of attention; together they have a total of 269 GoogleScholar citations (M = 30, Max = 91, Min = 5). Indeed, many of these citations are not specifically to the twinning idea, but reference other aspects of the work (e.g., the domains used). However, more recently, several papers have referenced their work. Weber *et al.* [64] claim a philosophic overlap with the Korean work in an ANN-CBR hybrid; yet the details of the feature-weighting used are not clear. Peng and Zhuang [65] propose a different feature-weighting scheme for an ANN-CBR twinning, that replaces feature-values of a case using the MLPs weights (but does not reference the Korean work). However, it is only in the last few years, that the Korean work has been seriously revisited. Biswas and colleagues [66–68] revisited the *sensitivity* measure and several limitations of earlier weighting schemes; they transform the MLP into an AND/OR graph from which weights are extracted for use in the CBR system. On applying this graph technique to several new domains, they find that it does better than other methods. Biswas *et al.* [68] also revisit global weighing-techniques in the context of class-imbalanced datasets, showing that a cost-sensitive learning algorithm displays improvements for such datasets. Finally, recently, the global-local distinction has been raised in XAI in deep learning [5, 6], without referencing the Korean work.

4.2   A Parallel Discovery in L.A.: Caruana *et al.* [20]

Around the same time as the Korean Group's work, another group working at UCLA reported an extension to an earlier system [69] that provided case-based explanations ([20] cited 33 times in GoogleScholar). Caruana *et al.* [20] describe a system for medical domains in which a "non-case-based learning method" (an MLP) could generate a distance metric over a training set, that could then be used to find an explanatory case that was most similar to a query-case. Caruana *et al.*'s MLP predicted pneumonia mortality and proposed case-based explanations based on a query-case's hidden-layer activation-vector (i.e., its latent features), by computing the Euclidean distance between the query-vector and all training-cases, thus enabling them to find the explanatory cases with the most similar latent features. The paper does not provide detailed results on the success of this method and neither does it report user trials, though it does discuss the issues surrounding how cases might be deployed to explain the ANN's predictions (for recent related work see [70]). Caruana *et al.*'s [20] feature-weighting method differs from those examined by the Korean group, that were based on *input space* weightings rather than *latent space* weightings (as well as being a local-weighting technique). This latent-space approach has not been pursued actively, perhaps, because it appears to be less transparent than input-space approaches (see [4]). Recently, in the deep learning literature reviews of XAI, [20] is regularly cited as *the* case-based explanation paper [7, 12, 71, 72], in the absence of references to the Korean work that, arguably, is more complete; a fact that, perhaps, indicates some discontinuities in the XAI literature.

4.3   An Irish Departure (Mid-2000s): Local Feature-Weighting Tests of Twins

The next major step in the development of the twinning idea came in the mid-2000s, from a group of Irish researchers, largely, at University College Dublin [73–77]. This group also saw a role for CBR in explaining the opaque, but accurate outputs of MLPs, arguing that "the use of actual training data, cases from the case base, as evidence in support of a particular prediction, is a powerful and convincing form of explanation" (p. 164, [75]). The Irish researchers, who did not cite the Korean work, proposed a new and intriguing local feature-weighting method [74–76]. Nugent and Cunningham [74] were concerned with capturing the function being computed by MLPs in the local region around a given query-case for a blood-alcohol dataset. So, they systematically perturbed the features of the query-case, queried labels for these perturbed cases from the MLP, and then built a linear model from the results of these tests. The coefficients of this linear model were then used to weight the k-NN search in the CBR system that shared its case-base with the MLP's training set. Nugent *et al.* [76, 77] also considered more complex use of cases than just providing nearest neighbors, by selecting *a fortori* cases; the idea being to use a case that is closest to the decision boundary for the query-case, which may not, necessarily, be the nearest neighbor. Finally, [73] did user tests to show that the retrieved cases have some explanatory value in these domains.

There are, at least, three significant contributions from this work: it explores (i) a very different approach to the computation of local feature-weights which, in contrast to the Korean local-weighting method, can be used for *post-hoc* explanations, (ii) a more complex scheme for selecting explanatory cases, beyond the simple use of nearest neighbors, (iii) it showed that this type of twin-system had some validity for human users by using case-based explanations over feature-based ones.

These five papers have received moderate attention in the literature; between them they have a total of 236 GoogleScholar citations (M = 47, max = 99, min = 8). However, few of these citations are about the twin-systems idea (i.e., often about user tests). Notably, though the linear-model idea has advanced significantly in the literature on interpretable classifiers [78, 79], few papers specifically cite this Irish work. For instance, the Local Interpretable Model-Agnostic Explanations (LIME) [79] technique also perturbs query-cases to build local linear models but does not cite [74]. Although, Olsson *et al.* [80] do, in a related approach to case similarity using logistic regression – the *principle of interchangeability* – and the notion of *local accuracy* to handle the identification of explanatory cases. More recently, there is some recognition of these papers in reviews [10, 81] but for the most part they are passed over [6, 7, 37, 72] with all CBR-solutions being attributed to Caruana *et al.* [20] and Kim, Rudin and Shah [21].

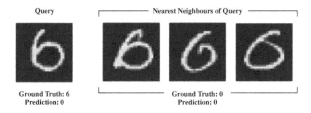

Fig. 3. In a CNN-CBR twin-system, the CNN classifies an image of the number "6" as "0". An explanation using nearest-neighbor cases from its CBR twin, shows the training-data used to model the CNN function in this area of the latent space was labelled as "0" but looks like "6"; so the model miss-classifies the query as it "looks like" these training cases (adapted from [4]).

4.4   Recent DNN-CBR Twinning

We have already established that there is a disconnect between more recent DNN research and the older twin-system literature in CBR. Yet, in the last two decades, a huge amount of work has been done on different ways to describe the functions of opaque ANN and ML systems. Recently, the focus of some of this work has shifted to the case-based explanation-by-example of DNNs. For instance, Chen *et al.* [82] and Li *et al.* [83] both build CBR into the DNN architecture itself, mainly to avoid the need for *post-hoc* explanations. Although these are not twin-systems, they do combine CBR and ANN techniques for the purpose of interpretability and explanations (though they fail to cite much of the previous work done). A review by Gilpin *et al.* [37] proposes that DNNs have often been explained by using simpler "proxy systems"[4] of which they identify four types: linear models [79], decision trees [84], automatic rule extraction [78], and saliency mapping [85–87]. Two of these approaches – linear models and saliency mapping – have resonances with the twin-systems literature reviewed here.

Linear Models: As in Nugent and Cunningham's work, a currently popular approach to explaining ANNs (and indeed any ML model) is to use local linear models built by perturbing an input in the neighborhood of a query. LIME [79] is *the* prime example of this approach, in finding relative feature-weightings for a given query-case. Recently, LIME has been used in comparative tests of several twin-systems ([4] influenced by Nugent and Cunningham) and was not appreciably better than global-weighting techniques (including the *sensitivity* method used by the Korean group).

Saliency Mapping: Another popular technique looks at the contribution of inputs to an ANN's output, deriving saliency maps by backpropagating contribution scores from a given activation in the network (usually in the output layer), to a previous layer (usually the input one). Amongst others, these methods include *Layer-wise Relevance Propagation* [85], *Integrated Gradients* [86], and *DeepLIFT* [87]. This *saliency mapping* is typically used to highlight important pixels in a CNN's classification of an image, however it has other uses and has recently been applied to MLP-CBR and CNN-CBR

---

[4] The proxy system is meant to behave similarly to the black-box system but is simpler for explanatory purposes (so, the CBR in ANN-CBR twin is one type of proxy model).

twin-systems (notably, with multiple fully connected dense layers using image transfer learning [88]) in comparative tests (see [4] and also Fig. 3).

Other DNN Options: There are also a handful of other DNN options that are arguably twin-systems though of quite a different ilk. Work on the extraction of prototypes from the analysis of DNNs have been cast as case-based approaches, though with a Bayesian aspect [21, 83]. These proposals look like a different type of twinning – Bayesian-CBR twins – that perhaps have other precursors in the CBR literature [89, 90]. Another approach tries to map the layers of a DNN onto particular exemplar cases using Deep k-Nearest neighbors (DkNN) [71, 91, 92]. However, it still remains to be seen whether these are to be accommodated as twin-systems.

5   Future Directions: Road Mapping

In the present paper, we have reviewed the history of how CBR has been used in a twinning fashion to explain the outputs of ANNs. The significance and importance of this survey is that it shows there are generalizations about XAI to be gleaned from the twin-systems approach. Such generalizations may help us avoid the current scattered fragmentary development of XAI solutions [5]. This review also suggests a research road map for future work in this area, along at least three paths:

- *Feature-Weighting Schemes.* It is clear that there is a large space of feature-weighting schemes that could be explored (especially, more recent ones in DNNs); this exploration needs to be done in a controlled and comparative fashion to determine which ones are best for which domains and tasks.
- *The Deployment of Cases.* CBR work has shown in twin-systems there are many different ways cases can be used for explanation (e.g., *a fortiori* usage, counter-factual cases, near misses, nearest unlike neighbors and so on; more needs to be done on these ideas in the context of ANNs, and especially DNNs).
- *The Embarrassment of User Testing.* In all the papers we examined we found less than a handful (i.e., <5) that performed any adequate user testing of the proposal that cases improved the interpretability of models; this gap needs to be rectified.

In conclusion, notwithstanding the citation gaps in the literature, it is clear that there are many fruitful directions in which the CBR-twin idea can be taken to answer the interpretability problems we currently face in XAI.

References

1. Adadi, A., Berrada, M.: Peeking inside the black-box: a survey on explainable artificial intelligence (XAI). IEEE Access **6**, 52138–52160 (2018)
2. Goodman, B., Flaxman, S.: European Union regulations on algorithmic decision-making and a "right to explanation". AI Mag. **38**(3), 50–57 (2017)
3. Wachter, S., Mittelstadt, B., Floridi, L.: Why a right to explanation of automated decision-making does not exist in the general data protection regulation. Int. Data Privacy Law **7**(2), 76–99 (2017)

4. Kenny, E.M., Keane, M.T.: Twin-systems to explain neural networks using case-based reasoning. In: Proceedings of the 28th International Joint Conference on Artificial Intelligence (IJCAI 2019), pp. 326–333 (2019)
5. Guidotti, R., Monreale, A., Ruggieri, S., Turini, F., Giannotti, F., Pedreschi, D.: A survey of methods for explaining black box models. ACM Comput. Surv. **51**(5), 93 (2018)
6. Doshi-Velez, F., Kim, B.: Towards a rigorous science of interpretable machine learning. arXiv preprint arXiv:1702.08608 (2017)
7. Lipton, Z.C.: The mythos of model interpretability. Queue **16**(3), 30 (2018)
8. Miller, T.: Explanation in artificial intelligence: insights from the social sciences. Artif. Intell. **267**, 1–38 (2019)
9. Abdul, A., Vermeulen, J., Wang, D., Lim, B.Y., Kankanhalli, M.: Trends and trajectories for explainable, accountable and intelligible systems: an HCI research agenda. In: Proceedings 2018 CHI Conference on Human Factors in Computing Systems, p. 582. ACM (2018)
10. Biran, O., Cotton, C.: Explanation and justification in machine learning: a survey. In: IJCAI 2017 Workshop on Explainable AI (XAI), vol. 8, p. 1 (2017)
11. Sørmo, F., Cassens, J., Aamodt, A.: Explanation in case-based reasoning–perspectives and goals. Artif. Intell. Rev. **24**(2), 109–143 (2005)
12. Johs, A.J., Lutts, M., Weber, R.O.: Measuring explanation quality in XCBR. In: Proceedings of ICCBR 2018, p. 75 (2018)
13. Tintarev, N., Masthoff, J.: A survey of explanations in recommender systems. In: 2007 IEEE 23rd International Conference on Data Engineering Workshop, pp. 801–810. IEEE (2007)
14. Harman, G.H.: The inference to the best explanation. Philos. Rev. **74**(1), 88–95 (1965)
15. Salmon, W.C.: Scientific Explanation and the Causal Structure of the World. Princeton University Press, Princeton (1984)
16. Van Fraassen, B.C.: The Scientific Image. Oxford University Press, Oxford (1980)
17. Keil, F.C.: Explanation and understanding. Ann. Rev. Psychol. **57**, 227–254 (2006)
18. Leake, D.B.: CBR in context: the present and future. In: Case-Based Reasoning: Experiences, Lessons, and Future Directions, pp. 3–30 (1996)
19. Leake, D., McSherry, D.: Introduction to the special issue on explanation in case-based reasoning. Artif. Intell. Rev. **24**(2), 103–108 (2005)
20. Caruana, R., Kangarloo, H., Dionisio, J.D., Sinha, U., Johnson, D.: Case-based explanation of non-case-based learning methods. In: Proceedings of the AMIA Symposium, p. 212. American Medical Informatics Association (1999)
21. Kim, B., Rudin, C., Shah, J.A.: The Bayesian case model: a generative approach for case-based reasoning and prototype classification. In: Advances in NIPs, pp. 1952–1960 (2014)
22. Pedreschi, D., Giannotti, F., Guidotti, R., Monreale, A., Ruggieri, S., Turini, F.: Meaningful explanations of Black Box AI decision systems. In: Proceedings of AAAI 2019 (2019)
23. Haykin, S.: Neural Networks, vol. 2. Prentice Hall, New York (1994)
24. Goodfellow, I., Bengio, Y., Courville, A.: Deep Learning. MIT Press, Cambridge (2016)
25. Kolodner, J.: Case Based Reasoning. Morgan Kaufmann, Burlington (2014)
26. Aamodt, A., Plaza, E.: Case-based reasoning: foundational issues, methodological variations, and system approaches. AI Commun. **7**(1), 39–59 (1994)
27. De Mantaras, R.L., et al.: Retrieval, reuse, revise and retention in CBR. Knowl. Eng. Rev. **20**(3), 215–240 (2006)
28. Sahin, S., Tolun, M.R., Hassanpour, R.: Hybrid expert systems: a survey of current approaches and applications. Expert Syst. Appl. **39**(4), 4609–4617 (2012)
29. Negnevitsky, M.: Artificial Intelligence. Pearson Education, London (2005)
30. Medsker, L.R.: Hybrid Neural Network and Expert Systems. Springer, Heidelberg (2012)
31. Szegedy, C., et al.: Intriguing properties of neural networks. arXiv preprint arXiv:1312.6199 (2013)

32. Krizhevsky, A., Sutskever, I., Hinton, G.E.: Imagenet classification with deep convolutional neural networks. In: Advances in NIPs, pp. 1097–1105 (2012)
33. Witten, I.H., Frank, E., Hall, M.A., Pal, C.J.: Data Mining: Practical Machine Learning Tools and Techniques. Morgan Kaufmann, Burlington (2016)
34. LeCun, Y., Bengio, Y., Hinton, G.: Deep learning. Nature **521**(7553), 436 (2015)
35. Olden, J.D., Jackson, D.A.: Illuminating the "black box". Ecol. Model. **154**(1–2), 135–150 (2002)
36. Selvaraju, R.R., Cogswell, M., Das, A., Vedantam, R., Parikh, D., Batra, D.: Grad-CAM: visual explanations from deep networks via gradient-based localization. In: Proceedings of the IEEE International Conference on Computer Vision, pp. 618–626 (2017)
37. Gilpin, L.H., Bau, D., Yuan, B.Z., Bajwa, A., Specter, M., Kagal, L.: Explaining explanations. arXiv preprint arXiv:1806.00069 (2018)
38. Zeiler, M.D., Fergus, R.: Visualizing and understanding convolutional networks. In: Fleet, D., Pajdla, T., Schiele, B., Tuytelaars, T. (eds.) ECCV 2014. LNCS, vol. 8689, pp. 818–833. Springer, Cham (2014). https://doi.org/10.1007/978-3-319-10590-1_53
39. He, K., Zhang, X., Ren, S., Sun, J.: Deep residual learning for image recognition. In: Proceedings of the IEEE Conference on Computer Vision and Pattern Recognition, pp. 770–778 (2016)
40. Erhan, D., Bengio, Y., Courville, A., Vincent, P.: Visualizing higher-layer features of a deep network. Univ. Montreal **1341**(3), 1 (2009)
41. Keane, M.T.: Analogical asides on case-based reasoning. In: Wess, S., Althoff, K.-D., Richter, M.M. (eds.) EWCBR 1993. LNCS, vol. 837, pp. 21–32. Springer, Heidelberg (1994). https://doi.org/10.1007/3-540-58330-0_74
42. Nunes, I., Jannach, D.: A systematic review and taxonomy of explanations in decision support and recommender systems. UMUAI **27**(3–5), 393–444 (2017)
43. Becker, L., Jazayeri, K.: A connectionist approach to case-based reasoning. In: Proceedings of the Case-Based Reasoning Workshop, pp. 213–217. Morgan Kaufmann (1989)
44. Thrift, P.: A neural network model for case-based reasoning. In: Proceedings of the Case-Based Reasoning Workshop, pp. 334–337. Morgan Kaufmann (1989)
45. Hilario, M., Pellegrini, C., Alexandre, F.: Modular integration of connectionist and symbolic processing in knowledge-based systems. C.de R. en Informatique de Nancy (1994)
46. Shin, C.K., Park, S.C.: Memory and neural network based expert system. Expert Syst. Appl. **16**(2), 145–155 (1999)
47. Shin, C.K., Yun, U.T., Kim, H.K., Park, S.C.: A hybrid approach of neural network & memory-based learning to data mining. IEEE Trans. Neural Netw. **11**, 637–646 (2000)
48. Shin, C.K., Park, S.C.: A machine learning approach to yield management in semiconductor manufacturing. Int. J. Prod. Res. **38**, 4261–4271 (2000)
49. Park, J.H., Shin, C.K., Im, K.H., Park, S.C.: A local weighting method to the integration of neural network and case based reasoning. In: Neural Networks for Signal Processing XI: Proceedings of the 2001 IEEE SPS Workshop, pp. 33–42. IEEE (2001)
50. Shin, C.K., Park, S.C.: Towards integration of memory based learning and neural networks. In: Pal, S.K., Dillon, T.S., Yeung, D.S. (eds.) Soft Computing in Case Based Reasoning, pp. 95–114. Springer, London (2001). https://doi.org/10.1007/978-1-4471-0687-6_5
51. Park, J.H., Im, K.H., Shin, C.K., Park, S.C.: MBNR: case-based reasoning with local feature weighting by neural network. Appl. Intell. **21**(3), 265–276 (2004)
52. Park, S.C., Kim, J.W., Im, K.H.: Feature-weighted CBR with NN for symbolic features. In: Huang, D.S., Li, K., Irwin, G.W. (eds.) ICIC 2006. LNCS, vol. 4113, pp. 1012–1020. Springer, Heidelberg (2006). https://doi.org/10.1007/11816157_123
53. Im, H., Park, S.C.: Case-based reasoning and neural network based expert system for personalization. Expert Syst. Appl. **32**(1), 77–85 (2007)

54. Ha, S.: A personalized counseling system using case-based reasoning with neural symbolic feature weighting (CANSY). Appl. Intell. **29**(3), 279–288 (2008)

55. Reategui, E.B., Campbell, J.A., Leao, B.F.: Combining a neural network with case-based reasoning in a diagnostic system. Artif. Intell. Med. **9**(1), 5–27 (1997)

56. Yang, B.S., Han, T., Kim, Y.S.: Integration of ART-Kohonen NN and CBR for intelligent fault diagnosis. Expert Syst. Appl. **26**(3), 387–395 (2004)

57. Rodriguez, Y., Garcia, M.M., De Baets, B., Morell, C., Bello, R.: A connectionist fuzzy case-based reasoning model. In: Gelbukh, A., Reyes-Garcia, C.A. (eds.) MICAI 2006. LNCS, vol. 4293, pp. 176–185. Springer, Berlin (2006). https://doi.org/10.1007/11925231_17

58. Amin, K., Kapetanakis, S., Althoff, K.-D., Dengel, A., Petridis, M.: Answering with cases: a CBR approach to deep learning. In: Cox, M.T., Funk, P., Begum, S. (eds.) ICCBR 2018. LNCS (LNAI), vol. 11156, pp. 15–27. Springer, Cham (2018). https://doi.org/10.1007/978-3-030-01081-2_2

59. Corchado, J.M., Rees, N., Lees, B., Aiken, J.: Data mining using example-based methods in oceanographic forecast models. In: IEE Colloquium on Knowledge Discovery and Data Mining (Digest No. 1998/310), p. 7-1. IET (1998)

60. Corchado, J.M., Lees, B.: A hybrid case-based model for forecasting. Appl. Artif. Intell. **15**(2), 105–127 (2001)

61. Fdez-Riverola, F., Corchado, J.M., Torres, J.M.: An automated hybrid CBR system for forecasting. In: Craw, S., Preece, A. (eds.) ECCBR 2002. LNCS (LNAI), vol. 2416, pp. 519–533. Springer, Heidelberg (2002). https://doi.org/10.1007/3-540-46119-1_38

62. Jothikimar, R., Shivakumar, N., Ramesh, P.S., Suganthan, Suresh, A.: Heart disease prediction system using ANN, RBF and CBR. Int. J. Pure Appl. Math. **117**(21), 199–217 (2017)

63. Kouser, R.R., Manikandan, T., Kumar, V.V.: Heart disease prediction system using artificial neural network, radial basis function and case based reasoning. J. Comput. Theoret. Nanosci. **15**(9–10), 2810–2817 (2018)

64. Weber, R., Proctor, Jason M., Waldstein, I., Kriete, A.: CBR for modeling complex systems. In: Muñoz-Ávila, H., Ricci, F. (eds.) ICCBR 2005. LNCS (LNAI), vol. 3620, pp. 625–639. Springer, Heidelberg (2005). https://doi.org/10.1007/11536406_47

65. Peng, Y., Zhuang, L.: A case-based reasoning with feature weights derived by BP network. In: Intelligent Information Technology Application, pp. 26–29. IEEE (2007)

66. Biswas, S.K., Sinha, N., Purakayastha, B., Marbaniang, L.: Hybrid expert system using case based reasoning and neural network for classification. Biol. Inspired Cogn. Archit. **9**, 57–70 (2014)

67. Biswas, S.K., Baruah, B., Sinha, N., Purkayastha, B.: A hybrid CBR classification model by integrating ANN into CBR. Int. J. Serv. Technol. Manag. **21**(4–6), 272–293 (2015)

68. Biswas, S.K., Chakraborty, M., Singh, H.R., Devi, D., Purkayastha, B., Das, A.K.: Hybrid case-based reasoning system by cost-sensitive neural network for classification. Soft. Comput. **21**(24), 7579–7596 (2017)

69. Cooper, G.F., et al.: An evaluation of machine-learning methods for predicting pneumonia mortality. Artif. Intell. Med. **9**(2), 107–138 (1997)

70. Caruana, R., Lou, Y., Gehrke, J., Koch, P., Sturm, M., Elhadad, N.: Intelligible models for healthcare: predicting pneumonia risk and hospital 30-day readmission. In: Proceedings of the 21th ACM SIGKDD International Conference on Knowledge Discovery and Data Mining, pp. 1721–1730. ACM (2015)

71. Papernot, N., McDaniel, P.: Deep k-nearest neighbours: Towards confident, interpretable and robust deep learning. arXiv preprint arXiv:1803.04765 (2018)

72. Mittelstadt, B., Russell, C., Wachter, S.: Explaining explanations in AI. In: Proceedings of Conference on Fairness, Accountability, and Transparency (FAT*-19) (2019)
73. Cunningham, P., Doyle, D., Loughrey, J.: An evaluation of the usefulness of case-based explanation. In: Ashley, K.D., Bridge, D.G. (eds.) ICCBR 2003. LNCS (LNAI), vol. 2689, pp. 122–130. Springer, Heidelberg (2003). https://doi.org/10.1007/3-540-45006-8_12
74. Nugent, C., Cunningham, P.: A case-based explanation system for black-box systems. Artif. Intell. Rev. **24**(2), 163–178 (2005)
75. Nugent, C., Cunningham, P., Doyle, D.: The best way to instil confidence is by being right. In: Muñoz-Ávila, H., Ricci, F. (eds.) ICCBR 2005. LNCS (LNAI), vol. 3620, pp. 368–381. Springer, Heidelberg (2005). https://doi.org/10.1007/11536406_29
76. Nugent, C., Doyle, D., Cunningham, P.: Gaining insight through case-based explanation. J. Intell. Inf. Syst. **32**(3), 267–295 (2009)
77. Doyle, D., Cunningham, P., Bridge, D., Rahman, Y.: Explanation oriented retrieval. In: Funk, P., González Calero, P.A. (eds.) ECCBR 2004. LNCS (LNAI), vol. 3155, pp. 157–168. Springer, Heidelberg (2004). https://doi.org/10.1007/978-3-540-28631-8_13
78. Andrews, R., Diederich, J., Tickle, A.B.: Survey and critique of techniques for extracting rules from trained artificial neural networks. Knowl.-Based Syst. **8**, 373–389 (1995)
79. Ribeiro, M.T., Singh, S., Guestrin, C.: Why should I trust you?: explaining the predictions of any classifier. In: Proceedings of the 22nd ACM SIGKDD International Conference on Knowledge Discovery and Data Mining, pp. 1135–1144. ACM (2016)
80. Olsson, T., Gillblad, D., Funk, P., Xiong, N.: Case-based reasoning for explaining probabilistic machine learning. Int. J. Comput. Sci. Inf. Technol. **6**(2), 87–101 (2014)
81. Zharov, Y., Korzhenkov, D., Shvechikov, P., Tuzhilin, A.: YASENN: Explaining Neural Networks via Partitioning Activation Sequences. arXiv preprint arXiv:1811.02783 (2018)
82. Chen, C., Li, O., Barnett, A., Su, J., Rudin, C.: This looks like that: deep learning for interpretable image recognition. arXiv preprint arXiv:1806.10574 (2018)
83. Li, O., Liu, H., Chen, C., Rudin, C.: Deep learning for case-based reasoning through prototypes: a neural network that explains its predictions. In: Thirty-Second AAAI Conference on Artificial Intelligence. AAAI (2018)
84. Zilke, J.R., Loza Mencía, E., Janssen, F.: DeepRED – rule extraction from deep neural networks. In: Calders, T., Ceci, M., Malerba, D. (eds.) DS 2016. LNCS (LNAI), vol. 9956, pp. 457–473. Springer, Cham (2016). https://doi.org/10.1007/978-3-319-46307-0_29
85. Bach, S., Binder, A., Montavon, G., Klauschen, F., Müller, K.R., Samek, W.: On pixel-wise explanations for non-linear classifier decisions by layer-wise relevance propagation. PLoS ONE **10**(7), e0130140 (2015)
86. Sundararajan, M., Taly, A., Yan, Q.: Axiomatic attribution for deep networks. In: Proceedings of the 34th International Conference on Machine Learning, vol. 70, pp. 3319–3328. JMLR. Org (2017)
87. Shrikumar, A., Greenside, P., Kundaje, A.: Learning important features through propagating activation differences. In: Proceedings of the 34th International Conference on Machine Learning, vol. 70, pp. 3145–3153. JMLR. Org (2017)
88. Zhang, C.-L., Luo, J.-H., Wei, X.-S., Wu, J.: In defense of fully connected layers in visual representation transfer. In: Zeng, B., Huang, Q., El Saddik, A., Li, H., Jiang, S., Fan, X. (eds.) PCM 2017. LNCS, vol. 10736, pp. 807–817. Springer, Cham (2018). https://doi.org/10.1007/978-3-319-77383-4_79
89. Myllymäki, P., Tirri, H.: Massively parallel case-based reasoning with probabilistic similarity metrics. In: Wess, S., Althoff, K.-D., Richter, M.M. (eds.) EWCBR 1993. LNCS, vol. 837, pp. 144–154. Springer, Heidelberg (1994). https://doi.org/10.1007/3-540-58330-0_83

90. Kofod-Petersen, A., Langseth, H., Aamodt, A.: Explanations in Bayesian networks using provenance through case-based reasoning. In: CBR Workshop Proceedings, p. 79 (2010)
91. Wallace, E., Feng, S., Boyd-Graber, J.: Interpreting Neural Networks with Nearest Neighbours. arXiv preprint arXiv:1809.02847 (2018)
92. Card, D., Zhang, M., Smith, N.A.: Deep weighted averaging classifiers. In: Proceedings of Conference on Fairness, Accountability & Trust, pp. 369–378. ACM, January 2019

Predicting Grass Growth for Sustainable Dairy Farming: A CBR System Using Bayesian Case-Exclusion and *Post-Hoc*, Personalized Explanation-by-Example (XAI)

Eoin M. Kenny[1,2,4(✉)], Elodie Ruelle[3,4], Anne Geoghegan[3,4],
Laurence Shalloo[3,4], Micheál O'Leary[3,4], Michael O'Donovan[3,4],
and Mark T. Keane[1,2,4]

[1] School of Computer Science, University College Dublin, Dublin, Ireland
[2] Insight Centre for Data Analytics, UCD, Dublin, Ireland
{eoin.kenny,mark.keane}@insight-centre.org
[3] Teagasc, Animal and Grassland Research, Fermoy, Ireland
{elodie.ruelle,anne.geoghegan,laurence.shalloo,
michael.oleary,michael.odonovan}@teagasc.ie
[4] VistaMilk SFI Centre, Fermoy, Ireland

Abstract. Smart agriculture has emerged as a rich application domain for AI-driven decision support systems (DSS) that support sustainable and responsible agriculture, by improving resource-utilization through better on-farm, management decisions. However, smart agriculture's promise is often challenged by the high barriers to user adoption. This paper develops a case-based reasoning (CBR) system called PBI-CBR to predict grass growth for dairy farmers, that combines predictive accuracy and explanation capabilities designed to improve user adoption. The system provides *post-hoc, personalized explanation-by-example* for its predictions, by using explanatory cases from the same farm or county. A key novelty of PBI-CBR is its use of Bayesian methods for case exclusion in this regression domain. Experiments report the tradeoff that occurs between predictive accuracy and explanatory adequacy for different parametric variants of PBI-CBR, and how updating Bayesian priors each year reduces error.

Keywords: CBR · Bayesian analysis · Smart agriculture ·
Case exclusion · XAI

1 Introduction

Although the promise of artificial intelligence (AI) in smart agriculture is usually advertised as increasing productivity, in the future it may become increasingly about improving sustainability [1, 2]. As climate change accelerates, what AI may actually deliver is a precision agriculture that allows farmers to measure, balance, and predict the outcomes of farm management-decisions in ways that mitigate the environmental impact of these activities. However, this future depends on the development of AI-enabled decision support systems (DSS) that are both predictively accurate (e.g., in predicting

© Springer Nature Switzerland AG 2019
K. Bach and C. Marling (Eds.): ICCBR 2019, LNAI 11680, pp. 172–187, 2019.
https://doi.org/10.1007/978-3-030-29249-2_12

grass growth), and explainable to the end user (i.e., farmers) to encourage adoption and usage. In this paper, an existing DSS called PastureBase Ireland (PBI) is extended by using case-based reasoning (CBR) techniques; the so-called PBI-CBR system. This new DSS predicts grass growth for dairy farmers and offers explanations designed to improve user adoption. As such, the system is an instance of eXplainable AI (XAI), providing *post-hoc, personalized explanation-by-example* for its predictions, based on location (using cases from the same or nearby farms). One key novelty of PBI-CBR is its use of what we refer to as *Bayesian Case-Exclusion,* which excludes outlier cases from the prediction process using prior beliefs about data distribution(s), reducing error and improving explanations. In the remainder of this introduction, the sustainability context for this work is briefly described, before outlining the structure of the paper.

1.1   Context: Agriculture, Sustainability and AI

Concerns about the impact of agriculture on climate change and the development of sustainable models are growing [2]. The agricultural sector and consumers are faced with varying views from climate change denial, to proposals that animal agriculture is responsible for 18–51% of greenhouse grass emissions [29, 30]. However, there is perhaps a middle ground that is exemplified by the work here.

Recently, an argument has emerged arguing for a quick move to sustainable farming systems [5]; the so-called *agroecology* perspective. For example, in the dairy sector this agroecology view has proposed a move to pasture-based systems, where animals are predominantly fed on grass outdoors rather than on meal and supplements indoors. The pasture-based proposal has the potential to be sustainable, in part, by using grass as a carbon sink and extending the grazing season (reducing slurry emissions) [11]. Furthermore, humans have limited capacity to digest grass, as it is a non-edible protein, so it is not consuming a food people could eat [28]. However, these initiatives depend on precision technology, using AI, to monitor variables such as climate and grass growth.

This paper considers a CBR system[1] that supplements an existing DSS used by several thousand Irish dairy farmers (i.e., PBI), which predicts grass growth in the coming week for a specific farm and offers personalized explanations (see Sects. 4 and 5). However, as we shall see, there are significant challenges in handling the data noise which arises in this domain, especially against the backdrop of increasing climate disruption. Finally, for the sake of brevity, note that we only consider the *retrieval* step of CBR for this current iteration of PBI-CBR.

1.2   PastureBase Ireland for Dairy Farmers (PBI)

Smart agriculture often depends upon providing new DSSs for farmers to aid them in making complex decisions about how to manage their farm for productivity and sustainability [2]. These systems have three main challenges. First, they must be predictively

---

[1] Several other approaches such as linear regression, neural networks, SVMs and tree algorithms were also tested alongside this CBR system. The CBR system's accuracy equalled or bettered these other systems.

accurate. Second, they need to be easy to use and interpretable for end users, to encourage adoption and continued use. Third, they need to be able to support decision making in the context of increasing climatic disruption, where the climate in past years may not be indicative of climate in future years. The present work extends an existing DSS called PBI used in the grass-fed, pasture-based dairy farming systems in Ireland.

PastureBase Ireland. Since 2013, Ireland's national agricultural research organization *Teagasc*, have provided PastureBase Ireland (PBI, https://pasturebase.teagasc.ie) as a grassland management system to provide information and advice for Irish dairy farmers. PBI has 6,000+ users out of ∼ 18,000 dairy farmers in Ireland. Among other features, the PBI database has weekly records of *grass covers* for individual farms from 2013 to present. A *grass cover* for a farm is principally, the amount of grass available on that farm for cows to eat; formally, it is a measure of biomass in grass on the farm above ground level or a height of 4 cm. PBI allows farmers to enter this grass cover data for each field/paddock of their farm in a given week, using their own measurements/ estimates, thus allowing them to budget grass-availability for their herd. Our system, PBI-CBR, uses the grass growth rates calculated by PBI from this grass cover data to predict grass growth rates on a farm from one week to the next, a critical part of the grass budgeting process. Note, farmers vary in how regularly they use PBI; there are ∼ 2,000 active users defined as those entering > 20 grass covers a year.

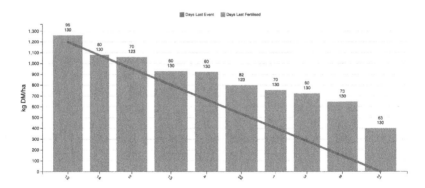

Fig. 1. A grass wedge as seen by farmer-users of PBI: The green columns represent each field/paddock on a farm, and the red line the *target pre-grazing yield* each paddock should be at before beginning rotational grazing. The y-axis is kilograms of dry matter per hectare, and the x-axis shows the farm's paddocks. The width of each paddock's green bar represents its total area. The *Days Last Event* number refers to when the paddock was last grazed. (Color figure online)

Feed Forecasting and Grass Wedges. Among other variables, the feed needs for a dairy herd depends on the size of one's farm, the size of the herd, and the status of the herd (e.g., lactating animals). PBI takes these variables and forecasts the feed needs for a farm. PBI accounts for both *rotational grazing* and *set stocking*, in which the farmer grazes certain paddocks while resting others (which may be grazed later or cut for silage). PBI allows farmers to modify variables such as rotation length and paddock status (e.g., is it currently being grazed), while producing a number of reports to show the effect of changing variables. Figure 1 shows all paddocks on a farm and the grass available in each paddock, measured in kilograms of dry matter per hectare (kg DM/ha; grass weight changes with moisture content, so dry weight is used). The red-line shows the *target pre-grazing yield* for each paddock, which can move up and down as the farmer changes variables (e.g., size of herd). If the red-line is below the top of a green-bar, then more grass is available than is currently needed (it could be cut for silage or meal supplementation reduced). However, if the red-line is above a green-bar then there is not enough grass to begin rotational grazing, and some meal supplementation may be required. These calculations are critical to the sustainability of the farming enterprise; stated simply, grass is inexpensive and meal supplements are the opposite. Also, meal requires transportation and possibly importation, so it entails increased carbon costs.

Grass Growth Prediction. Management decisions are largely based on grass growth, which varies based on soil/grass type, farming practices, climatic factors etc. In PBI, the farmer estimates a grass cover in paddocks and a calculation is done to determine the average growth rate since the previous grass cover. PBI-CBR aims to predict growth rates using machine learning (ML), by forecasting the growth-rate in the coming week using previous cases. Note, Teagasc currently uses a *mechanistic model* (a.k.a. a *first-principles* model) called the *Moorepark St. Gilles Grass Growth* model (MoST) that can predict growth-rates and continues to be tested [4]. However, key parameters of this model are not available for all farms (e.g., soil maps). A future system may combine PBI-CBR and MoST to make predictions, alternating between both models.

The PBI Dataset. We used the PBI dataset recorded from thousands of private farms in Ireland between 2013–2017. The primary feature of concern is the average grass growth rate for a farm since the last grass cover recorded, but location features (Farm ID-anonymized and County) are also important for explanation purposes. Ideally, to explain a prediction our system aims to provide an explanatory case from the same farm, but a case from a nearby farm in the same county is also acceptable. This was the advice given by the domain experts running the current system, although ultimately this proposal needs to be user tested.

1.3   Outline of Paper

Section 2 discusses noise in the PBI dataset used here and how a Bayesian approach is both useful and intuitive for case exclusion. Section 3 describes Experiment 1 (Expt. 1) which compares four systems on accuracy and explanatory success, and the tradeoff between both measures. Section 4 describes Expt. 2 which shows how updating priors using Bayesian analysis can improve prediction accuracy, possibly providing a means to deal with climate change in DSSs for this domain. Section 5 reviews relevant previous work in the area before final conclusions in Sect. 6.

2   Noise: The Gold-Standard and Working-Farm Datasets

We believe that this grass-growth domain is representative of the datasets and problems that AI will face in many smart agriculture contexts, especially in being *highly noisy*. The data is gathered by end users (farmers) and, as such, is understood to contain errors, miss-recordings, adjustments, and estimates. For example, some of the recordings in the dataset are based on physical measurements with a device, whereas others are estimates from visual inspections. This inherent noise has profound implications for how prediction and explanation need to be handled in this domain. On the one hand, we need a systematic way to remove possibly-poor cases. On the other hand, we need to keep as many cases as possible, because each additional case has potential to improve the system's accuracy and interpretability. Indeed, case exclusion could also affect the tradeoff between predictive accuracy and explanatory adequacy. Our solution to these noise issues is to use one dataset to clean another; to use a gold-standard dataset gathered under controlled conditions by researchers (with is idealized but noise-free) to clean a working-farm dataset gathered by farmers as part of their daily work (which is noisy). Technically, we use a gold-standard set of historical grass-growth measurements to give a prior belief about the distribution of grass growth each week, which in turn allows us to exclude cases from the working-farm case-base that may contain errors; what we call *Bayesian Case-Exclusion*. As we shall see, this solution seems to exclude noisy cases while retaining enough high-quality ones to maintain accurate predictions and explanations. Next, we describe these two datasets and how the working-farm case-base was built.

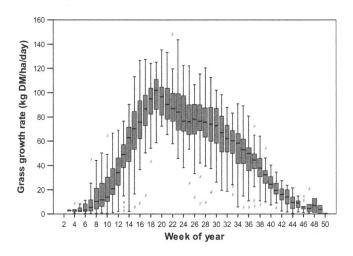

Fig. 2. The gold-standard dataset of grass growth measurements from 1982–2010 at Teagasc, Animal and Grassland Research and Innovation Centre, Moorepark, Fermoy, Co. Cork, Ireland [7], where the distribution of grass growth each week of the year is given as box plots.

2.1   A Gold-Standard Dataset: Teagasc Grass Growth (1982–2010)

The gold-standard dataset of grass-growth measurements we used covers 28 years of carefully-controlled, weekly measurements in which samples taken by researchers from the same pasture were cut, dried and weighted on a weekly basis at the Teagasc Moorepark Dairy Research Centre, Fermoy, Co. Cork (a major location for dairy farming). These measurements are somewhat idealized as they come from one location, which was not grazed (i.e., the livestock's impact – such as urine and trampling – on grass growth was excluded). However, they are very accurate and can thus serve as a good benchmark for determining outlier cases in the PBI dataset.

2.2   Case Definition and Case-Base Construction

The dataset used to construct the working-farm case-base came from the weekly grass covers entered by farmers in PBI. This dataset's growth rates were calculated using the grass covers recorded by farmers showing the estimate of grass available on a given farm for a given day. Some of these grass covers are known to be in-error; for example, often multiple entries are made on the same day, where the last entry of the day was the intended record. For the years 2013–2017, this dataset had 99,087 grass cover-records, that reduced to 92,635 when these same-day entries were removed. These grass cover-records are the raw data from which the cases used in PBI-CBR were generated to create the working-farm case-base.

Case Generation. Let a farm's data be $f = \{x_1, x_2, \ldots x_n\}$, where x_i is a grass cover-record for a single day, and n the total number of grass covers recorded (note the grass covers are in chronological order). The features of x_i used to generate a case (C_i) in the case-base are the average growth rate since the previous grass cover (gr), the week (wk), month (mth), and season ($seas$) in which the grass cover was recorded. Weather data (w_i) at the county level was scraped from Met Éireann (www.met.ie), and added as an average from x_i until x_{i+1}. The weather information in w_i is the maximum daily temperature ($maxt$), the average soil temperature 10 cm below the surface ($soilt$) on a given day, and the average global radiation ($grad$) in a given day. Finally, gr from x_{i+1} is also added to C_i as the target feature for prediction. Thus, a case is represented as:

$$C_i(x_i, w_i, x_{i+1}) = x_i(gr, wk, mth, seas), w_i(maxt, soilt, grad), x_{i+1}(gr) \qquad (1)$$

Case Base Construction. Taking the raw-data grass cover-records (N = 92,635) the cases as defined in (1) were constructed. However, given that the system has to predict one week ahead, only those cases where the target $x_{i+1}(gr)$ was recorded 5–9 days after x_i were included in the case base. Also, cases from January and December were excluded (as they tend to show zero growth), though they might be appropriate in a final deployed system. Finally, only those cases with accurate historical weather information until the next grass cover were considered (weather is a crucial factor in growth predictions). These steps resulted in a *working-farm case-base* of N = 20,760 cases for use in experimental tests. Note, in each system variant (except for the

Control) the number of cases in this *working-farm case-base* is reduced further by the respective method(s) used.

2.3   The Current Experiments

In the remainder of this paper, two experiments are reported that test several variants of the *Bayesian Case-Exclusion* idea. In Expt. 1, we examine what happens in this predictive CBR-system when cases are not excluded (*Control*), versus experimental systems in which we use the gold-standard dataset's distributions in different ways to modify or exclude cases (the *Exclude-2sd, Exclude-3sd and Transform-3sd* systems; see Sect. 3). These experimental system-variants examine performance when cases are *transformed* with reference to the gold-standard distributions or when cases are *excluded* a pre-defined number of standard deviations away from the means in the gold-standard distributions. The transformation system enables greater retention of cases, in turn helping with explanations. In Expt. 2, we explore *Adaptive Bayesian Case-Exclusion*, where priors derived from the gold-standard distributions are updated year-on-year, to see if performance improves (see Sect. 4).

3   Experiment 1: Bayesian Case-Exclusion

PBI-CBR is a CBR system for predicting grass growth, using the growth rates calculated from each farm. Two different datasets are used in the experiments, the gold-standard Teagasc data (1982–2010) and the PBI dataset (2013–2017), where the former is used to transform or exclude cases from the latter when making predictions for a particular farm in a given week of a given year. Hence, the gold-standard dataset is our "prior" belief (in Bayesian parlance), which is used to make probabilistic inferences in how to handle noise. In the working-farm case-base, the current week is used to predict one week ahead, allowing a farmer to make informed management decisions. In general, for this prediction, a mean squared error (MAE) of ≤ 10 kg DM/ha/day is sufficient. The main problem is the noise in the working-farm case-base, hence we use *Bayesian Case-Exclusion* to exclude outlier cases when making predictions. PBI-CBR also explains predictions using *post-hoc, personalized explanation-by-example* by referencing nearest neighboring cases from the same farm or county. So, the tests involve two measures: (i) *predictive accuracy*, as MAE for the growth-rate prediction measured in kg DM/ha/day, (ii) *explanatory success*, as the percentage of times nearest-neighbor cases are found from either the same-farm or same-county to the test-cases in the k nearest neighbors retrieved (a measure recommended by experts). However, it should be noted again that the "success" of these explanations is dependent on future user testing. Crucially, we tested four variants of the system:

- *Control*. A basic system that uses all the cases in the *working-farm case-base* (N = 20,760; see Sect. 2.2); this case base was built mostly from the PBI dataset (from 2013–17) and, accordingly, is quite noisy and has many outliers.

- *Exclude-2sd*. A Bayesian system that excludes cases two-standard deviations away from the weekly, mean growth-rates of the gold-standard dataset (see Fig. 2). The rationale being that grass growth in a given week approximates a normal distribution (verified by plotting thousands of growth rates in histograms) and using the properties of such a distribution can aid in making probabilistic assumptions for how to exclude cases. Formally, the data for growth rate (*GR*) in a given week across all years in the gold-standard dataset approximates $GR \sim N(\mu_g, \sigma_g^2)$, where N is a normal distribution with parameters μ_g and σ_g for the mean and standard deviation, respectively. All cases outside $\mu_g \pm 2\sigma_g$ are excluded (as well as other query-cases), thus excluding cases with $\sim 5\%$ probability of occurring. This step reduces the *working-farm case-base* by 42% (N = 12,042 cases).
- *Exclude-3sd*. This is identical to the *Exclude-2sd* system but $\mu_g \pm 3\sigma_g$ is used to exclude cases, thus excluding cases with $\sim 0.3\%$ probability of occurring. This reduces the *working-farm case-base* by 21% (N = 16,443 cases).
- *Transform-3sd*. This is a Bayesian system that transforms the growth-rates of cases using the gold-standard distributions. That is, the distribution of growth in a given week from the gold-standard dataset $[GR \sim N(\mu_g, \sigma_g^2)]$ is used to transform the growth-rate values of cases for the same week in the *working-farm case-base*, to fit to the parameters μ_g and σ_g^2. Formally, to transform the growth-rate (*gr*) in a grass cover x in any given week of the year we use:

$$y_{gr} = \left(x_{gr} - \mu\right) \times \frac{\sigma_g}{\sigma} + \mu_g \tag{2}$$

where x_{gr} is the growth rate in grass cover x, y_{gr} is the transformed growth rate of x_{gr}, μ and σ are the mean and standard deviation for the overall growth rate in that week in the *working-farm case-base*, respectively, and μ_g and σ_g are the mean and standard deviation for the overall growth rate in that week in the gold-standard dataset, respectively. The intuition being that the gold-standard dataset is closer to the ground-truth, hence if it is used to transform the growth rates (in the *working-farm case-base*), the overall deviation from the ground truth will reduce. Note, in this system cases that fall outside $\mu_g \pm 3\sigma_g$ after the transformation are still excluded, and, so, the *working-farm case-base* is reduced by 2% (N = 20,282 cases).

As we shall see, exclusion methods improve prediction accuracy, with varying levels of explanatory success. The transform system retains as many cases as possible, aiding accuracy and explanatory success. Indeed, there are indications that the transformed case-base is closer to the ground truth as the correlation of Pearson's r between *maxt* and *growth-rate* across all cases increases from $r = 3.92$ to $r = 5.11$ after transformation, reflecting known dependencies between temperature and grass-growth (<5 °C grass does not grow, from 5–10 °C it grows with temperature [4]).

3.1   Method: Procedure and Measures

For each system variant Monte Carlo cross-validation was used with 30 resampling iterations, each time taking 80/20% data for training and testing, respectively. An unweighted k-NN algorithm with Euclidean distance was used for case retrieval, with the averaged value of all nearest neighbors' target-growth-rates used as the prediction. Selected values of k ranging from 5-1000 were tested for each system variant to observe effects on prediction and explanatory outcomes. For each evaluation of k for each system, three measures were taken: (i) the MAE (ii) the *%Farm-Retrieval-Success*, the percentage of times the k-nearest-neighbors contained a case from the same farm as the query, and (iii) the *%County-Retrieval-Success*, the percentage of times the k-nearest-neighbors contained a case from the same county as the query.

3.2   Results and Discussion

Figure 3a shows the results of running the system variants – *Control, Exclude-2sd, Exclude-3sd, Transform-3sd* – for all values of k in three graphs, one for each measure: MAE, %Farm-Retrieval-Success (%FRS), and %County-Retrieval-Success (%CRS). Across all systems, MAE is worst for the lowest and highest k with some improvement in between ($k = 20–35$). With regard to %FRS all system variants are very similar, though success does change for different values of k. For all systems, %FRS is very poor for low values of k, but beyond $k = 50$ it rises to $\sim 80\%$; showing that only higher values of k deliver enough cases from the same farm to explain the predictions made. For all systems, %CRS is much better, as it starts high ($\sim 80\%$) for low values of k and rapidly reaches $\sim 100\%$; showing that finding explanatory cases for a prediction from the same county is a common occurrence. However, the differences between the system variants are, perhaps, more interesting.

Overall, the *Control* system, which includes all cases, does the worst; it never gets lower than a MAE of 15 kg DM/ha/day (recall, acceptable error is ≤ 10 kg DM/ha/day). Similarly, the two exclusion-systems – *Exclude-2sd* and *Exclude-3sd* – do not reach the acceptability threshold. Overall, *Bayesian Case-Exclusion* does much better than the *Control*, but only *Exclude-2sd* with $k = 35$ has the somewhat acceptable MAE of ~ 10.01 kg DM/ha/day. Overall, the *Transform-3sd* system is the best with a MAE < 10 kg DM/ha/day for all values of k (note, many current mechanistic models have MAEs of $\sim 10–20$ kg DM/ha/day, showing the potential for AI solutions in this domain).

Finally, the best system is *Transform-3sd*; in Fig. 3a, comparing the 1st and 2nd graphs, we can see the tradeoff between MAE and %FRS for all values of k. The 1st graph shows that the lowest error (MAE = 8.6 kg DM/ha/day) occurs at $k \sim 35$, but at this level %FRS is poor at $\sim 7\%$ (see 2nd graph). Accordingly, $k = 1000$ is required to improve %FRS to $\sim 85\%$. However, even at this value for k, an acceptable MAE is achieved (~ 9.8 kg DM/ha/day), making *Transform-3sd* the only system that successfully balances the tradeoff between accuracy and explanation. Note, with additional data from a given farm, it should be possible to improve this tradeoff even further.

4   Experiment 2: Updating Priors Year-on-Year

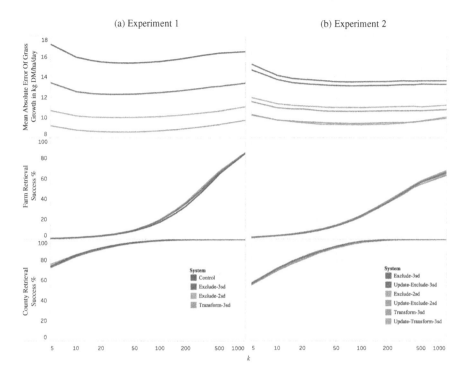

Fig. 3. The tradeoff between error and explanation. (a) Expt. 1 shows that as the value for k approaches 1000, more explanatory cases are retrieved, but the MAE for all systems also increases. *Transform-3sd* has the best MAE of ~ 8.6 kg DM/ha/day at $k \sim 35$, but same-farm explanatory success is low at $\sim 7\%$; however, at $k = 1000$, the tradeoff is balanced, with the MAE still acceptable and %FRS at $\sim 85\%$. (b) Expt. 2 shows MAE is improved for almost every update-variant, although the improvement in the transform-system is minimal; explanatory success and MAE are similar to Expt. 1, but poorer, likely due to less training data. Finally, note the log scale on the x-axis.

In Expt. 1 Bayesian exclusion or transformation of cases from the *working-farm case-base* gave improved performance. However, these systems exclude cases using parameters from the gold-standard dataset, gathered between 1982 and 2010. Recently climate change appears to be impacting the distribution for grass growth. For example, in the hot Irish summer of 2018 grass-growth stopped during July (normally it is ~ 100 kg DM/ha/day). In Expt. 1, this not considered, but Expt. 2 rectifies this by combining the two datasets to update Bayesian priors year-on-year by Bayesian analysis (e.g., see [3]) to estimate the unknown distributions of grass growth each week with a view to making predictions in 2017[2] (the final year's data). Hence, Expt. 2 has

---

[2] Predictions could only be made for 2017 because the earlier years of the PBI dataset (2013–2016) have too few cases, as the DSS was in its early years of adoption.

six versions of PBI-CBR, three systems from in Expt. 1 (*Exclude-2sd, Exclude-3sd, Transform-3sd*) and three variants of these in which priors were updated (*Update-Exclude-2sd, Update-Exclude-3sd, Update-Transform-3sd*). The updating procedure used is described next.

4.1   Updating Priors in Exclusion and Transformation Systems

To perform Bayesian updating, we take priors from the gold-standard dataset and then progressively use each year's data from the PBI-dataset to update them. First, take the gold-standard dataset and, binning all its data into weeks, for any given week, let the growth rate (*GR*) approximate a normal distribution $GR \sim N(\mu, \sigma^2)$, where μ and σ^2 are its mean and variance, respectively. In 2013, all the data for this week was processed into cases (see Sect. 3). Then, we proceed with *transformation* or *exclusion* methods on these cases depending on the system variant (as in Expt. 1), which gives the new data $D = \{C_1, C_2 \ldots C_n\}$ where n is the number of cases. Take the prior to be $\mu \sim N(\mu_0, \sigma_0^2)$, where the value σ_0 is initially chosen as 4^3 and μ_0 is initially chosen as μ. Here the value for σ^2 is assumed to remain fixed[4]. Bayes rule shows that the posterior (for a given week) is proportional to the *likelihood* times the *prior*, in addition, because σ^2 and σ_0^2 are known we can ignore the constant of proportionality and derive that the posterior μ_p is:

$$\mu_p \sim N\left(\frac{\sigma^2}{\sigma^2 + \sigma_0^2 n}\mu_0 + \frac{\sigma_0^2}{\sigma^2 + \sigma_0^2 n}n\bar{x}, \frac{\sigma^2\sigma_0^2}{\sigma^2 + \sigma_0^2 n} \right) \tag{3}$$

where $\bar{x}$ is the empirical mean of the growth rates in the cases of D, for a full derivation and explanation the reader is referred to [3]. Although in CBR the word "Bayesian" usually infers the use of Bayesian networks, in this experiment it is used in a more traditional sense and refers to the estimation of an unknown distribution (a.k.a. the *posterior*) of grass growth using a prior belief (a.k.a. the *prior*) and a sample of data from the new year (i.e., the *likelihood*).

Using Eq. (3) we update values for μ_0 and σ_0^2, the new value of μ_0 was then used to update the original μ from the gold-standard dataset, which was used with σ^2 (the fixed variance from the gold-standard dataset) to repeat the whole process in 2014 for the same week. This process is repeated for all weeks of each year until the end of 2016 when all training data was collected. The latest priors in each week were again used to exclude or transform cases in 2017 for evaluation[5]. All evaluations were carried out on 2017 because there was insufficient training data in previous years to ensure adequate evaluations (2017 has $\sim 40\%$ of usable cases), though the years prior to 2017 were all used in the year-on-year updating to acquire the training data.

---

[3] The relatively large value of 4 was chosen to represent that we are not highly certain of the validity of the gold standard prior mean when compared to a typical dairy farm pasture.

[4] The variance σ^2 wasn't adapted; if it changes it could lead to an unfair evaluation as *updated*-variants may differ a lot in the amount of data excluded compared to non-updated variants.

[5] Note, for *transform* methods some knowledge about a given week's data distribution would need to be inferred if we were doing this in a live-system for formula (2) to be used.

4.2   Case-Base Sizes After Transformation or Exclusion

Expt. 2 has six system variants, the *Exclude-2sd, Exclude-3sd* and *Transform-3sd* systems from Expt. 1, and matched versions of these systems, which used the updating methods described above called *Update-Exclude-2sd, Update-Exclude-3sd and Update-Transform-3sd.* In the updated variants, several aspects change, so the number of cases after transformation or exclusion vary slightly: *Exclude-2sd* (N = 12,042), *Update-Exclude-2sd* (N = 12,183), *Exclude-3sd* (N = 16,443), *Update-Exclude-3sd* (N = 16,379), *Transform-3sd* (N = 20,282), and *Update-Transform-3sd* (N = 20,120).

4.3   Method: Procedure and Measures

For each system variant the respective case base was split in a $\sim 60/40\%$ ratio of training and testing cases, respectively; the former coming from the PBI data from 2013–2016 and the latter from 2017. Crucially, note that results will be different from identical systems in Expt. 1 because of the different ratio for splits. For case retrieval, an unweighted k-NN was again used with Euclidean distance for selected values for k ranging from 5-1000. The same three measures were used as in Expt. 1: Mean Absolute Error (MAE), %Farm-Retrieval-Success (%FRS), and %County-Retrieval-Success (%CRS).

4.4   Results and Discussion

Figure 3b shows the results of running the six system variants – *Exclude-2sd, Exclude-3sd, Transform-3sd, Update-Exclude-2sd, Update-Exclude-3sd and Update-Transform-3sd* – for all values of k in three graphs, one for each measure: MAE, %FRS, and %CRS. In general, the shape of the results replicates many of the findings of Expt. 1.

Regarding MAE, as before the transformation-versions do better than the exclusion-versions, the error decreases in order from *exclude-3sd* to *exclude-2sd* to *transform-3sd*; $k = 75$ is optimal for all systems, doing better than the lower and higher values of k. Overall the MAE scores (and explanation-success scores) are not as good as in Expt. 1, perhaps, reflecting the different ratios in the training and testing splits (i.e., they were 80/20% in Expt. 1 and $\sim 60/40\%$ in Expt. 2); note, the evaluation dataset is reduced to one-year in Expt. 2 (i.e., 2017), whereas it is across all 5 years in Expt. 1 (2013–17).

Expt. 2 shows that systems with Bayesian updating (*Update-Exclude-2sd, Update-Exclude-3sd and Update-Transform-3sd*) do better than systems without updating (*Exclude-2sd, Exclude-3sd, Transform-3sd*) at nearly every value of k, though the improvements are relatively modest, particularly in the transform version (see Fig. 3b).

Regarding explanation measures (%FRS, %CRS) the overall curve-shapes are similar to those in Expt. 1, with maximum values being %FRS = 68% and %CRS = 100%, in contrast to %FRS = 85.94% and %CRS = 99.98% in Expt. 1. Acceptable tradeoffs for accuracy and explanation are achieved for both of the transform systems (*Transform-3sd, Update-Transform-3sd*) in that at $k = 1000$ the MAE is ~ 9.95 kg DM/ha/day with $\sim 67.5\%$ explanatory-success rate for same-farm cases in both systems. These systems would likely improve if training and testing splits were more favorable as in Expt. 1.

5   Related Work

This work impinges on many areas, though the most relevant literatures are arguably in case-based maintenance, Bayesian CBR, and explanation in CBR DSSs for smart agriculture. Here we review the relevant literature and discuss its relevance to this work.

Case base maintenance (i.e., case base editing/deleting/exclusion/inclusion etc.) is a notable area of research for the CBR community [19]. However, the most popular methods have tended to focus on classification [16, 20–25], as opposed to regression [17]. Redmond and Highley [17] did try to convert Edited Nearest Neighbors [22] to handle regression by assigning two hyperparameters for *agree* and *accept* thresholds, but they acknowledge that applying the classification algorithms to regression is difficult. Our method requires no hyperparameters, though it does require the specification of a prior(s). Furthermore, most of the literature on case base maintenance is concerned with deleting cases to optimize case-bases; here we have used the phrase "case exclusion" rather than "case deletion" because we believe it is important to retain cases for future use. For instance, cases deemed outliers with extreme environmental conditions may be useful if climate change results in these extreme conditions becoming common or more data becomes available (e.g., soil type) identifying them as non-outlier data.

Much work has been done using Bayesian methods in CBR systems. Nikpour *et al.* [8] used Bayesian posterior distributions to modify case descriptions and dependencies in a model, showing the capability of such an approach to increase similarity assessment. Moreover, the vast majority of work combining CBR and Bayesian methods has involved combining Bayesian Networks with CBR systems, for which there are many architectures and approaches [27]. However, beyond the combination of Bayesian methods and CBR, these systems have little in common with the present work, which uses prior distributions for case exclusion. The best algorithm for a particular problem regarding case base maintenance will likely always depend on the domain in question [19], but here we present a novel option.

XAI within CBR has been shown to be important in intelligent systems [9, 31, 32], with some consideration of smart agriculture [10]. Additionally, it has been argued that recommender systems should play a central part in smart agriculture [12], and CBR is a popular approach for such systems [6]. Pu and Chen [26] have conducted user studies showing that designers should build trusted interfaces into recommender systems due to the high likelihood users will return. As smart agriculture arguably requires a recommender component [12], and it suffers from a user retention issue; this is of particular relevance. Moreover, in understanding the effects of environmental changes, Cho *et al.* [12] note that global warming and pollution have made environmental and agricultural modelling difficult, thus suggesting the use of a recommender system to support users, but no specific instances are described. Moreover, Holt [14] suggested that CBR could be used to help farm management decisions. CBR gives a unique ability to offer intuitive exemplar-based explanations, and user studies have shown it potentially superior to rule-based explanations [13], frameworks have been proposed for CBR XAI [18], but to the best of our knowledge no instance in smart agriculture has been proposed until the present paper. Branting *et al.* [10] did use CBR in the agricultural advisory system CARMA (which also produces explanations), but it only forms part of the consultation process, whilst our solution appears to be the first pure CBR approach.

6 Conclusions and Future Research

We have shown that a CBR system can be used for decision support in dairy farming to predict a key aspect of the enterprise accurately, while also providing case-based explanations that are personalized for a specific farm. To deal with noise in the dataset, we have used historical distributions based on accurate research measurements to determine what cases should or should not be included in the predictive model (i.e., our Bayesian exclusion approach). Furthermore, we have shown that transforming key-attributes of cases based on a goal-standard distribution (that is closer to a ground truth) can improve accuracy, and that using Bayesian analysis for updating priors year-on-year also improves performance. By our knowledge, all of this work is novel.

These systems have the ability to improve the sustainability of grasslands for dairy farming into the future. Accordingly, for us, the key question for future research is whether these techniques can continue to deliver accurate predictions in the face of climate change. One would hope that these CBR systems can maintain predictive accuracy by selectively picking useful cases from historical datasets (e.g., as soon as the data is available, we plan to test PBI-CBR against the extreme weather of 2018). So, though we may experience significant climate shifts, there will always be a case somewhere in the historical record that can provide accurate predictions.

Acknowledgements. This publication has emanated from research conducted with the financial support of (i) Science Foundation Ireland (SFI) to the Insight Centre for Data Analytics under Grant Number 12/RC/2289 and (ii) SFI and the Department of Agriculture, Food and Marine on behalf of the Government of Ireland to the VistaMilk SFI Research Centre under Grant Number 16/RC/3835.

References

1. Gafsi, M., Legagneux, B., Nguyen, G., Robin, P.: Towards sustainable farming systems: effectiveness and deficiency of the French procedure of sustainable agriculture. Agric. Syst. **90**(1–3), 226–242 (2006)
2. Lindblom, J., Lundström, C., Ljung, M., Jonsson, A.: Promoting sustainable intensification in precision agriculture: review of decision support systems development and strategies. Precis. Agric. **18**(3), 309–331 (2017)
3. Murphy, K.P.: Conjugate Bayesian analysis of the Gaussian distribution. def, $1(2\sigma 2)$, p. 16 (2007)
4. Ruelle, E., Hennessy, D., Delaby, L.: Development of the Moorepark St Gilles grass growth model (MoSt GG model): a predictive model for grass growth for pasture based systems. Eur. J. Agron. **99**, 80–91 (2018)

5. Poux, X., Aubert, P.M.: An agroecological Europe in 2050: multifunctional agriculture for healthy eating (2018)
6. Tintarev, N., Masthoff, J.: A survey of explanations in recommender systems. In: 2007 IEEE 23rd International Conference on Data Engineering Workshop, pp. 801–810. IEEE (2007)
7. Hurtado-Uria, C., Hennessy, D., Shalloo, L., O'Connor, D., Delaby, L.: Relationships between meteorological data and grass growth over time in the south of Ireland. Irish Geogr. **46**(3), 175–201 (2013)
8. Nikpour, H., Aamodt, A., Bach, K.: Bayesian analysis in a knowledge-intensive CBR system. In: 22nd UK Symposium on Case-Based Reasoning. Miltos Petridis (2017)
9. Leake, D.B.: Evaluating Explanations: A Content Theory. Psychology Press, London (2014)
10. Branting, K., Hastings, J.D., Lockwood, J.A.: CARMA: a case-based range management advisor. In: IAAI, pp. 3–10, August 2001
11. Teagasc - 8/2/19: The Dairy Carbon Navigator. Improving Carbon Efficiency on Irish Dairy Farms. https://www.teagasc.ie/publications/2019/the-dairy-carbon-navigator.php
12. Cho, Y., Cho, K., Shin, C., Park, J., Lee, E.-S.: A recommend service based on expert knowledge model in agricultural environments. In: Lee, G., Howard, D., Kang, J.J., Ślęzak, D. (eds.) ICHIT 2012. LNCS, vol. 7425, pp. 189–194. Springer, Heidelberg (2012). https://doi.org/10.1007/978-3-642-32645-5_24
13. Cunningham, P., Doyle, D., Loughrey, J.: An evaluation of the usefulness of case-based explanation. In: Ashley, K.D., Bridge, D.G. (eds.) ICCBR 2003. LNCS (LNAI), vol. 2689, pp. 122–130. Springer, Heidelberg (2003). https://doi.org/10.1007/3-540-45006-8_12
14. Holt, A.: Understanding environmental and geographical complexities through similarity matching. Compl. Int. **7**, 1–16 (2000)
15. Bridge, D., Göker, M.H., McGinty, L., Smyth, B.: Case-based recommender systems. Knowl. Eng. Rev. **20**(3), 315–320 (2005)
16. Hart, P.: The condensed nearest neighbor rule (Corresp.). IEEE Trans. Inf. Theory **14**(3), 515–516 (1968)
17. Redmond, M.A., Highley, T.: Empirical analysis of case-editing approaches for numeric prediction. In: Sobh, T., Elleithy, K. (eds.) Innovations in Computing Sciences and Software Engineering, pp. 79–84. Springer, Dordrecht (2010). https://doi.org/10.1007/978-90-481-9112-3_14
18. Sørmo, F., Cassens, J., Aamodt, A.: Explanation in case-based reasoning–perspectives and goals. Artif. Intell. Rev. **24**(2), 109–143 (2005)
19. Smiti, A., Elouedi, Z.: Overview of maintenance for case based reasoning systems. Int. J. Comput. Appl. **975**, 8887 (2011)
20. Gates, W.: The reduced nearest neighbor rule. IEEE Trans. Inf. Theory **18**(3), 431–433 (1972)
21. Ritter, G., Woodruff, H., Lowry, S., Isenhour, T.: An algorithm for a selective nearest neighbor decision rule. IEEE Trans. Inf. Theory **21**, 665–669 (1975)
22. Wilson, D.L.: Asymptotic properties of nearest neighbor rules using edited data. IEEE Trans. Syst. Man Cybern. **2**, 408–421 (1972)
23. Guan, D., Yuan, W., Lee, Y.K., Lee, S.: Nearest neighbor editing aided by unlabeled data. Inf. Sci. **179**(13), 2273–2282 (2009)
24. Aha, D.W., Kibler, D., Albert, M.K.: Instance-based learning algorithms. Mach. Learn. **6**, 37–66 (1991)
25. Markovitch, S., Scott, P.D.: The role of forgetting in learning. In: Proceedings of the Fifth International Conference on Machine Learning. pp. 459–465 (1988)
26. Pu, P., Chen, L.: Trust-inspiring explanation interfaces for recommender systems. Knowl.-Based Syst. **20**(6), 542–556 (2007)

27. Bruland, T., Aamodt, A., Langseth, H.: Architectures integrating case-based reasoning and bayesian networks for clinical decision support. In: Shi, Z., Vadera, S., Aamodt, A., Leake, D. (eds.) IIP 2010. IAICT, vol. 340, pp. 82–91. Springer, Heidelberg (2010). https://doi.org/10.1007/978-3-642-16327-2_13

28. Johns, T.: Phytochemicals as evolutionary mediators of human nutritional physiology. Int. J. Pharmacognosy **34**(5), 327–334 (1996)

29. Steinfeld, H., et al.: Livestock's long shadow: environmental issues and options. Food & Agriculture Organization (2006)

30. Goodland, R., Anhang, J.: Livestock and climate change: What if the key actors in climate change are… cows, pigs, and chickens? (2009)

31. Keane, M.T., Kenny, E.M.: How Case Based Reasoning Explained Neural Networks: An XAI Survey of Post-Hoc Explanation-by-Example in ANN-CBR Twins. arXiv preprint arXiv:1905.07186 (2019)

32. Kenny, E.M., Keane, M.T.: Twin-systems to explain artificial neural networks using case-based reasoning: comparative tests of feature-weighting methods in ANN-CBR twins for XAI. In: Proceedings of the Twenty-Eight International Joint Conference on Artificial Intelligence (2019)

Learning Workflow Embeddings to Improve the Performance of Similarity-Based Retrieval for Process-Oriented Case-Based Reasoning

Patrick Klein$^{(\boxtimes)}$ 🆔, Lukas Malburg$^{(\boxtimes)}$ 🆔, and Ralph Bergmann$^{(\boxtimes)}$ 🆔

Business Information Systems II, University of Trier, 54286 Trier, Germany
{kleinp,malburgl,bergmann}@uni-trier.de
http://www.wi2.uni-trier.de

Abstract. In process-oriented case-based reasoning, similarity-based retrieval of workflow cases from large case bases is still a difficult issue due to the computationally expensive similarity assessment. The two-phase MAC/FAC ("Many are called, but few are chosen") retrieval has been proven useful to reduce the retrieval time but comes at the cost of an additional modeling effort for implementing the MAC phase. In this paper, we present a new approach to implement the MAC phase for POCBR retrieval, which makes use of the StarSpace embedding algorithm to automatically learn a vector representation for workflows, which can be used to significantly speed-up the MAC retrieval phase. In an experimental evaluation in the domain of cooking workflows, we show that the presented approach outperforms two existing MAC/FAC approaches on the same data.

Keywords: Process-oriented case-based reasoning ·
MAC/FAC retrieval · Graph embeddings

1 Introduction

As more and more workflows are supported and executed electronically, the amount of available data in process repositories as well as the procedural knowledge gathered through past problem-solving experience increases. Such workflows can represent business processes, scientific experiments, repair instructions, or activities from daily life such as cooking recipes. It is valuable to reuse this procedural knowledge since creating workflows from scratch is typically a complex and time-consuming task [18]. Process-Oriented Case-Based Reasoning (POCBR) [2,15] can be used for retrieving, reusing, revising, and retaining procedural experiential knowledge represented as workflows. A case base in POCBR specifies best-practice workflows that can be (re-)used in similar situations. A critical factor for the performance of a CBR system is the efficiency of case retrieval [7]. The retrieval time is of importance due to its impact on the user's system acceptance as well as being a requirement in time critical environments

ⓒ Springer Nature Switzerland AG 2019
K. Bach and C. Marling (Eds.): ICCBR 2019, LNAI 11680, pp. 188–203, 2019.
https://doi.org/10.1007/978-3-030-29249-2_13

where decisions need to be made within clearly defined time boundaries. However, obtaining an acceptable retrieval time is particularly difficult for POCBR, as cases in POCBR are usually represented by semantically labeled graphs leading to a similarity assessment that requires a kind of inexact sub-graph matching, which is computationally expensive [2,12,16].

POCBR research has addressed the issue of efficient retrieval [5,11,12,17] through a two-phase retrieval following the MAC/FAC ("Many are called, but few are chosen") [9] principle. The retrieval is divided into two phases: The first phase (MAC) utilizes a simplified and often knowledge-poor similarity measure for a fast pre-selection. The second phase (FAC) then applies the computationally intensive graph-based similarity measure to the results of the MAC phase. This method improves the retrieval performance, if the MAC stage efficiently selects a small number of relevant cases. However, there is a risk that the MAC phase reduces the retrieval quality, as it might disregard highly similar cases due to its simplified assessment of the similarity. As a consequence, the retrieval approach for the MAC phase must be designed very carefully. Today, existing approaches [5,12] are based on a manually designed simplified domain specific case representation as well as a related method for the pre-selection of cases, which leads to a significantly increased development effort for the CBR system. The cluster-based retrieval approach introduced by Müller and Bergmann [17] avoids this additional effort but only works well for case bases with a strong cluster structure.

The aim of this paper is to present a novel approach for the design of a MAC phase for POCBR, which automatically learns an appropriate simplified case representation in the form of *workflow embeddings*. Thereby, we avoid the manual domain modeling for the MAC phase. For learning workflow embeddings, we investigate the general-purpose neural embedding model *StarSpace* [22], which has shown impressive performance on many different tasks such as text classification, entity ranking, and also graph embeddings. We aim at applying StarSpace to entities, which are workflows described by sets of linked task and data nodes.

The next section introduces previous work on POCBR, including semantic workflow representation, similarity assessment, and MAC/FAC approaches for POCBR. In Sect. 3, we present our approach for learning workflow embeddings and their use for retrieval in the MAC phase. An experimental evaluation of our approach in the domain of cooking recipes is presented in Sect. 4. Finally, Sect. 5 concludes the results and discusses future work.

2   Foundations and Previous Work

Process-Oriented Case-Based Reasoning [2,15] deals with the integration of CBR with Process-Aware Information Systems (PAISs) [8]. An example of a certain type of PAISs are workflow management systems [8]. Using POCBR, workflow developers are supported with best-practice workflows from a case base during their development process. Thus, POCBR supports the development of workflows as an experience-based activity [2,15]. POCBR methods require an appropriate case representation for workflows as well as a similarity measure that assesses the suitability of a workflow for a new problem situation.

2.1   Semantic Workflow Representation

In general, workflows are used for the automation of a defined sequence of tasks that can exchange inputs and outputs in order to achieve an overarching corporate objective [10]. Therefore, the ordering of tasks is modeled through structures such as sequences, parallel (AND split/join) as well as alternative (XOR split/join) and loops, which result together in the so-called control-flow. Additionally, tasks consume inputs and produce outputs, both of which can be physical or virtual in nature, and, along with their relationship between tasks, they form the data-flow.

In order to represent workflows, we use semantically labeled directed graphs named *NEST* graphs by Bergmann and Gil [2]. A *NEST* graph is a quadruple $G = (N, E, S, T)$ where N is a set of nodes and $E \subseteq N \times N$ represents the edges between nodes. Semantic descriptions $S : N \cup E \to \Sigma$ can be used for semantic enrichment of nodes or edges. Semantic descriptions are based on a semantic meta data language Σ and are domain-dependent. Additionally, each node and edge is annotated with a type $T : N \cup E \to \Omega$. The set Ω is predefined for nodes (e.g., task and data nodes) and edges (e.g., control-flow and data-flow edges).

In this paper, we use the well-known cooking domain to illustrate our approach and to perform an experimental evaluation. Thus, workflows are cooking recipes, tasks represent cooking steps, and data items take the role of ingredients. Figure 1 shows an example of a *NEST* graph representing a simple sandwich recipe, consisting of three task nodes (cooking steps) and four data nodes (ingredients).

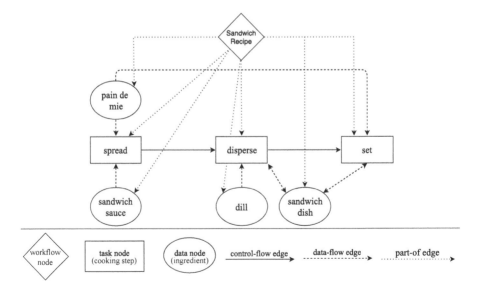

Fig. 1. Exemplary cooking workflow for a sandwich dish

In the cooking domain, the semantic meta data language is defined by taxonomic ontologies, one for ingredients and one for cooking steps. These ontologies are complete in the sense that all items occurring in the recipes are also included in the ontology. Figure 2 illustrates a fraction of the used ingredients taxonomy.

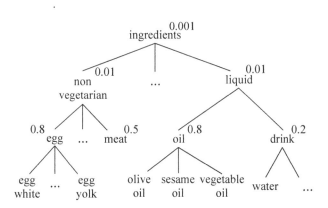

Fig. 2. Part of the data taxonomy showing the modeled similarity between ingredients

2.2 Similarity Assessment

Based on the *NEST* graph format, an assessment of the similarity between workflows through consideration of its constituents as well as the link structure is possible. Therefore, the local similarity of nodes and edges is defined based on the semantic meta data language $sim_\Sigma : \Sigma \times \Sigma \to [0,1]$. For task and data nodes, the taxonomies are used to derive their similarity using a taxonomic similarity measure that is based on the assignment of similarity values to the inner nodes of the taxonomy [4]. Building on that, the global similarity between a query workflow QW and a workflow from the case base CW is calculated by a type-preserving, partial, injective mapping function m from the nodes and edges of QW to those of CW. With respect to a mapping m, the local similarities are aggregated, leading to a similarity value $sim_m(QW, CW)$ based on which the overall workflow similarity $sim(QW, CW)$ is determined by the best possible mapping m as follows (see [2] for more details):

$$sim(QW, CW) = max\left\{sim_m(QW, CW) \mid \text{admissible mapping } m\right\} \quad (1)$$

Thus, computing the similarity between a query and a single case requires solving an optimization problem, for which we have proposed to use A^* search. Bergmann and Gil also developed a parallelized version of the A^* search, which is complete but still not sufficiently fast for large case bases.

2.3 MAC/FAC Retrieval for POCBR

To overcome the issue of long retrieval times in POCBR, two-phase MAC/FAC retrieval approaches for workflows have been introduced [5,11,12]. The major difficulty with MAC/FAC retrieval in general is the definition of the filter condition of the MAC stage. Since cases that are not selected by the MAC stage will not appear in the overall retrieval result, the completeness of the retrieval can be easily violated if the filter condition is too restrictive. Hence, retrieval errors, i.e., missing cases will occur. On the other hand, if the filter condition is less restrictive, the number of pre-selected cases may become too large, resulting in a low retrieval performance. To balance retrieval error and performance, the filter condition should be a good approximation of the similarity measure used in the FAC stage, while at the same time it must be efficiently computable to be applicable to a large case base in the MAC stage.

Bergmann and Stromer [5] addressed this problem by adding a feature-based domain specific case representation of workflows, which simplifies the original representation while maintaining the most important properties relevant for similarity assessment. The MAC stage then selects cases by performing a similarity-based retrieval using an appropriately modeled similarity measure. The number of cases selected in the MAC phase can be controlled by a parameter called *filter size FS*, i.e., the MAC stage retrieves the *FS*-most similar cases using feature-based retrieval. The choice of the filter size determines the behavior of the overall retrieval method with respect to retrieval speed and error in the following manner: the smaller the filter size, the faster the retrieval but the larger the retrieval error will become.

In order to avoid the additional modeling effort for the feature-based representation, Müller and Bergmann [17] developed a MAC/FAC approach that is based on the structuring of the case base into clusters of similar cases. Therefore, a binary cluster-tree is learned, which hierarchically partitions the case base into sets of similar cases. Traversing the cluster-tree allows finding clusters with cases similar to the query, thus reducing the number of required similarity computations. Again, a filter size parameter *FS* is used to determine the number of cases transferred to the FAC phase. This algorithm did not reach quality and retrieval speed of the feature-based MAC/FAC approach, but shows acceptable performance if the case base has a clear cluster structure.

3 Learning Workflow Embeddings for MAC Retrieval

We now present our approach for the design of a MAC phase for POCBR retrieval that avoids the manual construction of a simplified representation and a related similarity measure. The main idea is to automatically transform the original semantic workflow representation by use of an embedding method.

As workflows are represented as graphs, *graph embedding techniques* [6] are useful, as their main purpose is to address the complexity problems of many graph analytic methods by converting the graph data into a low dimensional

space. The transformation is performed such that the graph structural information and the graph properties are preserved as best as possible. Thus, we expect
that graph embeddings are also helpful for designing a MAC retrieval approach
for POCBR. There is already a large range of graph embedding methods reported
in the literature [6], which enable to learn embeddings for nodes, edges, or whole
graphs or sub-graphs. We decided to chose a recently proposed algorithm called
StarSpace [22], a general purpose neural embedding model, which provides an
efficient strong baseline on various tasks. In particular, it can be used for the
embedding of multi-relational graphs.

In the remainder of this section, we first introduce the general StarSpace
algorithm before we describe how we apply it to learn workflow embeddings
as simplified representations of POCBR cases. Afterwards, the straightforward
application during the MAC retrieval phase is explained.

3.1 Embedding Learning with StarSpace

StarSpace embeds entities of different types into a single, fixed-length dimensional space with the result of having entity representations that are comparable
to each other. It learns to rank a set of entities, documents, or objects given
a query entity, document, or object, where the query is not necessarily of the
same type as the items in the set [22]. As we will further see, this is an important
property explored in our application.

The basic concept of StarSpace is that it learns embeddings for entities that
consist of one or more features. As we will see, in our case a feature is a node, an
edge, or a whole workflow. The algorithm maintains a dictionary D of features
and assigns to each feature an embedding vector, which is stored in an embedding
matrix $F \in R^{|D| \times d}$, where $|D|$ is the number of features and d is the size of the
dimension of the embedding space. Each row F_i is the embedding for a feature
i. An entity a is embedded by a vector representing the sum of the embeddings
F_i of the features i it consists of so that $a = \sum_{i \in a} F_i$.

To learn the embeddings for each feature, StarSpace compares the similarity
of two entities a and b that are provided from the set of training data E^+ against
randomly generated entities, which constitute the set of negative examples E^-.
The goal is that entities, which are labeled as similar based on the training
data E^+ will be rated higher by a margin h than randomly generated samples
from E^-. During learning, the following ranking loss function L is minimized by
stochastic gradient descent:

$$\sum_{\substack{(a,b) \in E^+ \\ b^- \in E^-}} L^{batch}(sim(a,b), sim(a, b_1^-), \dots, sim(a, b_k^-)) \tag{2}$$

The similarity function sim can be cosine or dot product, while cosine usually
leads to better results. The margin ranking loss for $sim(a,b)$ and $sim(a, b^-)$ is
calculated by $max\{0, h - sim(a,b) + sim(a, b^-)\}$.

How exactly positive training examples are generated from the set E is task-specific. StarSpace provides a large number of options, which makes StarSpace a general-purpose embedding approach. Since a straightforward word embedding approach to transform the labels of the graph components into a bag of words leads to less promising results in our previous experiments, we apply the multi-relational knowledge graph approach for generating training examples. This option enables to learn embeddings of a graph represented as triples (h, r, t) consisting of a head concept h, a relation r, and a tail concept t.

3.2 Learning Embeddings for Workflow Graphs

In order to learn embeddings for workflow graphs, a workflow becomes an entity in StarSpace that is described by a set of tasks and data nodes, each of which are features for which an embedding vector is learned. Also the main workflow node (i.e., the top node in Fig. 1) is a feature of the workflow entity.

To construct the training input for StarSpace, all *NEST* graphs from the case base are serialized into a triple format similar to the *Turtle* notation for RDF graphs[1]. This representation fully represents the semantic workflow and can be used for learning embeddings by StarSpace. The following triples can be extracted from the workflow example depicted in Fig. 1:

Sandwich_Recipe	hasTask	spread
Sandwich_Recipe	hasInput	pain de mie
Sandwich_Recipe	hasInput	sandwich sauce
Sandwich_Recipe	hasTask	disperse
Sandwich_Recipe	hasInput	dill
Sandwich_Recipe	hasInput	sandwich dish
Sandwich_Recipe	hasTask	set

Moreover, we can add further triples that represent the connections between tasks and data nodes:

pain de mie	dataflow	spread
sandwich sauce	dataflow	spread
pain de mie	dataflow	set
dill	dataflow	disperse
sandwich dish	dataflow	disperse
disperse	dataflow	sandwich dish
sandwich dish	dataflow	set
set	dataflow	sandwich dish
spread	controlflow	disperse
disperse	controlflow	set

---

[1] https://www.w3.org/TR/2014/REC-turtle-20140225/.

Based on this data, we can use the method proposed by StarSpace to learn multi-relational graphs. For this purpose, each triple (h, r, t) results in two training examples: the left entity (h) is predicted based on the relation (r) and the right feature vector (t) and the right entity (t) is predicted based on the left (h) and the reverse relation $(\bar{r})$ feature vector. Since we are only interested in the prediction of workflow graphs based on their task and data nodes, we have adapted the training such that only triples with *hasTask* and *hasInput* relations are used and we set the main workflow node as the label to predict. In other words, by using the relation feature and the feature of the task or data node, we try to predict the workflow. The learning algorithm of StarSpace then optimizes our feature vectors accordingly.

As a result of the learning phase, StarSpace produces feature vectors for each node of each workflow in the case base. Please note that equivalent nodes in different cases (e.g., nodes referring to the same ingredient or the same cooking step) are represented only once, thus have the same embedding vector. However, each workflow in the case base has its own main workflow node. We use the embedding vector learned for this main workflow node as workflow embedding representation of the case to be used in the MAC phase of the POCBR retrieval.

3.3   Plausibility of Similarities of Learned Embeddings

As a side effect of the described learning approach, all items of the task and data ontology used as semantic annotation for the nodes in the case workflows also occur as features in the StarSpace dictionary and thus have an embedding vector attached. Consequently, their similarity can be assessed by using the similarity measure for which StarSpace performs its optimization, i.e., the cosine similarity. In order to check whether the resulting similarity values are plausible, we performed a spot-checking of selected ingredient pairs. We compared the similarity value resulting from the embedding with the similarity value determined using the ingredient taxonomy (see Fig. 2), i.e., the local node similarity measure used in the graph-based similarity measure of the FAC phase (see Sect. 2.1.). Table 1 illustrates selected similarity comparisons with value ranges adjusted to the interval $[0, 1]$. For all three pairs of ingredients, which all have a relatively high similarity according to the manually modeled similarity measure, the learned embedding similarity is also quite high, which is a first indication that both measures are inline with each other. This observation could also be made for many more similarity pairs that we have checked in a random fashion but certain negative examples could also be found. In total, however, this inspection is a first hint that the embedding approach is able to learn useful similarity knowledge.

Table 1. Comparing selected similarity values

Query	Result	Embedding Similarity	Taxonomic Similarity
Egg White	Egg Yolk	0.854	0.8
Coconut	Pineapple	0.705	0.6
Bananas	Strawberries	0.695	0.6

3.4 Embedding-Based Workflow Retrieval

Using the learned workflow embeddings, the implementation of the MAC retrieval stage is quite straightforward and similar to the approach used for the feature-based MAC/FAC approach [5]. Prior to retrieval, the workflow embeddings must be learned for each case in the case base in an offline-phase. MAC retrieval then simply performs a linear search for the FS-most similar cases using the workflow embedding representation and the similarity measure used by StarSpace, i.e., the cosine measure. Thus, the parameter FS is the filter size for the MAC phase.

In this process, however, one aspect is less obvious, i.e., how the embedding vector of the query is determined. Typically, a query is not an already existing workflow in the case base, but a new workflow or even only a partial workflow, just consisting of a small number of nodes and edges. Consequently, there is no workflow embedding vector for the main workflow node of the query available, as this node does not exist in the StarSpace directory. To construct the embedding vector for the query, we make use of the StarSpace property that all items are embedded in the same common embedding space; there is no difference between different types of features. Given this, we construct the embedding vector of the query using the bag of features approach, i.e., by adding the embedding vectors of the task and data nodes the query consists of. This can be done at least for those nodes that previously occurred in the case base and that are thus also present in the dictionary of StarSpace.

As an example, consider we would use the workflow from Fig. 1 as a query. The resulting query embedding q would be determined as follows:

$$q = F_{spread} + F_{disperse} + F_{set} + F_{pain\_de\_mie} + F_{sandwich\_sauce} + F_{dill}$$
$$+ F_{sandwich\_dish}$$

4 Experimental Evaluation

In this section, we present the evaluation setup and the results. To determine the suitability of our approach, we compare our results with the MAC/FAC retriever by Bergmann and Stromer [5] and with the results of the cluster-based retriever by Müller and Bergmann [17]. For this purpose, we implemented the presented approach in the POCBR component of the CAKE framework[2]. We used the StarSpace implementation available at GitHub[3] and integrated it into the CAKE framework. The StarSpace implementation is started via a command line statement.

4.1   Hypotheses

In our experimental evaluation, we investigate the following hypotheses:

H1 The embedding-based retriever provides at least as good results as the MAC/FAC retriever using the feature-based representation in the MAC phase [5].

H2 The embedding-based retriever achieves better results than the cluster-based retriever [17] for case bases without cluster structure.

The first hypothesis expresses the expectation that the embedding-based retriever is as good as the feature-based retriever although no manual modeling effort has been invested. This leads to a significant benefit, since manual knowledge modeling is often complex and time-consuming. The second hypothesis claims that the embedding-based retriever is independent of the case distribution and thus more universally applicable compared to the alternative approach, which also avoids manual modeling of the MAC phase.

4.2   Experimental Setup

The evaluation is conducted on a case base with 1529 case workflows and 200 query workflows, taken from the extraction of case workflows from cooking recipes from Allrecipes[4] by Schumacher et al. [20]. The case workflows in the case base and the query workflows are the same as those used in the evaluation of the cluster-based retriever – named as CB-I [17] – and the MAC/FAC retriever using the feature-based representation [5]. Each workflow case contains 11 nodes on average and a corresponding taxonomic ontology of 208 individual ingredients and 225 cooking preparation steps is used. For learning our embedding model, the 1529 case workflows are serialized into 18.169 triples that represent *part-of* (*hasTask* or *hasInput*) edges. For simplicity reasons, parallel (*AND*) and alternative (*XOR*) sequences are disregarded as training data. Since our embedding model that learned on the completed graph structure has performed slightly worse than those only learned on triples with *hasTask* and *hasInput* relations, we only discuss results for our best model learned on this smaller training data set.

As a starting point for hyper-parameter optimization, the default settings are used. We adjusted the values of the *margin* to 0.35, the embedding's *dimension*

---

[2] http://procake.uni-trier.de.

[3] https://github.com/facebookresearch/StarSpace.

[4] https://allrecipes.com/.

size to 200, the number of *training epochs* to 200, and the *similarity measure* to cosine. These changes are based on manual inspections of nearest neighbor results by using the projection of individual workflow cases from the case base and thus maximizing the similarity to the case itself and magnifying the selectivity to other workflow cases. The StarSpace learning phase only took approximately 4 min on a PC with an Intel i9-7900X CPU @3.30 GHz and 64 GB RAM running Linux Ubuntu.

In order to assess the validity of the two hypotheses, we evaluate the retrieval time (MAC+FAC phase) and the retrieval quality for various parameter combinations. For retrieval quality, we use the same quality criterion (see Formula 3) as in previous work by Müller and Bergmann [17]. For this purpose, it is examined if workflow cases from the set of the k-most similar case workflows ($MSC(QW, k)$) that are retrieved by the A^* parallel retriever (i.e., the gold standard without any MAC pre-selection), are also retrieved by the MAC/FAC retriever under investigation. If not all case workflows are contained in the *result list (RL)* of the corresponding MAC/FAC approach, the quality decreases proportional to the similarity of this workflow to the query. Thus, if a highly similar case is omitted, the negative impact on the quality is stronger than if a case with a low similarity is missing.

$$quality(QW, RL) = 1 - \frac{1}{|RL|} \cdot \sum_{CW \in \{MSC(QW,|RL|) \setminus RL\}} sim(QW, CW) \tag{3}$$

4.3   Experimental Results

We compared the three MAC/FAC approaches using different numbers of case workflows to be retrieved (k) and using different filter sizes (FS) for the MAC phase. We show the retrieval time in seconds and the quality value according to Formula 3. All results are average values over all queries.

Table 2 shows the results of the comparison of the feature-based retriever with our embedding-based approach. Each row in the table represents one particular parameter setting. Please note that the k-value is the same for both retrievers in a row (so both have to solve the same retrieval task) but the used filter size parameter is different and optimized for each of the two approaches. In particular, we have chosen the FS value for the embedding-based retriever in a way that the achieved retrieval quality is quite the same as what is achieved by the feature-based retriever. In addition, we illustrate the number of matches (*Hits*) without considering the corresponding rank. When we investigate the retrieval time, we can see that the embedding-based retriever is as fast as the feature-based retriever for larger values of k but clearly faster for small values of k. Thus, Hypothesis H1 is clearly confirmed. With respect to the complete A^* parallel retriever (its retrieval time is shown in the right column of Table 2), we achieve a speedup of a factor 2.3 to 10.8. When looking at the results in more detail, we can see that the embedding-based retriever requires a significantly higher filter size to achieve the same quality values. Thus, it does not approximate the FAC similarity as well as the feature-based retriever, which leads to more irrelevant cases among within the list of top-ranked cases resulting from the MAC phase.

However, the speed-up in MAC similarity assessment is so large that we can compensate this by affording a larger filter size, thereby shifting retrieval effort from the MAC phase to the FAC phase. Overall, this seems to be an effective approach in our experiments.

Table 2. Comparison of the feature-based and the embedding-based retriever

Feature Graph MAC-FAC					Embedding Graph MAC-FAC					A* Parallel
FS	k	Hits	Quality	Time	FS	k	Hits	Quality	Time	Time
5	5	2.37	0.79	0.166	25	5	2.44	0.79	0.083	
50	5	4.69	0.98	0.262	250	5	4.61	0.97	0.251	0.896
10	10	5.24	0.80	0.190	50	10	5.75	0.81	0.126	
80	10	9.48	0.98	0.336	300	10	9.26	0.97	0.340	0.982
25	25	14.37	0.82	0.259	100	25	15.77	0.84	0.228	
100	25	23.17	0.97	0.465	350	25	22.75	0.97	0.469	1.098

To compare the embedding-based retriever with the cluster-based retriever, we report the values for k, FS, quality, and retrieval time as published in [17]. To compensate for the improved hardware capabilities under which we measure the results of the embedding-based approach, we adjust the previously reported time values by a factor of 0.80556. This value is carefully determined based on the average improvement of the A^* parallel and the feature-based MAC/FAC retriever. The results are shown in Table 3. Please note that in this experiment, the filter size is the same for both approaches. The results clearly demonstrate that the embedding-based retriever outperforms the cluster-based approach in retrieval time and quality in all examined parameter settings. Since the case base used for evaluation throughout the whole experiment has no cluster structure, Hypothesis H2 is clearly confirmed.

Table 3. Overview of results between the cluster-based and the embedding-based approach

		Cluster Graph MAC-FAC		Embedding Graph MAC-FAC	
FS	k	Quality	Time	Quality	Time
10	10	0.60	0.199	0.65	0.066
50	10	0.70	0.261	0.81	0.126
100	10	0.77	0.321	0.89	0.181
25	25	0.61	0.254	0.69	0.101
50	25	0.67	0.300	0.75	0.166
100	25	0.74	0.371	0.84	0.228
50	50	0.65	0.338	0.74	0.169
100	50	0.72	0.420	0.81	0.308

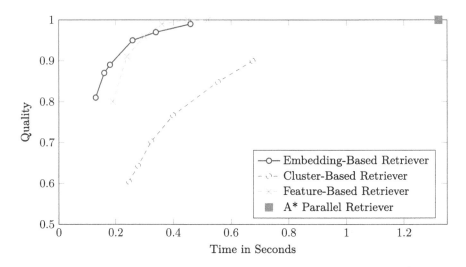

Fig. 3. Retrieval time and quality for all retrievers with $k = 10$

Finally, we summarize our results in a way that allows to compare all three MAC/FAC retrieval approaches. In Fig. 3, we present a plot that characterizes each retrieval approach as it directly relates retrieval time and retrieval quality for various values of *FS*. The value of k is fixed to 10 for this comparison. As shown by the plots, the embedding-based MAC/FAC retriever provides the highest quality in relation to retrieval time but does not fully reach the perfect quality value of 1 such as the feature-based approach. Hence, the speed advantage of the embeddings used to compensate for the poorer quality through increasing the filter size comes to an end at high-quality values because the marginal utility of an increased filter size diminishes. However, the loss of quality in this range is so small that it does not justify the effort involved in implementing a manually designed feature-based approach. Figure 3 also clearly shows the impressive advantage of the embedding-based retriever over the cluster-based approach.

5   Conclusion, Related and Future Work

We presented a new MAC/FAC approach for the retrieval of semantic workflows in POCBR, which is based on a novel general-purpose neural embedding approach. It enables to learn a vector representation for workflows that can be efficiently compared using the cosine similarity measure. The fact that this embedding approach is able to embed features of different kinds in the same embedding space allows us to efficiently determine also an embedding for a query workflow for which an embedding vector cannot be determined in advance. We could show that the presented approach achieves a performance, which is comparable to a feature-based MAC/FAC retrieval that works with manually modeled case representation and similarity measures for the MAC phase. As a result, an efficient

MAC phase is now available fully automatically by means of machine learning and comes at no cost (except for the offline computation time to be invested for training). The only previously available alternative approach to construct a MAC phase through learning is clearly outperformed in terms of retrieval time and quality.

To our knowledge, the use of neural embedding approaches for implementing the MAC phase in CBR has not yet been discussed in the literature before. Most similar is probably our work on retrieval of argumentation graphs [3] in which we use word embeddings as local similarity measures within a graph similarity measure but also as similarity measure for the MAC phase of retrieval. Not for MAC/FAC retrieval, but for case indexing in general, Metcalf and Leake [14] proposed several embedding techniques, include a knowledge graph embedding method in the domain of medical cases. In addition, the use of neural embedding approaches has been recently discussed in the CBR literature primarily for local similarity measures in textual CBR applications (see e.g., [1, 21]).

In future work, we aim to extend our experimental evaluation by using case bases from other domains and including workflow cases with higher complexity. Given the fact that the current embedding approach is not particularly designed to predict the modeled graph-based similarity measure, we still see potential for further improvements. Thus, we aim to investigate the idea to train a siamese network on top of an embedding network using the graph-based similarity values of case pairs from the case base. Furthermore, we propose to examine how the graph structure (e.g., data-flow and control-flow edges) and semantic annotations (e.g., amounts of ingredients) could be better considered during learning. An promising approach that will be considered in future work is presented by Li et al. [13]. Thereby, we hope to improve prediction of the current MAC phase, which would lead to further improvements in retrieval time and quality. In addition, we will explore the idea of an incremental MAC/FAC approach [19], which successively increases the filter size based on an analysis of the FAC similarity of the found cases.

Acknowledgments. This work is funded by the German Research Foundation (DFG) under grant No. BE 1373/3-3.

References

1. Amin, K., Kapetanakis, S., Althoff, K.-D., Dengel, A., Petridis, M.: Answering with cases: a CBR approach to deep learning. In: Cox, M.T., Funk, P., Begum, S. (eds.) ICCBR 2018. LNCS (LNAI), vol. 11156, pp. 15–27. Springer, Cham (2018). https://doi.org/10.1007/978-3-030-01081-2_2
2. Bergmann, R., Gil, Y.: Similarity assessment and efficient retrieval of semantic workflows. Inf. Syst. **40**, 115–127 (2014)
3. Bergmann, R., Lenz, M., Ollinger, S., Pfister, M.: Similarity measures for case-based retrieval of natural language argument graphs in argumentation machines. In: Barták, R., Brawner, K.W. (eds.) Proceedings of the Thirty-Two International Florida Artificial Intelligence Research Society Conference, FLAIRS 2019, Florida,

pp. 329–334. AAAI Press, 19–22 May 2019. https://dblp.uni-trier.de/rec/bibtex1/conf/flairs/BergmannLO019

4. Bergmann, R., Stahl, A.: Similarity measures for object-oriented case representations. In: Smyth, B., Cunningham, P. (eds.) EWCBR 1998. LNCS, vol. 1488, pp. 25–36. Springer, Heidelberg (1998). https://doi.org/10.1007/BFb0056319

5. Bergmann, R., Stromer, A.: MAC/FAC retrieval of semantic workflows. In: Boonthum-Denecke, C., Youngblood, G.M. (eds.) Proceedings of the Twenty-Sixth International Florida Artificial Intelligence Research Society Conference, FLAIRS 2013, Florida, 22–24 May. AAAI Press (2013). http://www.aaai.org/ocs/index.php/FLAIRS/FLAIRS13/paper/view/5834

6. Cai, H., Zheng, V.W., Chang, K.C.: A comprehensive survey of graph embedding: problems, techniques, and applications. IEEE Trans. Knowl. Data Eng. **30**(9), 1616–1637 (2018)

7. Dalal, S., Athavale, V., Jindal, K.: Case retrieval optimization of case-based reasoning through knowledge-intensive similarity measures. Int. J. Comput. Appl. **34**(3), 12–18 (2011)

8. Dumas, M., van der Aalst, W.M.P., ter Hofstede, A.H.M. (eds.): Process-Aware Information Systems: Bridging People and Software Through Process Technology. Wiley, Hoboken (2005)

9. Forbus, K.D., Gentner, D., Law, K.: MAC/FAC: a model of similarity-based retrieval. Cognit. Sci. **19**(2), 141–205 (1995)

10. Hollingsworth, D.: Workflow management coalition - the workflow reference model: Document Number TC00-1003 - Version 1.1 (1995)

11. Kendall-Morwick, J., Leake, D.B.: A study of two-phase retrieval for process-oriented case-based reasoning. In: Montani, S., Jain, L. (eds.) Successful Case-based Reasoning Applications-2. SCI, vol. 494, pp. 7–27. Springer, Heidelberg (2014). https://doi.org/10.1007/978-3-642-38736-4_2

12. Kendall-Morwick, J., Leake, D.B.: On tuning two-phase retrieval for structured cases. In: ICCBR-Workshop on Process-Oriented CBR, Lyon, pp. 25–34 (2012)

13. Li, Y., Gu, C., Dullien, T., Vinyals, O., Kohli, P.: Graph matching networks for learning the similarity of graph structured objects. In: Proceedings of the 36th International Conference on Machine Learning, ICML 2019, 9–15 June, California, Proceedings of Machine Learning Research, vol. 97, pp. 3835–3845. PMLR (2019)

14. Metcalf, K., Leake, D.B.: Embedded word representations for rich indexing: a case study for medical records. In: Cox, M.T., Funk, P., Begum, S. (eds.) ICCBR 2018. LNCS (LNAI), vol. 11156, pp. 264–280. Springer, Cham (2018). https://doi.org/10.1007/978-3-030-01081-2_18

15. Minor, M., Montani, S., Recio-García, J.A.: Process-oriented case-based reasoning. Inf. Syst. **40**, 103–105 (2014)

16. Montani, S., Leonardi, G.: Retrieval and clustering for supporting business process adjustment and analysis. Inf. Syst. **40**, 128–141 (2014)

17. Müller, G., Bergmann, R.: A cluster-based approach to improve similarity-based retrieval for process-oriented case-based reasoning. In: ECAI 2014–21st European Conference on Artificial Intelligence, pp. 639–644. IOS Press (2014)

18. Müller, G.: Workflow Modeling Assistance by Case-based Reasoning. Springer, Heidelberg (2018). https://doi.org/10.1007/978-3-658-23559-8

19. Schumacher, J., Bergmann, R.: An efficient approach to similarity-based retrieval on top of relational databases. In: Blanzieri, E., Portinale, L. (eds.) EWCBR 2000. LNCS, vol. 1898, pp. 273–285. Springer, Heidelberg (2000). https://doi.org/10.1007/3-540-44527-7_24

20. Schumacher, P., Minor, M., Walter, K., Bergmann, R.: Extraction of procedural knowledge from the web: a comparison of two workflow extraction approaches. In: Proceedings of the 21st World Wide Web Conference, pp. 739–747. ACM (2012)
21. Terada, E.: The writer's mentor. In: Proceedings of ICCBR 2018 Workshops Co-located with the 26th International Conference, ICCBR 2018, Sweden, July 9–12, pp. 229–233 (2018)
22. Wu, L.Y., Fisch, A., Chopra, S., Adams, K., Bordes, A., Weston, J.: Starspace: embed all the things! In: McIlraith, S.A., Weinberger, K.Q. (eds.) Thirty-Second AAAI Conference on Artificial Intelligence, pp. 5569–5577 (2018). https://dblp. uni-trier.de/rec/bibtex/conf/aaai/WuFCABW18

On Combining Case Adaptation Rules

David Leake[(✉)] and Xiaomeng Ye

School of Informatics, Computing, and Engineering, Indiana University,
Bloomington, IN 47408, USA
leake@indiana.edu, xiaye@iu.edu

Abstract. The case adaptation process in case-based reasoning is often
modeled as having two steps: enumerating differences between a new
problem and the problem part of a retrieved case and then applying
an adaptation rule for each difference. This model is sufficient when (1)
predefined adaptation rules exist for all differences the system encounters, and (2) adaptation rules are sufficiently independent that interactions are not a major issue. This paper presents an approach to handling
case adaptation when these assumptions fail. It proposes an approach,
RObust ADaptation (ROAD), that uses heuristics to guide multi-step
adaptations, with each adaptation chosen in the context of adaptations
applied previously. To reduce the potential for accumulated degradation
of solution quality from long adaptation chains, it performs incremental
retrieval of new source cases along the adaptation path, *resetting* the
partially modified case to the "ground truth" of existing cases when an
existing case is nearby. An evaluation supports the benefits of the model
and illuminates some tradeoffs.

Keywords: Adaptation rule interactions · Case adaptation ·
Differential adaptation · Adaptation path

1 Introduction

The case-based reasoning (CBR) *case adaptation* process modifies the solution of
a retrieved case to fit a new problem [10]. Because case adaptation provides the
flexibility to apply retrieved cases to new situations, case adaptation is fundamental to the performance of CBR systems. The case adaptation process is commonly driven by a list of differences between the new problem and the problem
of the retrieved case, with adaptation rules retrieved according to the differences
to adapt. Models of the adaptation process often assume that a suitable adaptation rule will be available for each difference the system may have to adapt,
making adaptation for each difference a one-step process. However, in practice
it may be infeasible to provide a CBR system with adaptation rules for every
possible difference. Generating adaptation rules by hand is costly, and it may be
difficult or impractical to anticipate or capture a sufficiently extensive rule set.
Another issue is the potential need to address multiple differences between old

© Springer Nature Switzerland AG 2019
K. Bach and C. Marling (Eds.): ICCBR 2019, LNAI 11680, pp. 204–218, 2019.
https://doi.org/10.1007/978-3-030-29249-2_14

and new problems. Rules may be tailored for specific differences, but the number of potential difference combinations aggravates the rule generation problem. In practice, CBR systems often address the problem of multiple differences by treating the effects of differences along particular problem features as independent and decomposing the difference set by generating a list of differences to adapt in turn, applying one rule per difference.

For numerical problems, the decomposition approach has been formalized by Fuchs et al. as *differential adaptation* [4], which models case adaptation as the application of a sequence of adaptation operators. This paper is in the spirit of that work, but adopts a search-based framework for adaptation generation that is appropriate for symbolic domains as well and can pursue multiple alternative adaptation paths. It proposes ROAD (RObust ADaptation), a search-based model of the adaptation rule application process that applies a flexible chaining process and integrates ongoing retrievals to help guard against the degradation of adaptation performance that can occur for long adaptation paths [9].

The paper begins by considering general issues for applying sequences of adaptation rules. It then proposes a search-based model, guided by similarity. It identifies a potential pitfall when similarity does not match adaptability (cf. [12]), and proposes addressing it by ongoing retrievals that "reset" the state of the adaptation sequence to a known case. Resetting also provides a potential solution to solution quality degradation over long chains. An evaluation illustrates the benefits of the multi-step approach and path resetting, while illustrating tradeoffs based on adaptation rule specificity and reach.

2   Models of Applying Multiple Case Adaptation Rules

We begin by presenting three models of adaptation rule combination: adaptation by multiple independent one-step adaptation rules, adaptation by multi-step adaptation paths, and ROAD—robust adaptation by dynamically adjusted and reset adaptation paths.

Adaptation by Multiple Independent One-Step Adaptation Rules: It is common for CBR systems to adapt a collection of differences by selecting adaptation rules for each one, applying each one, and combining the results. Given a set of independent features f_i of a case, and corresponding differences d_i between the value of the feature in the input problem and a retrieved case to adapt, and a collection of adaptation rules r_i, with each r_i applicable to difference d_i, the adaptation combination process combines the independent component results. For example, for a regression task (*e.g.* real estate appraisal), combination might simply sum the price effect of each difference.

The one-step approach assumes that there will be an adaptation rule suitable for each difference. However, this may not be the case. For example, if no sufficiently similar case is available, extensive adaptation may be required to bridge the distance between cases, with no predefined rules that bridge the gap. When a limited set of rules is available, a multi-step adaptation may be the only way to address certain feature differences, even for a single feature.

Fig. 1. Adapting from Solution A to Solution C through adaptation path (f, g).

Adaptation by Multi-step Adaptation Paths: Path-based adaptation discards the assumption that there will exist an adaptation rule that directly addresses a given difference. Instead, the system may have rules that incrementally provide the desired solution. In the cooking domain, to change a recipe for regular pancakes to buttermilk pancakes, a cook might first adapt to use buttermilk (by replacing the leavening with buttermilk and baking soda) and then, if buttermilk is not available, adapt by substituting regular milk and vinegar for the buttermilk. This generates an adaptation path, the composition of a sequence of adaptation rules resulting in intermediate solution states (Fig. 1).

An adaptation path is a sequence of triples (c_i, a_i, c_{i+i}), where c_i is a case to adapt, a_i is an adaptation rule, and c_{i+1} is the result of applying a_i to c_i. In our formulation, as in that of D'Aquin, Lieber, and Napoli [2] and Leake and Schack [8], the adaptation rule modifies both the solution and the problem description, to keep the problem and solution descriptions consistent for a new complete case. As this case does not correspond to a situation encountered in the world, we call it a *ghost case* [8]. The adaptation path retains provenance information about how each case in the sequence is derived, which could be used, e.g. for estimating result quality [9]. In an adaptation path, differences are not assumed to be independent; adaptations are performed in a sequence. Issues with interaction problems are handled when the CBR cycle revision step repairs the candidate solution [10].

In the adaptation path model of D'Aquin, Lieber, and Napoli, a similarity path is built first and then a corresponding adaptation path to modify the source solution to the target solution. Badra, Cordier, and Lieber [1] present an algorithm building an adaptation path by which the query is modified to match at least one case from the case base. Inspired by differential calculus, the differential adaptation model of Fuchs et al. [4] uses partial derivatives to make small variations in the problem and solution. Adaptation knowledge is generated by computing derivatives of every solution feature with respect to every problem feature. Differential adaptation can be applied to any numerical task domain provided that the dependencies between descriptors can be computed. Under differential adaptation, small-step adaptation is preferable to big-step adaptation as the former introduces less error using derivatives. This assumption motivates their multi-step differential adaptation.

Robust Adaptation by Dynamically Adjusted and Reset Adaptation Paths: The ROAD approach to adaptation paths is based on three principles.

First, rather than dividing difference identification and adaptation into separate steps, after every adaptation rule application it re-assesses differences and selects the next adaptation rule to apply. Incremental choices of each adaptation rule are made in the context of the previous adaptations. Second, ROAD can simultaneously pursue multiple potential adaptation paths, enabling exploiting alternative adaptations and comparing competing alternatives. Third, ROAD's adaptation process is coupled to the case base: As adaptation proceeds, if an incremental solution is similar to an existing case, ROAD can "reset" the adaptation path to proceed from the existing case. This substitution effectively restarts solution generation from a known solution in the case base, reducing the required adaptation. In the common situation of imperfect adaptation knowledge, this is expected to increase solution quality.

3   Issues for Adaptation Paths and ROAD

Compared to applying single adaptations, adaptation path composition increases adaptation flexibility and coverage. However, moving from single adaptations to adaptation paths complicates multiple parts of the case adaptation process:

1. Estimating adaptability: CBR commonly uses similarity as a proxy for adaptability. This assumption has been questioned [12], leading to methods for adaptation-guided retrieval [3,7,11,12]. However, when adaptation could in principle involve long sequences of adaptation steps, estimating adaptation cost could require consideration of many possible alternatives, making adaptation-guided retrieval extremely expensive.
2. Guiding path construction: If many alternative adaptation sequences could be considered, how to select them becomes important for adaptation efficiency and quality (e.g., if some rules are known to be more reliable than others or shorter paths are more desirable for quality [9]).
3. Terminating unpromising paths: When a path is judged to be unpromising (e.g., because its length suggests risk of excessive quality loss or explainability), it may be appropriate to terminate; heuristics are needed to determine when to terminate.
4. Handling adaptation interactions: If adaptations may interact, the choice of adaptations must be sensitive to side-effects of previous adaptations. In principle, in fully understood domains, it would be possible to use standard AI methods to manage such interactions. However, CBR is often applied to domains that are poorly understood or imperfectly formalized. An adaptation approach practical for such domains must use other methods. This paper presents heuristic-based methods.

The ROAD model of dynamically adjusting and resetting adaptation paths raises additional issues:

1. Determining a strategy for exploring alternative adaptation paths: When multiple paths are pursued, any search strategy could be used to manage the choice of partial paths to extend (e.g., best-first, breadth-first).

Algorithm 1. ROAD Overall Procedure

Input:
CB: case base
target: the input query
settings parameters: from section 5.1
rules: adaptation rules

Output: a prediction for the solution of the target case
$sourceCases \leftarrow intialRetrieval(CB, target, k)$
$paths = new\ priorityQueue()$
$paths.addAll(createPathsFrom(sourceCases))$

$finishedPaths = []$ ▷ building k paths
while not $paths.isEmpty()$ **do**
    $p \leftarrow paths.pop()$
    $resetIfDecay(resetByDecay, CB, p)$
    $resetIfConflict(resetByConflict, CB, target, p, paths, finishedPaths)$
    $step(path, rules)$
    if $done(path, maxLen)$ **then**
        $finishedPaths.add(path)$
    else
        $paths.add(path)$

$solutions = []$
for all $path$ in $finishedPaths$ **do**
    $solutions.add(path.head.solution)$
$prediction \leftarrow average(solutions)$
return $prediction$

2. Reconciling path intersections: If two adaptation paths lead to generating the same (or a nearby) ghost case, various strategies could be used for merging those results in the final path, such as selecting the single "best" path leading to the point of overlap or retaining both paths.
3. Resolving path conflicts and path reinforcements: If the overlapping ghost cases have different (similar) solutions, the conflict could be evidence for reducing (increasing) path reliability.
4. Determining when to reset the search: As described in Sect. 4.5, when the ghost cases in a path are similar to stored cases it may be beneficial to reset the path by replacing the ghost case with a new retrieval. This requires determining criteria for resetting the search.

4   ROAD's Strategies for Building Adaptation Paths

ROAD builds adaptation paths by a search process. Given a target problem p, the process begins by retrieving one or more most similar cases as a starting point.

The starting point may be reset by new retrievals as the process progresses. The goal is to generate an adaptation path from the initial case to one applicable to the new situation, while minimizing a cost function $f : P \rightarrow \mathbb{R}^+$. The function f could reflect criteria such as minimizing solution error, minimizing path length (as a proxy for minimizing error, or for increasing explainability of the solution derivation), minimizing adaptation cost (i.e., path construction cost), or domain-specific criteria such as, in case-based planning, minimizing the execution cost of the generated plan. For example, for numerical domains, Fuchs et al. [4] propose building the adaptation path by hill climbing using the derivative to reflect the relationship between the variation of problem features and the variation of solution values.

ROAD uses four categories of strategies, for: (1) avoiding prohibited regions, (2) extending a single adaptation path, (3) initializing and pursuing multiple paths, and (4) preserving the reliability of paths.

4.1 Avoidance of Prohibited Regions (Dead Zones)

ROAD supports the avoidance of generation of ghost cases in certain regions of the problem space. We call these regions *dead zones*. For example, in the housing price prediction domain, local housing codes may prohibit houses in a certain area. Rather than allowing case adaptation to hypothesize ghost cases in that area, ROAD's can be provided with a test function to reject ghost cases there. This has two motivations. First, for an adaptation rule to hypothesize a case in such a region shows that the rule is missing portions of the relevant context; thus it might be expected to be less reliable in that region, making it reasonable to favor a path with all steps within the realm of possibility. Second, if the CBR system will present the adaptation path to the user as an explanation for its result, presenting an impossible intermediate step might undermine confidence in the explanation.

We note that this prohibition is not required and might sometimes be undesirable. For example, for a domain in which solutions are hard to generate but easy to evaluate, the path might not be important to trust, and enabling adaptation to hypothesize impossible ghost cases might lead to creative solutions.

4.2 Extending a Single Adaptation Path

An adaptation path is extended by adapting the case at the head of the path, using an adaptation rule, and appending the triple of the original case, the rule, and the ghost case generated by adaptation to the path. Because multiple adaptation rules may apply to a single case, heuristics are needed to select the adaptation rule to apply. Our implementation of ROAD searches by a modified best-first process, favoring adaptation rules that result in cases most similar to the target case.

Inspired by Fuchs et al. [4], ROAD extends paths by a best-first process, but it differs in three respects. First, Fuchs et al. only addressed regression tasks, requiring the generation of differential numerical adaptation operators, with the sequence of chosen operators always generating the correct adaptation. ROAD, addresses adaptation for non-regression as well as regression tasks, and does not assume that a best-first process will necessarily produce the best path. Second, ROAD supports the simultaneous pursuit of multiple paths, with back-tracking as needed and with the ability to block consideration of portions of the candidate adaptation space (dead zones). Third, as described below, rather than simply applying the sequence of chosen adaptations, ROAD monitors the ongoing results of partial adaptations to potentially re-start adaptation from a nearby case.

When extending a path, ROAD first retrieves all adaptation rules applicable to the case at the head of the path. The rules are then applied, with the results ranked by applying the cost function (e.g., closeness of the problem of the new ghost case to the target problem). The rule with the best result is chosen and applied to the head (ties are broken arbitrarily). If the result has already been considered along an adaptation path, or if the result is impossible in the task domain (falls in a dead zone), then the next best rule is chosen.

4.3   Heuristics for Initializing and Pursuing Multiple Paths

We considered three methods for the initial retrieval of source cases to use to initialize paths: (1) using 1-nn to retrieve one case and start one path from it, (2) using k-nn to retrieve k cases and starting one path from each of the k cases, and (3) using a k-nn to retrieve k cases but starting paths from a subset of those k cases selected for diversity. We expect (2) to perform better than (1) as combining solutions from multiple paths averages out errors in individual paths. We expect (3) to increase efficiency by reducing the number of paths considered by (2), with little quality loss because similar cases result in less independent paths–they involve similar adaptation steps toward the target–and less potential benefit from an ensemble of solutions. However, a tradeoff is that we would expect substantially non-similar cases to result in lower performance. Selecting the diverse cases from the k nearest neighbors balances similarity and diversity. ROAD uses method (3).

ROAD stores all paths in a priority queue. In every iteration, the path with the highest priority is removed from the queue, extended, assigned a new priority, and added back to the queue. Using a priority queue ensures that computational resources are shared among all paths. In ROAD, the priority of a path is defined as the inverse of the path length. Alternatively, the priority can be defined by the distance between the head and the target case, or confidence of a path.

4.4   Path Termination and Resetting as a Heuristic for Maintaining/Increasing Path Reliability

One source of confidence in the results of case-based reasoning is that conclusions are derived from similar prior cases known to have correct solutions. In imperfectly understood domains, adaptation rules are generally imperfect, so adaptation is not guaranteed to generate correct solutions. Repeated application of unreliable adaptation rules may compound errors; in some contexts, reliability

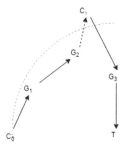

Fig. 2. Illustration of path resetting

may be estimated by assigning a confidence decay rate to each adaptation rule and combining the decay effects of multiple rules [9].

ROAD uses several simple methods to reduce the chance of compound error. First, its greedy search process is aimed at finding a short adaptation path to the target (though as with any greedy search, the shortest path is not guaranteed). ROAD can limit adaptation path length. It also uses a novel method, path resetting, described in the following section, to bring the speculative path from adaptations back to case knowledge grounded in experience.

4.5   Resetting a Ghost Case with a Nearby Retrieval

ROAD's resetting process is triggered when heuristics suggest a potential solution quality problem. Resetting moves the focus from the current ghost case to a similar case in the case base. We note that changing to this case can result in a shorter adaptation path from a prior case, even when the previous case most similar to the source case was retrieved initially. Figure 2 presents an intuitive example. In the figure, C_0 and C_1 are cases in the case base, with C_0 the initial retrieval to solve target case T. The solid arrows represent adaptation steps, with G_1, G_2 and G_3 ghost cases produced by adaptation.

Because C_1 is farther away from T than C_0 (the dashed arc), C_0 is initially used as the source case for an adaptation path. By greedy search, C_0 is adapted to G_1. The path further expands to G_2. Notice that even though C_1 is farther from T than C_0, it is closer to G_3 than C_0. The solution of T can be predicted by either continuing adapting the path $C_0 - G_1 - G_2$, which may have accumulated error due to imperfect adaptation rules, or "resetting" the path to C_1 (with C_1's correct solution) and extending from there. The dashed arrow represents the reset to continue the path from C_1.

ROAD triggers resetting in two conditions:

- *Cumulative quality decay:* Input cases are associated with a perfect reliability score; a decay function decreases reliability based on each adaptation rule applied, with reset triggered when reliability drops below a threshold.
- *Conflicting adaptation paths:* When the ghost cases at the heads of two adaptation paths are within a preset similarity threshold but their solution dif-

ference exceeds a difference threshold, the conflict puts their reliability in question. Because ROAD allows multiple adaptation paths to start from different (nearby) cases, ROAD only resets if the new solution difference exceeds their original solution difference. ROAD resets the path whose head is farthest from the target case.

5   Evaluation

Our evaluation addressed the following questions:

1. How does the use of single adaptation paths vs. multiple adaptation paths affect solution accuracy?
2. What are the tradeoffs between one-step adaptation and multi-step adaptation, for varying degrees of case base sparsity around a target, and how is this affected by rule specificity?
3. How do rule specificity and locality affect the ROAD's performance?
4. How does resetting paths affect performance compared to not resetting?
5. How does the combined effect of multiple adaptation paths, multi-step adaptation, and resetting affect solution accuracy?

5.1   Experimental Design

Task and Data: We tested the performance of ROAD for an automobile price prediction task, using the Kaggle automobile dataset [6]. The original dataset contains 205 cases, each with 26 features. We removed the first two features because they concern insurance risk rather than attributes of a car and cases with missing features, leaving 193 cases. Every case has 10 nominal features and 13 numerical features, plus *price* as the solution.

Adaptation Rule Generation: Case adaptation rules were generated automatically for this task using a method based on the case difference heuristic [5], with specifics as described in Schack and Leake [8].

   Their method generated adaptation rules that were considered applicable only when all nominal features in the source case used for learning were also present in the case to adapt. Such rules have high specificity (which corresponds to low generality). An alternative approach is to generate rules whose applicability is based on a smaller set of features, or even just one feature, making them more generally applicable (low specificity). Our rule generation procedure can be tuned to record only a subset of the features. A parameter *rule specificity* (*rspec*) governs the fraction of the original features retained in the adaptation rule (both in antecedents and in the features to adapt). The specific features for a given rule are chosen at random.

Similarity Criteria: Similarity of nominal features is 1 if the features are identical and otherwise 0; all numerical features are assigned equal weight and are normalized by the range of feature values.

Controlling for Case Base Sparseness: In general, the need for case adaptation depends on the density of the case base: When the case base is dense, with a high probability of nearby similar cases for any input problem, adaptation will be less crucial than when the case base is sparse and distant cases must be brought to bear. To study this effect, experiments were done both with the original case base, and with varying levels of case deletion in the neighborhood of the target case. Testing each target case, 0, 10, 25, 50, 75, 100, 125, and 150 cases around the target are removed from the case base to simulate situations where an initially retrieved source case is at a certain distance from the target.

Experimental Parameters: Parameters regulate adaptation rule generation and adaptation path building for each experiment. Performance is measured by the relative error of *price*. Each experiment is carried out 100 times, with 10-fold cross validation for each trial. Cases are considered solved when either (1) no adaptation results in a ghost case closer to the target, or (2) the path length limit is reached. Settings for all experiments are listed in Table 1. Experimental conditions are determined by the following parameters (1–4 affect system behavior; 5–7 affect adaptation rule generation):

1. **k**: Number of cases retrieved as path starting points (also k in baseline k-nn)
2. **MaxLen**: Maximum allowed path length
3. **ResetByDecay**: reliability threshold for resetting path due to reliability decay. In the experiments, the decay is simply a constant value subtracted from the initial reliability
4. **ResetByConflict**: Enables/disables resetting when paths disagree
5. **Rcount**: the number of rules to generate
6. **Rspec**: The level of rule specificity. For example, $rspec = 0.1$ if 10% of all feature differences are included in the antecedents of the rules.
7. **RuleGenDist**: the distance between pairs of cases generating rules. For example, if $ruleGenDist = 0.1$, rules are generated from cases whose difference is less than 10% of the maximum possible difference. Small $ruleGenDist$ values correspond to generated rules covering only small intercase differences.

5.2   Experimental Results

Question 1: Single Adaptation Path vs. Combined Results of Multiple Adaptation Paths: To assess the value of combining the results of multiple adaptation paths from multiple starting points, we compared results when starting adaptations from a single case (experiments #3 and #5) and from five cases (#1 and #8), adapting each, and averaging the solutions, for two levels of rule specificity. Figure 3 shows that using multiple adaptation paths lowered error at both levels of specificity.[1]

---

[1] In the experimental comparisons, when cases have been removed, all differences are statistically significant ($p < .05$), except for: Fig. 4(a) when more than 75 cases removed, Fig. 4 (b) when more than 25 have been removed, Fig. 5(b), which has no significant difference, and Fig. 6 (b), when fewer than 150 cases removed.

Table 1. Experimental parameter settings

#	k	maxLen	resetByDecay	resetByConflict	rcount	rspec	ruleGenDist
1	5	9	true	true	300	1.0	1.0
2	5	9	true	false	300	1.0	1.0
3	1	9	true	false	300	1.0	1.0
4	1	1	true	false	300	1.0	1.0
5	1	9	true	false	300	0.5	1.0
6	1	9	false	false	300	0.5	1.0
7	1	1	false	false	300	0.5	1.0
8	5	9	true	true	300	0.5	1.0
9	1	9	true	false	300	1.0	0.2
10	1	1	true	false	300	1.0	0.2
11	1	9	true	false	300	0.8	0.2
12	1	1	true	false	300	0.8	0.2
13	5	9	false	true	300	0.5	1.0
14	5	9	false	false	300	0.5	1.0

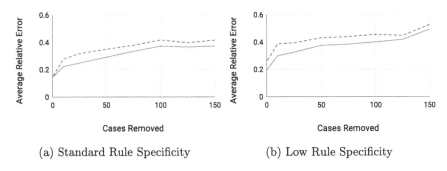

(a) Standard Rule Specificity   (b) Low Rule Specificity

Fig. 3. Error rate with increasing case base sparsity near target. Dashed red is one-path adaptation; solid blue is five-path adaptation. (Color figure online)

Question 2: Tradeoffs of One-Step Adaptations and Adaptation Paths:

To assess the tradeoffs between one-step adaptations and adaptation paths, we compared the performance of ROAD for one-step adaptation (experiment #4) and adaptation limited to nine adaptation steps (experiment #3), with adaptation rules reflecting the standard configuration of case difference heuristic rule generation (all differences reflected in the rules, for a rule set including both rules for short ($ruleGenDist = 0.2$) and long distance adaptations ($ruleGenDist = 1.0$)). As shown in Fig. 4(a), ROAD with the longer path length limit has lower error than ROAD with one-step adaptation. It might be expected that with a lower limit on distance covered, ROAD with longer path length limit

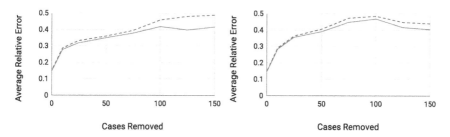

(a) Standard Rule Generation Distribu- (b) Local Rule Generation Distribution
tion

Fig. 4. Error rate with increasing case base sparsity near target. Dashed red is one-step adaptation; solid blue has max of nine adaptation steps. (Color figure online)

will generate a longer path, which reduces path reliability and decreases benefit over one-step adaptation.

To illustrate the effect of rule distance coverage on the benefit of path-based adaptation, Fig. 4(b) compares experiments #9 ($maxLen = 9$) and #10 ($maxLen = 1$) for a lower distance maximum ($ruleGenDist = 0.2$). Results are similar, with slightly higher error for rules with lower distance coverage when the region around the target case is less sparse. This suggests that the compounding of error from rules with less distance coverage may decrease the benefit of ROAD with long paths. For both levels of rule specificity, the error rate with adaptation paths increases more slowly than with one-step adaptation.

One interesting phenomenon observed in Fig. 4(b) is that for experiment #9, the error rate increases as the number of cases removed increases from 0 to 100, but decreases when the number of cases removed is 125 and 150. As more cases are removed, the initial retrieved cases become farther from the target case. Adaptation paths have to cover a larger distance (measured by similarity measure) to produce a ghost case close to the target case, and are therefore more prone to decay in reliability. As even more cases are removed, the initial retrieved case may reside on the boundary of the case space while the target case is in the middle. In such situations, an adaptation path often has only one direction to work toward the target case, therefore leading to slightly less error. This phenomenon can be observed in experiments #9, #10, #11, #12. A common characteristic of these experiments is that $ruleGenDist$ is set to 0.2, meaning the rules only cover small feature differences.

Question 3: Sensitivity of Result Quality to Adaptation Rule Specificity and Locality: To assess the dependence of ROAD on adaptation rule characteristics, we tested the effect of two factors: rule specificity and locality.

Rule Specificity: To test the effect of rule specificity on the benefit of longer paths compared to one-step adapations, we compared experiments #5 ($maxLen = 9$) and #7 ($maxLen = 1$), both of which use low specificity rules ($rspec = 0.5$) but allow non-local rules ($ruleGenDist = 1.0$). Non-local rules enable one-step

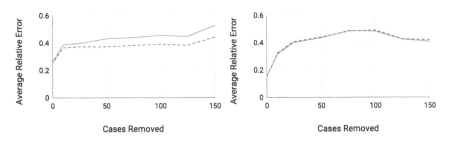

(a) Standard Rule Generation Distribu-(b) Local Rule Generation Distribution,
tion, Low Rule Specificity   High Rule Specificity

Fig. 5. Error rate with increasing case base sparsity near target. Dashed red is one-step adaptation; solid blue has max of nine adaptation steps. (Color figure online)

adaptations to bridge long problem differences. Results are shown in Fig. 5(a). Here one-step adaptations outperform longer paths, which contrasts with Fig. 4, which shows ROAD with long paths consistently outperforms one-step adaptations for more specific rules. Our explanation is that decreasing rule specificity broadens rule applicability, and therefore increases average path length as more applicable rules are available to apply. For example, the average path length is 2.5 when $rspec = 1.0$, and 7.3 when $rspec = 0.5$. Decreasing rule specificity also decreases rule accuracy, and longer paths compound error. This nullifies the benefit of longer paths seen in the previous experiment.

Rule Locality: In our tests, the rule generation distribution parameter determines the maximum distance between the pair of cases from which an adaptation is generated, determining rule locality. To illustrate the effect of rule locality, we compared experiment #11 ($maxLen = 9$) and #12 ($maxLen = 1$) in Fig. 5(b), for $rspec = 0.8$ and $ruleGenDist = 0.2$. $ruleGenDist$ is set to 0.2 to favor generation of local adaptation rules, while $rspec$ of 0.8 increases rule reliability. A local rule generation distribution limits one-step adaptation, because single rules cover shorter distances than paths with multiple rules. Here the performance of ROAD closely matches that of one-step adaptation for all sparsity levels.

Question 4: Effect of Path Resetting: We proposed two triggers for adaptation path resetting: (1) When a cumulative measure of path reliability drops below a threshold, or (2) when the solutions of two paths disagree. We conducted experiments to assess the benefit of resetting under each strategy. Figure 6(a) shows the error as a function of increased sparsity near the adaptation target, with and without resetting at the decay threshold (experiments #5 and #6). Resetting results in a substantial drop in error. Figure 6(b) shows the error as a function of increased sparsity near the adaptation target, with and without resetting for conflicts (experiments #13 and #14). Here resetting benefits as well, but less uniformly.

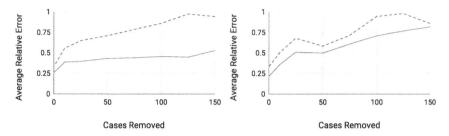

(a) No Resetting vs. Resetting by Decay (b) No Resetting vs. Resetting by Conflict

Fig. 6. Error rate with increasing case base sparsity near target. Dashed red is without resetting; solid blue is with resetting. (Color figure online)

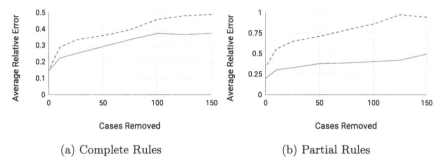

(a) Complete Rules   (b) Partial Rules

Fig. 7. Error rate with increasing case base sparsity near target. Dashed red is single case retrieval with at most one adaptation; solid blue is ROAD with combined methods. (Color figure online)

Question 5: Combined Effect: A final question is the combined effect when the individual aspects of ROAD are combined. Figure 7 compares accuracy for basic adaptation—a single adaptation path, at most one adaptation applied, no resetting—with ROAD using five paths, a maximum of nine adaptation steps, and both resetting mechanisms. It compares these for two types of adaptation rules generated from random case pairs, complete rules (rSpec = 1 and rule-GenDist = 1), in Fig. 7(a), and partial rules (rSpec = 0.5 and ruleGenDist = 1) in Fig. 7(b). (These respectively compare the results of experiments #1 and #4, and #6 and #8; note differing scales because rule characteristics heavily influence accuracy.) The experiments show substantial benefits for the full ROAD configuration, especially with partial rules.

6 Conclusion

This paper presented ROAD, a model of robust adaptation. ROAD uses heuristics to guide generation of multi-step adaptation paths and resets points along the adaptation path to nearby previous cases when path reliability decays or when multiple adaptation paths suggest conflicting results. The resetting process is a novel way to help ground uncertain adaptation in the real experience of the case base. Evaluations support the benefit of the approach, and especially the benefit of resetting. The evaluation demonstrated, as expected, that the benefit of ROAD depended on case base sparsity: If cases are available near a target, there is less need for extensive adaptation. Future work includes extending testing to additional domains and rule sets, to further study the effects of case base and adaptation rule characteristics. It also includes refining the heuristics used to guide the adaptation process, for example, by alternative search strategies and bidirectional search, and further investigation of "dead zones," regions in which no real cases exist but for which ghost cases might be hypothesized.

References

1. Badra, F., Cordier, A., Lieber, J.: Opportunistic adaptation knowledge discovery. In: McGinty, L., Wilson, D.C. (eds.) ICCBR 2009. LNCS (LNAI), vol. 5650, pp. 60–74. Springer, Heidelberg (2009). https://doi.org/10.1007/978-3-642-02998-1_6
2. D'Aquin, M., Lieber, J., Napoli, A.: Adaptation knowledge acquisition: a case study for case-based decision support in oncology. Comput. Intell. **22**(3/4), 161–176 (2006)
3. Díaz-Agudo, B., Gervás, P., González-Calero, P.A.: Adaptation guided retrieval based on formal concept analysis. In: Ashley, K.D., Bridge, D.G. (eds.) ICCBR 2003. LNCS (LNAI), vol. 2689, pp. 131–145. Springer, Heidelberg (2003). https://doi.org/10.1007/3-540-45006-8_13
4. Fuchs, B., Lieber, J., Mille, A., Napoli, A.: Differential adaptation: an operational approach to adaptation for solving numerical problems with CBR. Knowl.-Based Syst. **68**, 103–114 (2014)
5. Hanney, K., Keane, M.T.: Learning adaptation rules from a case-base. In: Smith, I., Faltings, B. (eds.) EWCBR 1996. LNCS, vol. 1168, pp. 179–192. Springer, Heidelberg (1996). https://doi.org/10.1007/BFb0020610
6. Kaggle: Automobile dataset (2017). data retrieved from Kaggle. https://www.kaggle.com/toramky/automobile-dataset
7. Leake, D., Kinley, A., Wilson, D.: Linking adaptation and similarity learning. In: Proceedings of the Eighteenth Annual Conference of the Cognitive Science Society, pp. 591–596. Lawrence Erlbaum, Mahwah (1996)
8. Leake, D., Schack, B.: Exploration vs. Exploitation in case-base maintenance: leveraging competence-based deletion with ghost cases. In: Cox, M.T., Funk, P., Begum, S. (eds.) ICCBR 2018. LNCS (LNAI), vol. 11156, pp. 202–218. Springer, Cham (2018). https://doi.org/10.1007/978-3-030-01081-2_14
9. Leake, D., Whitehead, M.: Case provenance: the value of remembering case sources. In: Weber, R.O., Richter, M.M. (eds.) ICCBR 2007. LNCS (LNAI), vol. 4626, pp. 194–208. Springer, Heidelberg (2007). https://doi.org/10.1007/978-3-540-74141-1_14
10. López de Mántaras, R., et al.: Retrieval, reuse, revision, and retention in CBR. Knowl. Eng. Rev. **20**(3), 215–240 (2005)
11. Nouaouria, N., Boukadoum, M.: Case retrieval with combined adaptability and similarity criteria: application to case retrieval nets. In: Bichindaritz, I., Montani, S. (eds.) ICCBR 2010. LNCS (LNAI), vol. 6176, pp. 242–256. Springer, Heidelberg (2010). https://doi.org/10.1007/978-3-642-14274-1_19
12. Smyth, B., Keane, M.: Adaptation-guided retrieval: Questioning the similarity assumption in reasoning. Artif. Intell. **102**(2), 249–293 (1998)

Semantic Textual Similarity Measures for Case-Based Retrieval of Argument Graphs

Mirko Lenz(✉)[iD], Stefan Ollinger[iD], Premtim Sahitaj[iD],
and Ralph Bergmann[iD]

Business Information Systems II, University of Trier, 54296 Trier, Germany
info@mirko-lenz.de, {ollinger,sahitaj,bergmann}@uni-trier.de
http://www.wi2.uni-trier.de

Abstract. Argumentation is an important sub-field of Artificial Intelligence, which involves computational methods for reasoning and decision making based on argumentative structures. This paper contributes to case-based reasoning with argument graphs in the standardized Argument Interchange Format by improving the similarity-based retrieval phase. We explore a large range of novel approaches for semantic textual similarity measures (both supervised and unsupervised) and use them in the context of a graph-based similarity measure for argument graphs. In addition, the use of an ontology-based semantic similarity measure for argumentation schemes is investigated. With a range of experiments we demonstrate the strengths and weaknesses of the various methods and show that our methods can improve over our previous work. Our code is publicly available on GitHub.

Keywords: Argument graph similarity · Semantic textual similarity · Argument retrieval

1 Introduction

Argumentation is an increasingly important sub-field of Artificial Intelligence (AI). It involves various computational methods for reasoning and decision making, which are not only based on individual facts, but on coherent argumentative structures. The German special research program *Robust Argumentation Machines* (RATIO)[1] aims at developing new methods for extracting arguments from documents as well as new semantic models and ontologies for the deep representation of arguments which allows argument-based reasoning for various kinds of real-world problem solving. The major challenge is the development of so-called *argumentation machines* [27], which are specialized in reasoning with arguments. An argumentation machine could find supporting and opposing arguments for a user's topic or it could synthesize new arguments for an upcoming,

---

[1] http://www.spp-ratio.de/home/.

© Springer Nature Switzerland AG 2019
K. Bach and C. Marling (Eds.): ICCBR 2019, LNAI 11680, pp. 219–234, 2019.
https://doi.org/10.1007/978-3-030-29249-2_15

not yet well explored topic. Thereby it could support researchers, journalists, and medical practitioners in various tasks, overcoming the very limited support provided by traditional search engines used today.

In the ReCAP project [6], which is part of the RATIO program, we aim at combining methods from case-based reasoning (CBR), information retrieval (IR), and computational argumentation (CA) to contribute to the foundations of argumentation machines. In previous work [5], we developed an initial version of a similarity measure for arguments represented as argument graphs [7] for the purpose of case-based argument retrieval. This similarity measure was inspired by our own previous work on process-oriented CBR (POCBR), in which the similarity of graphs is assessed that represent semantically annotated workflows [4]. Argument graphs, however, are largely based on textual representations of claims and premises and thus require the use of textual similarity measures, thereby pushing this work closer to the sub-field of textual CBR [35]. While in our previous work, we only apply a standard word embedding technique for the assessment of local textual similarities, the aim of this paper is to explore a larger range of new approaches for semantic textual similarity measures (both supervised and unsupervised) used in the context of a graph-based similarity measure for argument graphs. In addition the use of an ontology-based semantic similarity measure for argumentation schemes is investigated.

Next, we present the foundations and related work in the field. Section 3 introduces our general approach for argument graph similarity as well as the spectrum of semantic textual similarity measures and the argumentation scheme similarity, which are the major contributions of this paper. The various methods and selected combinations of them are systematically evaluated in Sect. 4. Finally, Sect. 5 concludes the paper.

2   Foundations and Related Work

In the field of CA, an argument consists of a set of premises and a claim together with a rule of inference which concludes the claim from the premises. A premise can support or oppose a claim as well as an inference step. Together premises, claims, and inference steps form an argument graph. The Argument Interchange Format (AIF) standardizes such a graph representation for arguments [15] to be used in CA. In Fig. 1 an example of an argument graph in AIF format is given. Claims and premises are represented as information nodes (I-nodes), depicted as rectangular boxes which are related to each other via scheme nodes (S-nodes), depicted as rhombuses. In the example there are two arguments for a claim related to health insurance. The opposing argument has a single premise, whereas the supporting argument has two distinct premises. Argumentation schemes, corresponding to archetypical forms of arguments, are annotated as types of an argument. Here, the supporting argument has a type of *Position to Know*. The opposing argument has the type *Default Conflict*. There are many different argumentation schemes which cover diverse facets of argumentation [34], such as *Argument from Positive Consequence*, *Argument from Expert Opinion*, or *Argument from Cause to Effect*.

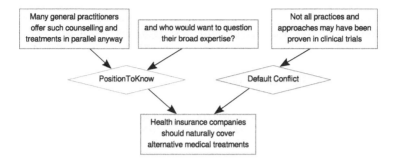

Fig. 1. Example of an argument graph in AIF format from the Microtexts corpus [25]

While argument mining methods [22] aim at converting natural language argumentative texts into such argument graphs, our work aims at supporting the reasoning with such graphs. Several formal argumentation frameworks currently exist which are based on formal logic, but we believe that they are of limited use for future argumentation machines that reason with real-world arguments [12]. Thus, we propose CBR as it does not require a complete and consistent domain theory and is able to make use of vague information. Thereby, we continue the traditional path for the use of textual CBR [35] in the context of argumentation for legal reasoning [3,11] and aim at linking it with ideas from POCBR and novel semantic text similarity approaches.

Existing work on CBR for legal argumentation is based on a model of legal argument. Cases are represented based on hierarchically structured factors or issues [29], which are used during similarity-based retrieval. A factor is similar to an argument or premise. The similarity of two arguments is defined by the commonalities and differences of the factors. CATO extends those argument graphs with intermediate factors, forming a factor hierarchy [1]. Branting [11] proposes case-based adaptation in legal reasoning by reusing and adapting justifications to create new arguments. Interestingly, similar ideas have been recently established in the field of CA such as the "recycling" of arguments for synthesis of claims [8].

3 Argument Graph Retrieval Using Semantic Textual Similarities

We now describe our approach to the representation of cases in the form of semantically labeled argument graphs, we recapitulate the basic approach for similarity assessment [5], and introduce the main enhancements by semantic textual similarity measures and the argumentation scheme ontology.

3.1 Argument Graph Representation

We developed a case representation using argument graphs, which is based on the graph representation of AIF. It is similar to text reasoning graphs [31] for representing causal information, but argument graphs contain in addition semantic information in different forms. Formally, an argument graph is a semantically labeled directed graph and represented as a tuple $A = (N, E, \tau, \lambda, t)$ [4]. N is the set of nodes and $E \subseteq N \times N$ is the set of directed edges connecting two nodes. $\tau : N \rightarrow \mathcal{T}$ assigns each node a type and $\lambda : N \rightarrow \mathcal{L}$ assigns each node a semantic description from a language $\mathcal{L}$. $t \in \mathcal{L}$ describes the overall topic of the argument represented in the graph. The types $\mathcal{T}$ follow the AIF standard [15] so that a node can either be an I-node with natural language propositional content or an S-node characterized by the respective argumentation scheme. The mapping function λ is used to link a semantic representation to a node. For an I-node n, $\lambda(n)$ is the original textual representation (possibly after the application of traditional pre-processing such as stopword removal) together with a semantic representation of this text in the form of a vector, produced by a sentence encoder (see Sect. 3.3). For an S-node n, $\lambda(n)$ corresponds to an argumentation scheme identifier, from an ontology of argumentation schemes constructed following the classification as proposed by Walton [33]. The argumentation scheme ontology is further used to define a local similarity measure for comparing two S-nodes, as described in Sect. 3.4. Finally, the overall topic t of an argument graph corresponds to the concatenated textual contents of all I-nodes as well as their semantic vector representation.

For retrieval, a case base of argument graphs is assumed, which could result from argument mining or from the manual transformation of text corpora. In our work, we also consider a query to be an argument graph or a fraction of it. In particular, a query can also consist only of a single I-node.

3.2 Argument Graph Similarity and Retrieval

The general principle of argument graph similarity and retrieval introduced by Bergmann et al. [5] has been adopted from POCBR [4] and follows the local-global principle [28]. The global similarity is computed from local node and edge similarities. The local node similarity $\text{sim}_N(n_q, n_c)$ of a node n_q from the query argument graph QA and a node n_c from the case argument graph CA is computed as follows:

$$\text{sim}_N(n_q, n_c) = \begin{cases} \text{sim}_I(n_q, n_c), & \text{if } \tau(n_q) = \tau(n_c) = \text{I-node} \\ \text{sim}_S(n_q, n_c), & \text{if } \tau(n_q) = \tau(n_c) = \text{S-node} \\ 0, & \text{otherwise} \end{cases} \qquad (1)$$

Approaches for concrete I-node and S-node similarity functions sim_I and sim_S are the main contribution of this paper and introduced in the next subsections. The similarity of two edges $\text{sim}_E(e_q, e_c)$ is the average of the similarities of their endpoints l and r respectively:

$$\text{sim}_E(e_q, e_c) = 0.5 \cdot (\text{sim}_N(e_q.l, e_c.l) + \text{sim}_N(e_q.r, e_c.r)) \qquad (2)$$

To construct a global similarity value, an admissible mapping m is applied which maps nodes and edges from QA to CA, such that only nodes of the same type

(I-nodes to I-nodes and S-nodes to S-nodes) are mapped. Edges can only be mapped if the nodes they link are mapped as well by m. For a given mapping m let sn_i be the node similarities $\text{sim}_N(n_i, m(n_i))$ and se_i the edge similarities $\text{sim}_E(e_i, m(e_i))$. The similarity for a query graph QA and a case graph CA given a mapping m is the normalized sum of the node and edge similarities.

$$\text{sim}_m(QA, CA) = \frac{sn_1 + \cdots + sn_n + se_1 + \cdots + se_m}{n_N + n_E} \tag{3}$$

Finally, the similarity of QA and CA is the similarity of an optimal mapping m, which can be computed using an A^* search [4].

$$\text{sim}(QA, CA) = \max_m \{\text{sim}_m(QA, CA) \mid m \text{ is admissible}\} \tag{4}$$

For similarity-based retrieval of argument graphs from a case base a linear retrieval approach should be avoided due to unacceptable retrieval times caused by the complexity of A^* search as well as the complexity of the involved node similarity measures. Thus, we propose a MAC/FAC *(many are called, but few are chosen)* approach [17], which divides the retrieval into an efficient pre-filter stage (MAC phase) and the subsequent FAC phase, in which only the a few filtered cases are assessed using the complex similarity measure. We proposed a MAC/FAC approach for argument graphs in which the MAC phase is implemented as a linear similarity-based retrieval of the cases based only on the semantic similarity of the topic vector t [5]. The filter selects the k most similar cases, which are passed over to the FAC phase which implements the ranking by a linear assessment of the cases using the graph-based similarity as described above.

3.3 Semantic Textual Similarity Measures for I-Node Similarity

The quality of the overall similarity assessment heavily depends on the applied node similarity measures. In our previous work we only employed Word2vec Skip-gram [23] embeddings aggregated with an arithmetic mean and compared with a cosine similarity. In this paper we investigate a larger, more diverse set of novel methods for semantic textual similarity based on neural networks. The approaches include unsupervised word and sentence embeddings and their combination as well as supervised sentence embeddings which are trained on a large amount of training data. There are also other methods available like SIF [2] or Skip-Thought vectors [19] which however will not be evaluated here.

Unsupervised Word Embeddings. Word embeddings are distributed representations of words, which means each word is associated with a word vector. Word vectors capture the semantics of a word, in the sense that similar words have similar word vectors. Word embedding models are trained on textual data in an unsupervised manner. The models rely on the distributional hypothesis, namely that words in similar contexts share meaning.

Word2vec Skip-gram (WV) [23] trains word vectors based on the prediction of context words. The model architecture employs a softmax classificator and maximizes the log likelihood of the word vectors based on (word, context) pairs. Words appearing in similar contexts have therefore similar word vectors. For performance reasons the softmax is replaced by either a hierarchical softmax or an alternative negative sampling objective [24]. The fastText (FT) embedding [9] is based on the Skip-gram model. In addition it uses subword information as each word is represented as the sum of its character n-grams together with the word. A vector for n-grams is learned which allows to build word representations for previously unseen words. GloVe (GL) [26] learns word vectors from global corpus statistics directly, in contrast to Skip-gram's context window approach. An objective function based on ratios of co-occurrence probabilities is maximized.

In order to assess the similarity of I-nodes, the individual embeddings of the words in the node's text have to be aggregated to an overall node embedding, based on which the similarity can be assessed, e.g. by a cosine measure. Traditional unsupervised aggregation methods for this task include arithmetic mean ($\overline{x}_a$), median ($\overline{x}_m$), geometric mean ($\overline{x}_g$), min pooling ($\min x$), max pooling ($\max x$) and p-means ($\overline{x}_p$) [30].

Unsupervised Sentence Embeddings. Sentence embedding methods are an alternative approach that can be applied to assess the similarity of the I-nodes based on their text. As they work on sentences rather than on words, no aggregation is needed. The Distributed Memory Model of Paragraph Vectors (DV) [21] is such a method trained similarly to word2vec's CBOW model [23], but with an additional vector representing the sentence as a whole. Embeddings for previously unseen sentences are inferred by backpropagation on the paragraph vector keeping all other parameters fixed.

Supervised Sentence Embeddings. While the previous embedding approaches are purely unsupervised, several approaches exist which aim at improving the similarity assessment including to a certain degree also supervised learning, thereby accepting the additional effort caused by labeled training data. InferSent [16] is one such approach trained on the Stanford Natural Language Inference corpus [10]. During training a shared BiLSTM encodes two sentences and the encoded sentence pair is further enhanced with additional features, such as the absolute difference of both sentences and their element-wise product, before it is passed through a feed-forward network for classification. After training the BiLSTM yields a 4096 dimensional vector for a sentence. Universal Sentence Encoder [13] trains a sentence encoder on multiple unsupervised and supervised tasks. The transformer-based variant (USE-T) uses a transformer encoder [32]. Deep Average Network-based Universal Sentence Encoder (USE-D) uses a Deep Average Network encoder [18] instead, which averages unigram- and bigram-embeddings and passes the averaged value through a feed-forward network. The output of both networks is a 512 dimensional vector, representing a sentence.

Combining Different Embeddings. The various embeddings just described can also be combined, following the idea that each embedding type captures different kinds of information [30]. The concatenation of two embeddings A and B is denoted by A $\oplus$ B, resulting in an embedding with dimension dim(A) + dim(B). For example WV $\oplus$ FT is the concatenation of WV and FT embeddings.

Similarity Measures for I-Nodes. In order to assess the similarity of I-nodes, a similarity measure is required which compares the computed embedding vectors of the nodes. Traditionally, the cosine similarity is used in semantic textual similarity tasks, but various alternative approaches exist. The MaxPool-Jaccard approach applies the fuzzy Jaccard index to max pooled word embeddings and has recently demonstrated a significant benefit in semantic textual similarity tasks [37]. In addition, the DynaMax-Jaccard approach was proposed, which is a completely unsupervised and non-parametric similarity measure that dynamically extracts and max-pools good features.

Finally, I-node similarity can be computed using the Word Mover's Distance (WMD) [20] which computes the distance of two sentences by a mapping between the word embeddings of the sentences. An optimal mapping is found by taking into account the distances of the words in a word embedding space, so that each word in one sentence needs to travel the lowest distance to the words in the other sentence.

Please note that WMD, DynaMax and MaxPool do not operate on the representation of the whole node text but on the representation of the individual words. As such they combine aggregation and similarity assessment.

3.4   Ontology-Based Similarity Measure for S-Node Similarity

We now introduce an approach with which we aim to improve the similarity assessment of argument graphs by considering the semantics of the argumentation schemes used in the S-nodes of the graph. In our previous work [5] we only used two different schemes and an exact match similarity. We now introduce a more fine grained representation and created an ontology consisting of 38 argumentation schemes which are arranged in a taxonomy based on a classification of argumentation schemes [33]. Figure 2 shows an excerpt of this ontology.

The similarity between two schemes can then be computed by using an edge-count based approach. Wu and Palmer introduce a similarity measure sim_{wp} that considers the depth of schemes S_1, S_2 and the closest common predecessor scheme S_x of S_1 and S_2. The Wu and Palmer similarity of two argumentation schemes S_1 and S_2 [36] is given by

$$\text{sim}_{wp}(S_1, S_2) = \frac{2N_x}{N_1 + N_2} \tag{5}$$

where N_1, N_2 and N_x describe the depth of the schemes S_1, S_2 and S_x respectively in terms of edges from root element to scheme. Wu and Palmer similarity

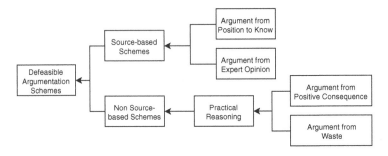

Fig. 2. Excerpt from the argumentation scheme ontology based on a classification of argumentation schemes [33]

assumes that schemes located deeper in the ontology are more specific and therefore more similar.

4   Experimental Evaluation

Given the various approaches described so far for I-node similarity as well as the advanced approach for S-node similarity, we now want to experimentally evaluate the benefit of them. Thus, we performed a systematic evaluation to test how well the various approaches are able to retrieve and rank cases in a way that is in line with the assessment of a human expert. Our code is publicly available on GitHub[2].

4.1   Hypotheses

The following four hypotheses are subject of this evaluation and relate to the quality of the ranking produced by the argument graph similarity.

- **H1:** The simple approach based on WV embeddings, mean aggregation, and cosine similarity as investigated in previous work [5] can be improved by some of the newly investigated methods.
- **H2:** The concatenation of embeddings achieves a higher quality than a single embedding.
- **H3:** Supervised sentence embeddings achieve a higher quality than unsupervised embeddings.
- **H4:** The use of argumentation scheme similarity improves the quality.

4.2   Experimental Setup

For the evaluation we rely on various pre-trained word embeddings and sentence encoder models. Word2vec GoogleNews[3] vectors are trained on the Google News

---

[2] https://github.com/MirkoLenz/ReCAP-Argument-Graph-Retrieval.

[3] https://code.google.com/archive/p/word2vec/.

dataset on about 100B tokens. GloVe[4] is trained on the Common Crawl dataset on 840B tokens. fastText[5] vectors are trained on Wikipedia and Common Crawl. The dimensionality of all word embeddings is 300. The Paragraph Vector model was trained by us on the english Wikipedia dump with 1M tokens. The Universal Sentence Encoder models[6,7] are trained on multiple unsupervised and supervised tasks, such as predicting context sentences [19] and classification on the SNLI corpus. InferSent[8] is trained on the SNLI corpus as well. We evaluate the model in version 1.

For the evaluation of the retrieval we choose the annotated corpus of argumentative microtexts [25] following the work in [5]. This corpus consists of 112 argument graphs with a total of 576 I-nodes and 443 S-nodes. For this paper, we refined these argument graphs by introducing a more fine-grained annotation of the S-nodes by argumentation schemes based on the ontology developed. These refinements were made by two students who were experienced in AIF and the OVA+ modelling tool[9]. For our evaluation, we used the same 24 queries from 6 topics as in our previous work. However, due to the introduction of the more detailed argumentation scheme representation, a new reference ranking was needed. It was produced by the same experienced students who refined the representation of the cases.

In our experiments, we used various combinations of the similarity methods proposed for retrieval of cases. In each experiment all 24 queries are used and the resulting $k = 10$ most similar cases are considered. We assessed the relevance of the cases (i.e. whether a case deals with the same topic as the query) as well as the ranking of the cases. Thereby, the similarity measures are evaluated by means of various metrics. Precision (P) measures the fraction of relevant cases retrieved within the set of 10 most similar cases. Due to the size of the reference rankings in our experiment, the upper limit for P achievable is 0.717. Recall (R) measures the fraction of relevant cases retrieved. P and R are set-based,i.e., the ranking quality itself is not assessed.

Average Precision (AP) measures the quality of the ranking by averaging the precision at all relevant positions. Thus, AP is the area under the precision-recall curve. A high AP value indicates that relevant elements are ranked high as well.

Normalized Discounted Cumulative Gain (NDCG) assesses that elements with a high relevance appear early in the ranking. NDCG is computed as the normalized sum of all relevance values in the result giving lower positions in the ranking less weight. It is noteworthy that for NDCG non-relevant elements in the ranking have no influence on the metric.

Correctness (CR) and Completeness (CP) [14] explicitly assess how well the ordering of the ranking produced by similarity matches the ordering of the refer-

---

[4] https://nlp.stanford.edu/projects/glove/.
[5] https://fasttext.cc/.
[6] https://tfhub.dev/google/universal-sentence-encoder-large/3.
[7] https://tfhub.dev/google/universal-sentence-encoder/2.
[8] https://github.com/facebookresearch/InferSent.
[9] http://ova.arg-tech.org/.

ence ranking. CP measures the percentage of pairs of the reference ranking that are included the produced ranking. CR measures concordance/disconcordance of the orderings of those pairs, resulting in a value from $[-1, 1]$ with higher values indicating higher concordance. It is important to note that CR values are only meaningful if also the CP value is high. Thus we only interpret CR values if the CP value is above 0.9.

In the following we always report values averaged over all queries. In addition, we show the average retrieval time in seconds on a 2014 MacBook Pro 15" with a 2.8 GHz Intel Core i7 processor and 16 GB RAM.

4.3   Results and Discussion

In the following experiments the similarity measures for I-nodes are evaluated. Only stopword removal is consistently performed in all conditions as this was the most successful pre-processing approach in our previous work. The S-node similarity measure is evaluated lastly.

The first experiment evaluates WV embeddings together with the cosine measure as in our previous work, but using various aggregation functions. The results are shown in Table 1, while the abbreviations are used as introduced in Sects. 3.3 and 4.2.

Table 1. Results of retrieval with WV and cosine using different aggregation functions.

Aggregation method	Time	P	R	AP	NDCG	CR	CP
$\overline{x}_a$	10.625	**0.692**	**0.965**	**0.924**	0.834	0.106	**0.956**
$\overline{x}_m$	11.021	0.675	0.943	0.903	0.844	**0.139**	0.907
$\overline{x}_g$	6.311	0.017	0.021	0.003	0.053	–	0.0
$\min x$	10.515	0.604	0.84	0.798	0.856	0.171	0.744
$\max x$	12.687	0.588	0.827	0.786	**0.866**	0.127	0.696
$\overline{x}_2$	9.663	0.65	0.908	0.836	0.825	0.14	0.846
$\overline{x}_3$	8.059	0.654	0.913	0.876	0.849	0.146	0.876
$\overline{x}_5$	9.932	0.608	0.853	0.805	0.84	0.064	0.734
$\overline{x}_{10}$	8.831	0.575	0.81	0.746	0.843	0.174	0.676
$\overline{x}_{1000}$	6.969	0.479	0.667	0.584	0.802	0.184	0.471
$\overline{x}_a \oplus \overline{x}_m$	11.676	**0.692**	**0.965**	0.923	0.835	0.115	0.948
$\overline{x}_a \oplus \overline{x}_m \oplus \overline{x}_2 \oplus \overline{x}_3$	11.52	**0.692**	**0.965**	0.918	0.841	0.116	0.952

Arithmetic mean performs best regarding the unranked measures P and R and also w.r.t. AP. For the ranked measures, max pooling led to the best NDCG value, but for a significantly lower recall. Median achieves best results for the ranked measure correctness among all aggregations with a completeness above 0.9. We systematically evaluated also concatenations of aggregation functions without being able to outperform the individual methods. The two best concatenation results are shown in the last two rows of Table 1.

Next we evaluate all unsupervised and supervised embedding methods using cosine similarity. Word embeddings are aggregated using arithmetic mean. The concatenation of the unsupervised embeddings have also been evaluated systematically, while only the best results are reported. Table 2 shows the results.

Table 2. Results of retrieval with different embedding methods using cosine and arithmetic mean.

Embedding method	Time	P	R	AP	NDCG	CR	CP
DV	12.132	0.675	0.942	0.888	0.854	0.148	0.9
FT	11.312	0.675	0.943	0.909	0.847	0.141	0.914
GL	12.201	0.667	0.929	0.876	0.809	0.044	0.897
WV	10.625	0.692	0.965	0.924	0.834	0.106	0.956
DV $\oplus$ WV	11.737	0.696	0.97	0.934	0.855	0.097	0.958
DV $\oplus$ FT $\oplus$ WV	12.489	0.671	0.938	0.905	0.846	0.085	0.904
InferSent	40.125	0.683	0.95	0.915	**0.864**	**0.184**	0.908
USE-D	8.92	0.704	0.982	0.951	0.841	0.099	0.977
USE-T	13.785	**0.713**	**0.994**	**0.972**	0.848	0.12	**0.992**

All methods achieve a high recall and completeness. Among the unsupervised methods WV achieves the best results w.r.t. the unranked measures. DV and concatenations including DV achieve the best ranked results NDCG and CR. The supervised methods further improve the results. USE-D and USE-T yield the highest P, R, and AP scores. InferSent was best w.r.t. the ranked measures, but was even worse than WV concerning P and R. This indicates that supervised methods can actually learn useful signals for semantic textual similarity. Hypothesis H3 can thus be accepted.

The impact of the similarity measure on the retrieval quality was evaluated next. WV with arithmetic mean is used as sentence embeddings. Table 3 shows the results for the different similarity measures.

Table 3. Results of retrieval with WV while using different similarity measures.

Similarity method	Time	P	R	AP	NDCG	CR	CP
Cosine	10.625	**0.692**	**0.965**	0.924	0.834	0.106	**0.956**
DynaMax-Jaccard	12.276	**0.692**	0.964	**0.934**	**0.877**	**0.274**	0.936
MaxPool-Jaccard	9.725	0.417	0.58	0.548	0.846	*0.34*	*0.365*
WMD	88.377	0.683	0.953	0.913	0.859	0.226	0.932

Table 4. Results of retrieval with different embedding methods using DynaMax-Jaccard.

Method	Time	P	R	AP	NDCG	CR	CP
DV	11.241	0.633	0.888	0.842	0.867	0.286	0.767
FT	10.497	**0.696**	**0.971**	0.93	0.868	0.256	**0.956**
GL	12.664	0.688	0.959	0.917	0.868	0.217	0.918
WV	12.276	0.692	0.964	**0.934**	0.877	0.274	0.936
DV $\oplus$ FT	13.236	0.688	0.959	0.914	0.869	0.277	0.918
DV $\oplus$ GL	14.077	0.667	0.931	0.891	**0.883**	0.274	0.812
DV $\oplus$ WV	13.58	0.667	0.929	0.893	0.862	0.258	0.85
FT $\oplus$ GL	16.772	0.675	0.943	0.907	0.872	0.278	0.899
FT $\oplus$ WV	12.17	**0.696**	0.970	0.924	0.862	0.27	0.943
GL $\oplus$ WV	15.25	0.679	0.949	0.903	0.867	0.192	0.853
DV $\oplus$ FT $\oplus$ GL	17.902	0.663	0.926	0.886	0.881	0.328	0.815
DV $\oplus$ FT $\oplus$ WV	15.251	0.692	0.964	0.922	0.875	**0.307**	0.943
DV $\oplus$ GL $\oplus$ WV	15.55	0.679	0.947	0.905	0.877	0.304	0.872
FT $\oplus$ GL $\oplus$ WV	16.721	0.671	0.938	0.897	0.873	0.264	0.832
DV $\oplus$ FT $\oplus$ GL $\oplus$ WV	18.551	0.679	0.948	0.908	0.868	0.222	0.888

Cosine and DynaMax-Jaccard perform comparably well only on the unranked measures, while DynaMax-Jaccard significantly improves the ranking results NDCG and CR compared to cosine. WMD achieves nearly comparable results, but leads to very high retrieval times and is thus not competitive. MaxPool-Jaccard is very poor on R and CP and thus not useful.

Since the DynaMax-Jaccard similarity performed best we evaluated the various unsupervised embeddings methods and their combinations again. The supervised methods could not be evaluated here, since DynaMax-Jaccard works on word embeddings and supervised methods yield sentence embeddings. The results are shown in Table 4.

Interestingly FT embeddings perform now best w.r.t. the unranked measures and are also very high in AP. Overall concatenations are able to improve the ranking quality. DV $\oplus$ FT $\oplus$ WV yields the strongest CR and very high NDCG and can even improve over the results of the supervised methods using the cosine measure. Therefore hypothesis H2 can be accepted at least for DynaMax-Jaccard. It is noteworthy that all metrics show slightly higher values than for cosine, especially correctness and NDCG (compare Tables 2 and 4). This indicates that DynaMax-Jaccard generally leads to an improved ranking.

The use of argumentation schemes for retrieval is evaluated next. Supervised embeddings are compared using cosine, unsupervised embeddings using DynaMax-Jaccard. Concerning S-node similarity, three variants are included: no S-node similarity (always 1), exact match similarity using the argumentation scheme labels at the S-nodes and the ontology similarity (see Sect. 3.4). Table 5

Table 5. Results of retrieval with including argumentation scheme similarity and selected embedding methods

Embedding	Schemes	Time	AP	NDCG	CR	CP
USE-T	No	13.785	**0.972**	**0.848**	0.12	0.992
USE-T	Exact Match	15.541	0.925	0.843	**0.136**	0.992
USE-T	Onto. Sim	14.127	0.938	0.847	0.132	0.992
WV	No	12.276	**0.934**	**0.877**	**0.274**	0.936
WV	Exact Match	13.659	0.906	0.853	0.161	0.936
WV	Onto. Sim	10.677	0.902	0.851	0.174	0.936
DV $\oplus$ FT $\oplus$ WV	No	15.251	**0.922**	**0.875**	**0.307**	0.943
DV $\oplus$ FT $\oplus$ WV	Exact Match	19.83	0.908	0.859	0.252	0.943
DV $\oplus$ FT $\oplus$ WV	Onto. Sim	27.096	0.905	0.861	0.216	0.943

Table 6. Evaluation of the approach used in [5] compared to the best new methods.

Method	Time	P	R	AP	NDCG	CR	CP
Paper [5]	10.625	0.692	0.965	0.924	0.834	0.106	0.956
USE-T	13.785	**0.713**	**0.994**	**0.972**	0.848	0.12	**0.992**
WV	12.276	0.692	0.964	0.934	**0.877**	0.274	0.936
WV $\oplus$ FT $\oplus$ DV	15.251	0.692	0.964	0.922	0.875	**0.307**	0.943

presents the results. Since the argumentation schemes are used only in the FAC phase only the ranked metrics are affected and reported.

For USE-T, the use of the argumentation scheme labels slightly improves the ranking CR. The ontology-based similarity measures does not lead to an improvement for any embedding, it even worsens the ranking results. Thus, hypothesis H4 has to be rejected.

To come to a concluding assessment of hypothesis H1, we compare the three best methods, USE-T, WV, and DV $\oplus$ FT $\oplus$ WV against the approach in [5] (see Table 6). Again the supervised embedding is compared using cosine similarity and unsupervised embeddings using DynaMax-Jaccard. Argumentation schemes are not used.

All three methods clearly improve on the baseline. USE-T has best P, R, AP as well as CP and can reach also near the precision limit of 0.717. WV achieves very good results with minimal complexity. DV $\oplus$ FT $\oplus$ WV achieves the best CR score. Hypothesis H1 can thus be clearly accepted. Concerning the retrieval time, the new best methods are clearly more time consuming (up to 50 %), but we consider this as acceptable given the resulting quality improvements.

5   Conclusion and Future Work

In this work we investigated new methods from semantic textual similarity for improved case-based argument retrieval and demonstrated significant improvements over our own previous results [5]. Unsupervised word embeddings and concatenations achieve a good ranking quality using the DynaMax-Jaccard similarity measure and can improve clearly on the cosine similarity measure. Supervised methods achieve the best results using the unranked metrics and the highest completeness measures. The similarity measures for argumentation schemes cannot further improve these results. A possible reason could be that the use of schemes yields in too many constraints when performing the graph mapping and thus impairing the results.

In future work we want to improve the ranking quality of supervised methods as well as explore more advanced ontological similarity measures by automatic linking with domain specific ontologies. Another line of work would be to extend the argument retrieval task to new benchmark corpora and in particular corpora in German language. A big challenge is addressing semantic similarity for the German language as most recent methods have been mainly investigated and optimized for the English language. Additionally, we will look at reducing the computational complexity of the mapping algorithm, especially the A* search. Finally, we intend to move further on to the adaptation of argument graphs by transferring compositional adaptation methods from POCBR.

Acknowledgments. This work was funded by the German Research Foundation (DFG), project 375342983.

References

1. Aleven, V.: Teaching Case-Based Argumentation Through a Model and Examples. Ph.D. thesis, University of Pittsburgh (1997)
2. Arora, S., Liang, Y., Ma, T.: A simple but though baseline for sentence embeddings (2017)
3. Ashley, K.D.: Modelling Legal Argument: Reasoning with Cases and Hypotheticals. Ph.D. thesis, University of Massachusetts (1988)
4. Bergmann, R., Gil, Y.: Similarity assessment and efficient retrieval of semantic workflows. Inf. Syst. **40**, 115–127 (2014). https://doi.org/10.1016/j.is.2012.07.005
5. Bergmann, R., Lenz, M., Ollinger, S., Pfister, M.: Similarity measures for case-based retrieval of natural language argument graphs in argumentation machines. In: Proceedings of the 32nd International Florida Artificial Intelligence Research Society Conference, FLAIRS 2019, Sarasota, Florida, USA. AAAI-Press (2019)
6. Bergmann, R., Schenkel, R., Dumani, L., Ollinger, S.: ReCAP - information retrieval and case-based reasoning for robust deliberation and synthesis of arguments in the political discourse. In: Proceedings of the Conference "Lernen, Wissen, Daten, Analysen", LWDA, vol. 2191. CEUR-WS.org (2018)
7. Bex, F., Prakken, H., Reed, C.: A formal analysis of the AIF in terms of the aspic framework. In: Proceedings of COMMA, pp. 99–110. IOS Press (2010)

8. Bilu, Y., Slonim, N.: Claim synthesis via predicate recycling. In: Proceedings of 54th Annual Meeting of the Association for Computational Linguistics (ACL) (2016)
9. Bojanowski, P., Grave, E., Joulin, A., Mikolov, T.: Enriching word vectors with subword information (2016). https://arxiv.org/abs/1607.04606
10. Bowman, S.R., Angeli, G., Potts, C., Manning, C.D.: A large annotated corpus for learning natural language inference (2015). https://arxiv.org/abs/1508.05326
11. Branting, K.: A reduction-graph model of precedent in legal analysis. Artif. Intell. **150**(1), 59–95 (2003)
12. Caminada, M., Wu, Y.: On the limitations of abstract argumentation. In: Proceedings of the 23rd Benelux Conference on Artificial Intelligence (2011)
13. Cer, D., et al.: Universal sentence encoder (2018). http://arxiv.org/abs/1803.11175
14. Cheng, W., Rademaker, M., De Baets, B., Hüllermeier, E.: Predicting partial orders: ranking with abstention. In: Balcázar, J.L., Bonchi, F., Gionis, A., Sebag, M. (eds.) ECML PKDD 2010. LNCS (LNAI), vol. 6321, pp. 215–230. Springer, Heidelberg (2010). https://doi.org/10.1007/978-3-642-15880-3_20
15. Chesñevar, C., et al.: Towards an argument interchange format. Knowl. Eng. Rev. **21**(4), 293–316 (2006)
16. Conneau, A., Kiela, D., Schwenk, H., Barrault, L., Bordes, A.: Supervised learning of universal sentence representations from natural language inference data (2017). https://arxiv.org/abs/1705.02364
17. Forbus, K.D., Gentner, D., Law, K.: MAC/FAC - A model of similarity-based retrieval. Cognit. Sci. **19**(2), 141–205 (1995)
18. Iyyer, M., Manjunatha, V., Boyd-Graber, J., Daumé III, H.: Deep unordered composition rivals syntactic methods for text classification (2015). https://doi.org/10.3115/v1/P15-1162
19. Kiros, R., et al.: Skip-thought vectors (2015). https://arxiv.org/abs/1506.06726
20. Kusner, M.J., Sun, Y., Kolkin, N.I., Weinberger, K.Q.: From word embeddings to document distances. In: Proceedings of the 32nd ICML, vol. 37, pp. 957–966. JMLR.org (2015)
21. Le, Q., Mikolov, T.: Distributed representations of sentences and documents. In: Proceedings of the 31st ICML, vol. 32, pp. II-1188–II-1196. JMLR.org (2014)
22. Lippi, M., Torroni, P.: Argument mining from speech: detecting claims in political debates. In: Proceedings of 13th AAAI Conf on Artificial Intelligence. AAAI Press (2016)
23. Mikolov, T., Chen, K., Corrado, G., Dean, J.: Efficient estimation of word representations in vector space (2013). https://arxiv.org/abs/1301.3781
24. Mikolov, T., Sutskever, I., Chen, K., Corrado, G., Dean, J.: Distributed representations of words and phrases and their compositionality (2013). https://arxiv.org/abs/1310.4546
25. Peldszus, A., Stede, M.: An annotated corpus of argumentative microtexts. In: First European Conference on Argumentation, Portugal, Lisbon (2015)
26. Pennington, J., Socher, R., Manning, C.: Glove: global vectors for word representation. In: Proceedings of EMNLP (2014). https://doi.org/10.3115/v1/D14-1162
27. Reed, C., Norman, T.J. (eds.): Argumentation Machines, New Frontiers in Argument and Computation, Argumentation Library, vol. 9. Springer, Dordrecht (2004). https://doi.org/10.1007/978-94-017-0431-1
28. Richter, M.M., Weber, R.O.: Case-Based Reasoning - A Textbook. Springer, Heidelberg (2013). https://doi.org/10.1007/978-3-642-40167-1
29. Rissland, E.L., Ashley, K.D., Branting, K.: Case-based reasoning and law. Knowl. Eng. Rev. **20**(3), 293–298 (2005)

30. Rücklé, A., Eger, S., Peyrard, M., Gurevych, I.: Concatenated p-mean word embeddings as universal cross-lingual sentence representations (2018). https://arxiv.org/abs/1803.01400
31. Sizov, G., Öztürk, P., Štyrák, J.: Acquisition and reuse of reasoning knowledge from textual cases for automated analysis. In: Lamontagne, L., Plaza, E. (eds.) ICCBR 2014. LNCS (LNAI), vol. 8765, pp. 465–479. Springer, Cham (2014). https://doi.org/10.1007/978-3-319-11209-1_33
32. Vaswani, A., et al.: Attention is all you need. In: Advances in Neural Information Processing Systems, vol. 30, pp. 5998–6008 (2017)
33. Walton, D., Macagno, F.: A classification system for argumentation schemes. Argum. Comput. **6**(3), 219–245 (2015)
34. Walton, D., Reed, C., Macagno, F.: Argumentation Schemes. Cambridge University Press, Cambridge (2008)
35. Weber, R.O., Ashley, K.D., Brüninghaus, S.: Textual case-based reasoning. Knowl. Eng. Rev. **20**(03), 255–260 (2005)
36. Wu, Z., Palmer, M.: Verbs semantics and lexical selection. In: Proceedings of the 32nd Annual Meeting on Association for Computational Linguistics (1994)
37. Zheleniak, V., Savkov, A., Shen, A., Moramarco, F., Flann, J., Hammerla, N.Y.: Don't settle for average go for the max: fuzzy sets and max-pooled word vectors. In: International Conference on Learning Representations (2019)

An Approach to Case-Based Reasoning Based on Local Enrichment of the Case Base

Yves Lepage[1(✉)] and Jean Lieber[2(✉)]

[1] Waseda University, IPS, 2-7 Hibikino, Kitakyushu 808-0135, Japan
yves.lepage@waseda.jp
[2] Université de Lorraine, CNRS, Inria, LORIA, 54000 Nancy, France
jean.lieber@loria.fr

Abstract. This paper describes an approach to case-based reasoning by which the case base is enriched at reasoning time. Enrichment results from the local application of variations to seed cases: new hypothetical cases are created which get closer and closer to the target problem. The creation of these hypothetical cases is based on structures associated to the problem and solution spaces, called variation spaces, that enable defining a language of adaptation rules. Ultimately reaching the target problem (exactly or nearly) allows the system to deliver a solution. Application of the proposed approach to machine translation shows behind state-of-the-art, but promising results.

Keywords: Analogical reasoning · Case base enrichment ·
Case-based machine translation · Case-based reasoning

1 Introduction

In machine learning, some techniques are used to enrich the training set in order to improve the accuracy of a learning system. This is called data augmentation. It can be done using general transformations (flipping, rotating, etc. images when this does not affect the class in image classification, see for example Taylor and Nitschke [17]), adding some controlled noise (like Gaussian noise on images, see Hussain et al. [7]), or by analogical reasoning (see, e.g. Couceiro et al. [2]).

While case-based reasoning (CBR, Richter and Weber [16]) usually requires less data than most current machine learning techniques, the enrichment of the case base can be useful. In this paper, we propose to perform the enrichment of the case base in a "case-based way": the case base is enriched with new source cases that are "around" the target problem. This constitutes a local enrichment

The authors want to thank the anonymous reviewers for their detailed comments. They have tried to do their best to take them into account. The first author is supported by JSPS Grant-In-Aid 18K11447: "Self-explainable and fast-to-train example-based machine translation using neural networks".

K. Bach and C. Marling (Eds.): ICCBR 2019, LNAI 11680, pp. 235–250, 2019.
https://doi.org/10.1007/978-3-030-29249-2_16

of the case base, computed on-line, at the time of problem-solving. The case base is enriched with the help of learned adaptation rules: given a source case, another source case closer to the target problem is generated thanks to such a rule. The process can be repeated, though each application of a rule may degrade the generated case, in the sense that it is less and less likely to be a licit case.

To deal with this issue, the notion of penalized case is introduced: this is a hypothetical case, whose likelihood to be licit is characterized by a number that penalizes the re-usability of the case. Therefore, the enriched case base is a set of penalized cases: the cases to be reused are the ones which offer the best compromise between the similarity to the target problem and their penalties.

Section 2 presents general definitions. The notions of penalized cases are presented in Sect. 3. The notions of variations between problems and between solutions are defined in Sect. 4, with some strong assumptions that make the general ideas easier to understand. Section 5 is the core of the paper: it presents the approach for local enrichment of the case base. It presents a first approach consisting in generating from the case base all the penalized cases with a penalty under a given threshold. Since this approach leads to a combinatorial explosion of the case base, a more practical approach is presented afterwards. They show how case-based translation (Lepage and Lieber [13]) can be performed thanks to the approach of local enrichment of the case base. The results are demonstrative and promising.

2   Preliminaries

2.1   Notations and Assumptions on Case-Based Reasoning

Let $\mathcal{P}$ and $\mathcal{S}$ be two given sets, called the *problem space* and the *solution space*. A *problem* is by definition an element x of $\mathcal{P}$ and a *solution*, an element y of $\mathcal{S}$. A relation $\rightsquigarrow$ on $\mathcal{P} \times \mathcal{S}$ is assumed to exist and $x \rightsquigarrow y$ is read "x has for solution y" or "y is a solution to x". A case is a pair $(x, y) \in \mathcal{P} \times \mathcal{S}$ such that $x \rightsquigarrow y$. The *case base* CB is a finite and nonempty set of cases. A *source case* is an element (x^s, y^s) of CB. CBR aims at solving a new problem x^t called the *target problem* with the help of the case base. It usually consists of the following steps:

- Retrieval (aka "retrieve" in Aamodt and Plaza [1]) selects a subset of CB;
- Adaptation (aka "reuse") proposes a *plausible* solution y^t to x^t, using the retrieved source cases;
- Learning (aka as "revise" and "retain") consists in validating/correcting y^t (for example, with the help of a human expert) and in storing the newly formed case (x^t, y^t) in the case base, if this storage is judged appropriate.

It is worth noting that, for many applications, y^t is only a plausible solution: CBR often functions as a hypothetical reasoning process whose use is motivated by the incompleteness of the knowledge of the relation $\rightsquigarrow$. The term "hypothetical case" stands for any pair $(x, y) \in \mathcal{P} \times \mathcal{S}$, though this notion is generally used when a solution y to x is plausibly inferred. By contrast, if (x, y) is a case—and thus, $x \rightsquigarrow y$—it is called a *licit case* to stress its certainty.

In some applications, other knowledge containers are used (Richter and Weber [16]): the domain knowledge, the retrieval knowledge, and the adaptation knowledge (AK). These four knowledge containers are interrelated. In particular, there are studies on learning AK using CB presented further.

2.2   Analogies

An analogy is a quaternary relation on a set $\mathcal{U}$ denoted by $A : B :: C : D$ for $(A, B, C, D) \in \mathcal{U}^4$, that is to be read "$A$ is to B as C is to D". It satisfies the following postulates (for any $(A, B, C, D) \in \mathcal{U}^4$):[1]

(**reflexivity of conformity**) $A : B :: A : B$;
(**symmetry of conformity**) if $A : B :: C : D$ then $C : D :: A : B$;
(**exchange of the means**) if $A : B :: C : D$ then $A : C :: B : D$.

A *ratio* is an expression of the form $P : Q$ ("P is to Q"), the relation $::$ ("as") is called conformity. Thus, analogy is a conformity of ratios.[2]

Classical examples of analogies are as follows:

(**geometrical analogy**) Here the ratio is division, conformity is equality, and $\mathcal{U} = \mathbb{R} \setminus \{0\}$. This analogy is defined by $A : B :: C : D$ if $\frac{A}{B} = \frac{C}{D}$.
(**arithmetic analogy**) Here the ratio is subtraction, conformity is equality, and $\mathcal{U} = \mathbb{R}$. This analogy is defined by $A : B :: C : D$ if $A - B = C - D$.
(**analogy on tuples**) If an analogy is defined on each set $\mathcal{U}_i$ $(1 \leq i \leq n)$ and $\mathcal{U} = \mathcal{U}_1 \times \mathcal{U}_2 \times \ldots \times \mathcal{U}_n$ then the following analogy can be defined on $\mathcal{U}$: $A : B :: C : D$ if for every i, $A_i : B_i :: C_i : D_i$.
(**analogies on strings**) Let dist be the LCS distance.[3] The Parikh vector of a string is the tuple of the number of occurrences of each character in the string. A ratio $P : Q$ between two strings P and Q can be defined as the difference of their Parikh vectors (arithmetic analogies on tuples) plus the LCS distance between them. However, this ratio does not entail the exchange of the means because $\text{dist}(A, B) = \text{dist}(C, D)$ does not imply $\text{dist}(A, C) = \text{dist}(B, D)$ in general. To define an analogy it is necessary to state: $A : B :: C : D$ if $A : B = C : D$ and $A : C = B : D$.

An analogical equation is an expression of the form $A : B :: C : y$, where y is the unknown. Solving it amounts to find the set of y such that the analogy holds. It may have 0, 1 or several solutions, depending on the type of analogies: for geometrical analogies on $\mathbb{R} \setminus \{0\}$ and arithmetic analogies on $\mathbb{R}$, every analogical equation has exactly one solution. By contrast, with analogies on strings as defined above, an analogical equation may have 0, 1 or several solutions.

---

[1] Some authors consider that analogy requires additional postulates (Lepage [10]). However, only these three postulates are used in this paper.

[2] When conformity is an equivalence relation, reflexivity and symmetry are straightforward. But conformity is not necessarily an equivalence relation.

[3] The LCS ("longest common subsequence") distance is an edit distance based on the character insertion and character deletion edit operations, with a cost of 1 for both. In other terms, if P and Q are two strings and L is the LCS of P and Q, then $\text{dist}(P, Q) = (|P| - |L|) + (|Q| - |L|)$.

3   Penalized Cases and Penalized Case Bases

The enrichment of the case base presented here is based on hypothetical reasoning: it generates hypothetical cases $(\mathbf{x}, \mathbf{y})$, i.e., the assertion "$\mathbf{x} \rightsquigarrow \mathbf{y}$" is uncertain. Thus, a hypothetical case is less trustworthy than a licit case (such as a source case), and thus, the former has to be penalized in the reasoning. Two hypothetical cases have different penalties if one of them is more uncertain than the other. The notion of penalty as a way to model uncertainty is introduced for this purpose. The penalties are associated to hypothetical inferences: the more uncertain the inference, the higher the inference cost, that is the additional penalty associated with the inference. Finally, the notion of penalized case is introduced: they are triples $(\mathbf{x}, \mathbf{y}, \pi)$ where $\mathbf{x} \in \mathcal{P}$, $\mathbf{y} \in \mathcal{S}$ and π is a penalty. An enriched case base is actually a set of penalized cases, i.e., a penalized case base.

3.1   Uncertainty and Penalties

Hypothetical reasoning leads to uncertain results. In this paper, uncertainty of an event is measured by an *uncertainty penalty* (or, simply, a *penalty*) $\pi \in [0, \infty]$ such that the higher π is, the less certain the event is. The penalty of an event that is certain is $\pi = 0$. The penalty of an impossible event is $\pi = \infty$.[4]

Remark 1: If two penalties π^1 and π^2 are associated to the same event, with $\pi^1 < \pi^2$, the lower penalty—associated to the higher certainty—is kept. In other words, if π is associated to an event e, then every $\pi' \geq \pi$ can also be associated to e.

3.2   Cost Associated to a Hypothetical Inference

Let φ^0 be a piece of knowledge whose uncertainty (to be consistent with the real world) has an uncertainty penalty π^0. From φ^0, a new piece of knowledge φ^1 can be produced by a hypothetical inference **hypo**. Since **hypo** is hypothetical, it adds some uncertainty, thus an uncertainty penalty π^1 associated to φ^1 can be computed as $\pi^1 = \pi^0 + \mathbf{c}$ where $\mathbf{c} > 0$ measures the additional uncertainty of **hypo**. **c** is assumed to be computed on the basis of **hypo** by a function called cost: $\mathbf{c} = \mathrm{cost}(\mathbf{hypo})$.[5] In summary:

---

[4] If uncertainty is modeled thanks to a probability measure, it is possible to associate to a probability $P \in [0; 1]$ a penalty $\pi = -\log P \in [0; \infty]$. If uncertainty is thought of as a measure of the gap to consistency with the real world, it is possible to associate a distance to it. Then, by definition, licit cases have a penalty of 0. The representation of uncertainty by penalties is chosen in this paper for generality of expression.

[5] Once again, costs could be associated to probabilities: $\mathrm{cost}(\mathbf{hypo})$ could be defined by $-\log P(\varphi^2 \mid \varphi^1)$, where $P(\varphi^2 \mid \varphi^1)$ is the probability of φ^2 being true given that φ^1 is. But they can also be associated to distance. $\mathrm{cost}(\mathbf{hypo})$ could be defined as $\mathrm{dist}(\varphi^1, \varphi^2)$ which expresses the additional uncertainty on φ^2 when inferred from φ^1.

> **if** π^0 is an uncertainty penalty associated to φ^0
> **and** φ^1 is inferred from φ^0 by the hypothetical inference hypo   (1)
> **then** $\pi^1 = \pi^0 + \text{cost}(\text{hypo})$ is an uncertainty penalty associated to φ^1

Now, if φ^0, φ^1, ..., φ^n are pieces of knowledge, φ^0 being certain (it can be associated to an uncertainty penalty $\pi^0 = 0$) and φ^i being produced by a hypothetical inference hypo_i $(1 \leq i \leq n)$, then, according to (1), the uncertainty penalty $\pi^n = \sum_{i=1}^{n} \text{cost}(\text{hypo}_i)$ can be associated to φ^n.

In particular, let us consider the simple approach to CBR consisting in retrieving a source case $(\mathbf{x}^s, \mathbf{y}^s)$ and in reusing (without modification) $\mathbf{y}^s$ as a plausible solution to the target problem $\mathbf{x}^t$ $(\mathbf{y}^t = \mathbf{y}^s)$. If retrieval is based on a distance function dist, then it is reasonable to assume that the higher $\text{dist}(\mathbf{x}^s, \mathbf{x}^t)$ is, the more uncertain the assertion $\mathbf{x}^t \rightsquigarrow \mathbf{y}^t$ is, and the higher the cost of this inference. For this reason, $\text{dist}(\mathbf{x}^s, \mathbf{x}^t)$ can be used to measure the cost of this inference: this is how the distance function is interpreted and used in the rest of the paper.

Remark 2: If an uncertain piece of knowledge φ is inferred by two hypothetical inferences, leading to two penalty values π^1 and π^2, then, following Remark 1, φ is associated to the penalty $\min(\pi^1, \pi^2)$.

3.3   Penalized Cases

In this paper, every hypothetical case $(\mathbf{x}, \mathbf{y}) \in \mathcal{P} \times \mathcal{S}$ is either a licit case $(\mathbf{x} \rightsquigarrow \mathbf{y})$ or not $(\mathbf{x} \not\rightsquigarrow \mathbf{y})$: there is no gradual distinction between licit and illicit cases. By contrast, a hypothetical case is more or less certain to be a licit case. So, hypothetical cases should be preferred on the basis of their respective chances of being licit. A *penalized case* is a triple $(\mathbf{x}, \mathbf{y}, \pi)$ with $(\mathbf{x}, \mathbf{y})$, a hypothetical case and π, a penalty measuring the uncertainty that $\mathbf{x} \rightsquigarrow \mathbf{y}$. The estimation of π is made on the basis of an inference that was applied to generate the hypothetical case $(\mathbf{x}, \mathbf{y})$. If $\pi = 0$, then $(\mathbf{x}, \mathbf{y}, \pi) = (\mathbf{x}, \mathbf{y}, 0)$ is assimilated to the case $(\mathbf{x}, \mathbf{y})$.

A *penalized case base* PCB is a finite set of penalized cases $(\mathbf{x}^s, \mathbf{y}^s, \pi^s) \in \mathcal{P} \times \mathcal{S} \times [0, \infty]$. In particular, CB is a penalized case base with penalties set to 0.

When a penalized case base PCB is used, instead of a classical case base, how does it affect the CBR process? An answer is to take into account the penalties of the case by adding them to the cost of the inference. For example, since dist is interpreted as a cost of the simple CBR inference based on the retrieval of a single source case and reusing it as such (cf. Sect. 3.2), this approach to CBR consists in selecting the $(\mathbf{x}^s, \mathbf{y}^s, \pi^s) \in$ PCB which minimizes $\text{dist}(\mathbf{x}^s, \mathbf{x}^t) + \pi^s$.

4   Problem and Solution Variations

4.1   Definitions

Intuitively, the variation from a problem $\mathbf{x}^i$ to a problem $\mathbf{x}^j$, denoted by $\overrightarrow{\mathbf{x}^i \mathbf{x}^j}$ in this paper, encodes the information necessary to transform $\mathbf{x}^i$ into $\mathbf{x}^j$. More formally, a triple $(\Delta\mathcal{P}, +, \overrightarrow{\cdot\,})$ called the *problem variation space* and satisfying the following postulates is assumed to exist:

(i) $(\Delta\mathcal{P}, +)$ is a commutative group. Its identity element is denoted by $\overrightarrow{0}$; the inverse of $\overrightarrow{u} \in \Delta\mathcal{P}$ is denoted by $-\overrightarrow{u}$.[6]

(ii) $\overrightarrow{\cdot\,}$ is an onto mapping $(\mathbf{x}^i, \mathbf{x}^j) \in \mathcal{P}^2 \mapsto \overrightarrow{\mathbf{x}^i \mathbf{x}^j} \in \Delta\mathcal{P}$: for each $\overrightarrow{u} \in \Delta\mathcal{P}$, there exists $(\mathbf{x}^i, \mathbf{x}^j) \in \mathcal{P}^2$ such that $\overrightarrow{u} = \overrightarrow{\mathbf{x}^i \mathbf{x}^j}$.

(iii) For every $\mathbf{x}^i, \mathbf{x}^j, \mathbf{x}^k \in \mathcal{P}$, $\overrightarrow{\mathbf{x}^i \mathbf{x}^j} + \overrightarrow{\mathbf{x}^j \mathbf{x}^k} = \overrightarrow{\mathbf{x}^i \mathbf{x}^k}$.

(iv) For each $\mathbf{x}^i \in \mathcal{P}$ and $\overrightarrow{u} \in \Delta\mathcal{P}$, there exists *at most* one $\mathbf{x}^j \in \mathcal{P}$ such that $\overrightarrow{\mathbf{x}^i \mathbf{x}^j} = \overrightarrow{u}$. This $\mathbf{x}^j$, *when it exists*, is denoted by $tr_{\overrightarrow{u}}(\mathbf{x}^i)$ (*tr* stands for "translation", borrowing the term from the field of vector spaces).

From these postulates, the following properties can be deduced:

$$\overrightarrow{\mathbf{x}^i \mathbf{x}^j} = \overrightarrow{0} \qquad \text{iff} \qquad \mathbf{x}^i = \mathbf{x}^j \tag{2}$$

$$-\overrightarrow{\mathbf{x}^i \mathbf{x}^j} = \overrightarrow{\mathbf{x}^j \mathbf{x}^i} \tag{3}$$

$$\text{if } \overrightarrow{\mathbf{x}^1 \mathbf{x}^2} = \overrightarrow{\mathbf{x}^3 \mathbf{x}^4} \text{ then } \overrightarrow{\mathbf{x}^1 \mathbf{x}^3} = \overrightarrow{\mathbf{x}^2 \mathbf{x}^4} \tag{4}$$

for every $\mathbf{x}^i, \mathbf{x}^j, \mathbf{x}^1, \mathbf{x}^2, \mathbf{x}^3, \mathbf{x}^4 \in \mathcal{P}$.

There are many ways of defining $\Delta\mathcal{P}$ and the mapping $\overrightarrow{\cdot\,}$. They depend partly upon the problem space $\mathcal{P}$. For example:

- If $\mathcal{P}$ is an affine space of dimension n on $\mathbb{R}$, $\Delta\mathcal{P}$ can be the vector space $\mathbb{R}^n$ associated with $\mathcal{P}$: $\overrightarrow{\mathbf{x}^i \mathbf{x}^j} = (\mathbf{x}^j_1 - \mathbf{x}^i_1, \mathbf{x}^j_2 - \mathbf{x}^i_2, \ldots, \mathbf{x}^j_n - \mathbf{x}^i_n)$ and $tr_{\overrightarrow{u}}$ is the translation operator of vector $\overrightarrow{u}$. This example explains the notations chosen in this paper.

- More generally, if $\mathcal{P}$ is defined by attribute-value pairs, the problem of defining $\overrightarrow{\mathbf{x}^i \mathbf{x}^j}$ can be reduced to the problem of defining the variation from $\mathbf{x}^i$ to $\mathbf{x}^j$ for each attribute. This is considered in particular in d'Aquin et al. [4].

In the same way and with the same notations, a solution variation space $(\Delta\mathcal{S}, +, \overrightarrow{\cdot\,})$ can be defined.

4.2 Adaptation Knowledge Learning Expressed in Terms of Variations

The seminal paper of Hanney and Keane [6] presents the main principles of the AK learning issue. They are reformulated below, thanks to the notions of variations introduced above.

---

[6] $(\Delta\mathcal{P}, +)$ being a commutative group means that $\Delta\mathcal{P}$ is a set, that $+$ is an associative and commutative operation on $\Delta\mathcal{P}$, and that every $\overrightarrow{u} \in \Delta\mathcal{P}$ has an inverse element $-\overrightarrow{u}$ (meaning $\overrightarrow{u} + (-\overrightarrow{u}) = \overrightarrow{0}$).

From the case base, the multiset[7]

$$\text{TS} = \left\{\!\!\left\{ \left(\overrightarrow{x^i x^j}, \overrightarrow{y^i y^j}\right) \;\middle|\; (x^i, y^i), (x^j, y^j) \in \text{CB, with } x^i \neq x^j \right\}\!\!\right\}$$

is computed. This multiset is used in the training of a supervised learning process (the inputs of the examples are the $\overrightarrow{x^i x^j}$, the outputs are the $\overrightarrow{y^i y^j}$). The learned model is used as adaptation knowledge.

Several studies have followed this scheme, and a few examples are given below. In Craw et al. [3], a variety of learning techniques are used, in particular decision tree induction and ensemble learning. In d'Aquin et al. [4], frequent closed itemset extraction in used. The expert interpretation enables to produce adaptation rules to be added to AK. In Jalali et al. [8], an ensemble approach provides adaptation rules for a nominal representation (feature-value pairs, where values are categories).

An example of adaptation learning approach suited for discrete representations is as follows. First, a triple $(\overrightarrow{u}, \overrightarrow{v}, c) \in \Delta\mathcal{P} \times \Delta\mathcal{S} \times [0, \infty[$ such that $\overrightarrow{u} \neq \overrightarrow{0}$ can be seen as an adaptation rule (for (x, y, π), a penalized case and $x^t \in \mathcal{P}$):

if $\overrightarrow{xx^t} = \overrightarrow{u}$ **then** $y^t = tr_{\overrightarrow{v}}(y)$ is a plausible solution of x^t, with penalty $\pi + c$

(recall that $y^t = tr_{\overrightarrow{v}}(y)$ iff $\overrightarrow{yy^t} = \overrightarrow{v}$). Among the $(\overrightarrow{u}, \overrightarrow{v}, c) \in \Delta\mathcal{P} \times \Delta\mathcal{S} \times [0, \infty[$ ($\overrightarrow{u} \neq \overrightarrow{0}$), the ones that are selected are the ones which are the most supported by the training set. More formally, let $\text{supp}(\overrightarrow{u}, \overrightarrow{v})$ (the *support* of the ordered pair $(\overrightarrow{u}, \overrightarrow{v})$) be the multiplicity of $(\overrightarrow{u}, \overrightarrow{v})$ in TS. In other terms:

$$\text{supp}(\overrightarrow{u}, \overrightarrow{v}) = \left| \left\{ ((x^i, y^i), (x^j, y^j)) \in \text{CB}^2 \;\middle|\; \overrightarrow{x^i x^j} = \overrightarrow{u}, \overrightarrow{y^i y^j} = \overrightarrow{v} \right\} \right|$$

Hence, the adaptation knowledge learning process consists in computing the pairs $(\overrightarrow{u}, \overrightarrow{v})$ such that their support is above a given threshold τ_{supp}. It is assumed here that $\tau_{\text{supp}} \geq 2$. The value of the support is used on the basis of the following heuristics: the higher the support is, the less the application of the adaptation rule adds uncertainty. Therefore, a value c is computed thanks to a decreasing function $f : \mathbb{N} \setminus \{0, 1\} \to \mathbb{R}$ by $c = f(\text{supp}(\overrightarrow{u}, \overrightarrow{v}))$. For our experiments, we have chosen $f(n) = 1/n$. Finally, the rule $(\overrightarrow{u}, \overrightarrow{v}, c)$ is added to AK.

With this adaptation knowledge learning approach, it is noteworthy that for each learned adaptation rule $(\overrightarrow{u}, \overrightarrow{v}, c)$ there is another learned adaptation rule $(-\overrightarrow{u}, -\overrightarrow{v}, c)$. Indeed, $\left(\overrightarrow{x^i x^j}, \overrightarrow{y^i y^j}\right)$ occurs in the multiset TS with the same multiplicity as $\left(\overrightarrow{x^j x^i}, \overrightarrow{y^j y^i}\right)$.

---

[7] A multiset is denoted with double braces; for example $M = \{\!\{a, a, b, c, c, c\}\!\}$ contains a with multiplicity 2, b with multiplicity 1 and c with multiplicity 3. Thus the cardinality of M is $|M| = 2 + 1 + 3 = 6$.

4.3   Variation-Based Analogies

Let us consider the relation on problems defined, for $A, B, C, D \in \mathcal{P}$, as follows:

$$A : B :: C : D \quad \text{if} \quad \overrightarrow{AB} = \overrightarrow{CD}$$

It satisfies the postulates of analogy. An analogy on $\mathcal{S}$ can be defined likewise.

Therefore, using the problem variation space and the solution variation space, an analogy on $\mathcal{P}$ and an analogy on $\mathcal{S}$ can be built, and thus, the approach to CBR based on the following principle (called *extrapolation* in Lieber et al. [14] and used in Lepage and Denoual [12] and Lepage and Lieber [13]) can be applied:

- Given a target problem $\mathbf{x}^t$, a triple $((\mathbf{x}^1, \mathbf{y}^1), (\mathbf{x}^2, \mathbf{y}^2), (\mathbf{x}^3, \mathbf{y}^3)) \in \mathsf{CB}^3$ is retrieved such that $\mathbf{x}^1 : \mathbf{x}^2 :: \mathbf{x}^3 : \mathbf{x}^t$ in the problem space (i.e., $\overrightarrow{\mathbf{x}^1\mathbf{x}^2} = \overrightarrow{\mathbf{x}^3\mathbf{x}^t}$).
- Then, the analogical equation $\mathbf{y}^1 : \mathbf{y}^2 :: \mathbf{y}^3 : y$ in the solution space is solved, and the solution of this equation is given as a plausible solution to $\mathbf{x}^t$ (in the framework of the postulates given below, this solution, when it exists, is unique, and verifies $y = tr_{\overrightarrow{v}}(\mathbf{y}^3)$ with $\overrightarrow{v} = \overrightarrow{\mathbf{y}^1\mathbf{y}^2}$).

4.4   Analogy-Based Variations

In the domain of strings, it is possible to define a vector corresponding to a ratio as follows (using the notion of ratios introduced in Sect. 2.2): $\overrightarrow{AB} = A : B$. However, as mentioned in Sect. 2.2, the definition of a ratio alone does not make an analogy. In such a domain, we impose for $(A, B, C, D) \in \mathcal{P}^4$:

$$\overrightarrow{AB} = \overrightarrow{CD} \quad \text{if} \quad A : B :: C : D, \text{ i.e., if } A : B = C : D \text{ and } A : C = B : D$$

In other words, in such a domain, we implement the extrapolation approach mentioned at the end of Sect. 4.3 by restricting ourselves to the use of variations such that both $\overrightarrow{AB} = \overrightarrow{CD}$ and $\overrightarrow{AC} = \overrightarrow{BD}$ hold at the same time. This ensures that the postulates of analogy are verified for the used variations.

5   CBR by Local Enrichment of the Case Base

The enrichment of the case base consists in adding to the original case base CB some (penalized) cases inferred from CB. The inferences considered in this paper consist in applying the learned adaptation rules $(\overrightarrow{u}, \overrightarrow{v}, c) \in \mathsf{AK}$. In theory, all the penalized cases that can be so inferred can enrich CB: this is considered in Sect. 5.1. However, this leads usually to a penalized case base that is too large. In Sect. 5.2, a local enrichment is proposed that consists in adding to CB penalized cases that are "around" the target problem.

5.1 Theoretical View: Global Enrichment of CB

The principle of global enrichment of CB is simple: it consists in computing all the penalized cases (x, y, π) that can be inferred from CB by application of one or several adaptation rules, with the constraint $\pi \leq \tau_{\text{penalty}}$, where τ_{penalty} is a given threshold.

The size of the enriched case base, PCB, can be estimated as follows. Let $p = |\text{AK}|$ and $d = \lfloor \tau_{\text{penalty}} / \min\{c \mid (\overrightarrow{u}, \overrightarrow{v}, c) \in \text{AK}\} \rfloor$. If no hypothetical case (x, y) is generated twice in the process, then $|\text{PCB}| = |\text{CB}| \times \frac{p^{d+1}-1}{p-1}$, assuming $p \neq 1$. Therefore, $|\text{PCB}| / |\text{CB}| = \mathcal{O}(p^d)$. For example, using $p = 10$ adaptation rules having the same cost $c = 1$, if $\tau_{\text{penalty}} = 3$, the size of PCB is about a thousand times the size of CB. This illustrates the fact that this global enrichment of the case base approach produces a case base whose size is, for most CBR applications, too large, which motivates the local enrichment of the case base.

5.2 Practical View: Local Enrichment of CB

The principle of local enrichment is based first on the choice of *seed cases*, i.e., cases from CB that are chosen to produce penalized cases to be added to the case base. If (x^s, y^s) is a seed case, then a penalized case (x, y, π) is produced by a gradient descent starting from $(x^s, y^s, 0)$, by decreasing $\text{dist}(x, x^t) + \pi$, each step corresponding to the application of an adaptation rule. The penalized case (x, y, π) to be added to the case base thus constitutes a local optimum of the set of cases generated from the seed case (x^s, y^s).

The selection of the set of seed cases SC can be done following several strategies, such as the following ones:

– The simplest strategy consists in taking all the source cases: SC = CB. This has the advantage of simplicity, but may lead to considerable growth of the case base (the enriched case base size, $|\text{PCB}|$, will be between $|\text{CB}|$ and $2|\text{CB}|$).
– If the size of the case base is too large already, only a few additional cases should be added and the following solutions can be proposed:
   • Choose SC by a random sampling from CB;
   • Choose SC as the k nearest neighbors of x^t, e.g. according to dist.

6 Applications

In the remainder of the paper, we present two applications of local enrichment of the case base during case-based reasoning. Both examples create strings in a second domain (the solution space) that correspond to strings in a first domain (the problem space). The first application is a theoretical example: the problem space and the solution space are formal languages. The second application is actual machine translation: the problem space and the solution space are actual natural languages: French and English.

6.1 Machine Translation of Formal Languages

The first example shows how local enrichment of the case base can be used to translate from a regular language into a context-free language. The languages we use are the prototypical examples of these families of formal languages, i.e., the problem space and the solution space are $\mathcal{P} = \{(ab)^n \mid n \in \mathbb{N}\}$ and $\mathcal{S} = \{A^n B^n \mid n \in \mathbb{N}\}$, respectively. Let us suppose that our case base contains only the three smallest nonempty members of each of these languages: $CB = \{(ab, AB), (abab, AABB), (ababab, AAABBB)\}$.

From such a case base, in the problem space, one variation with a support greater than 1 is extracted. It corresponds to the ratio $ab : abab = abab : ababab$.[8] This variation in the problem space corresponds to a variation in the solution space: $AB : AABB = AABB : AAABBB$. Of course, these variations have their corresponding inverse variations in the problem and solution spaces.

An actual trace of the system is given in Fig. 1 for the translation of the string $(ab)^6$. We choose to select all cases in the case base as seed cases, i.e., $SC = CB$. The seed cases are sorted by distance to the target problem. Their LCS distance to the target problem is given by δ in Fig. 1. Starting from the problems in the seed cases, applying the variation has the effect of enriching the case base with cases of the form $((ab)^n, A^n B^n)$ from $n = 4$ to 6, one after another. This is indeed an induction over n for $(ab)^n$ and $A^n B^n$ simultaneously in both spaces.

During enrichment, the distance from the new source problems to the target problem decreases down to 2. The distance of 0 is not mentioned, as it means that the new source problem is indeed the target problem, for which a solution has been found.

Such an example can be easily amended to translate from a regular language into a context-sensitive language (like $\{A^n B^n C^n \mid n \in \mathbb{N}\}$), or a context-free language into a context-sensitive language. Changing the direction of translation is also possible: from context-free to regular, etc.

6.2 Machine Translation of Natural Languages

The second application deals with machine translation of natural languages. We use French–English as the language pair and data from the Tatoeba Corpus[9] as our bilingual corpus. There are important remarks to make on this domain.

Nature of the Data: strings of characters. This implies again that variations are defined as in Sect. 4.4. The case base consists of sentence pairs which are in a translation relation. We retain sentences of less than 10 words in length and select 90 % of them for training and the other 10 % for testing. This makes 109,390 sentence pairs in total in the training set. The average length of a sentence in French is 6.9 ± 1.8 words and 6.6 ± 1.6 in English. Such sentence pairs are illustrated in Fig. 2. Notice that the sentences are lowercased and tokenized.

---

[8] Note that this is the equality of two ratios. Of course, it is also an analogy by itself ($ab : abab :: abab : ababab$), but this is not what is meant here.

[9] https://tatoeba.org/ and http://www.manythings.org/anki/.

PROBLEM SPACE	SOLUTION SPACE	CASE BASE

CASE BASE

Initial case base = Seed cases

$(ab)^1$	A^1B^1
$(ab)^2$	A^2B^2
$(ab)^3$	A^3B^3

PROBLEM SPACE — Source problems in seed cases

$\delta = 6 \quad (ab)^3$
$\delta = 8 \quad (ab)^2$
$\delta = 10 \quad (ab)^1$

Enrichment

PROBLEM SPACE	SOLUTION SPACE	CASE BASE
$(ab)^1:(ab)^2 :: (ab)^3:\mathbf{(ab)^4}$	$A^1B^1:A^2B^2 :: A^3B^3:\mathbf{A^4B^4}$	$\mathbf{(ab)^4}\quad \mathbf{A^4B^4}$
$(ab)^2:(ab)^3 :: (ab)^3:\mathbf{(ab)^4}$	$A^2B^2:A^3B^3 :: A^3B^3:\mathbf{A^4B^4}$	$\mathbf{(ab)^4}\quad \mathbf{A^4B^4}$
$(ab)^1:(ab)^2 :: (ab)^2:(ab)^3$	$A^1B^1:A^2B^2 :: A^2B^2:A^3B^3$	$(ab)^3\quad A^3B^3$
$(ab)^2:(ab)^3 :: (ab)^2:(ab)^3$	$A^2B^2:A^3B^3 :: A^2B^2:A^3B^3$	$(ab)^3\quad A^3B^3$
$(ab)^1:(ab)^2 :: (ab)^1:(ab)^2$	$A^1B^1:A^2B^2 :: A^1B^1:A^2B^2$	$(ab)^2\quad A^2B^2$
$(ab)^2:(ab)^3 :: (ab)^1:(ab)^2$	$A^2B^2:A^3B^3 :: A^1B^1:A^2B^2$	$(ab)^2\quad A^2B^2$

New source problem

$\delta = 4 \quad \mathbf{(ab)^4}$

PROBLEM SPACE	SOLUTION SPACE	CASE BASE
$(ab)^1:(ab)^2 :: \mathbf{(ab)^4}:\mathbf{(ab)^5}$	$A^1B^1:A^2B^2 :: \mathbf{A^4B^4}:\mathbf{A^5B^5}$	$\mathbf{(ab)^5}\quad \mathbf{A^5B^5}$
$(ab)^2:(ab)^3 :: \mathbf{(ab)^4}:\mathbf{(ab)^5}$	$A^2B^2:A^3B^3 :: \mathbf{A^4B^4}:\mathbf{A^5B^5}$	$\mathbf{(ab)^5}\quad \mathbf{A^5B^5}$

New source problem

$\delta = 2 \quad \mathbf{(ab)^5}$

PROBLEM SPACE	SOLUTION SPACE	CASE BASE
$(ab)^1:(ab)^2 :: \mathbf{(ab)^5}:\mathbf{(ab)^6}$	$A^1B^1:A^2B^2 :: \mathbf{A^5B^5}:\mathbf{A^6B^6}$	$\mathbf{(ab)^6}\quad \mathbf{A^6B^6}$
$(ab)^2:(ab)^3 :: \mathbf{(ab)^5}:\mathbf{(ab)^6}$	$A^2B^2:A^3B^3 :: \mathbf{A^5B^5}:\mathbf{A^6B^6}$	$\mathbf{(ab)^6}\quad \mathbf{A^6B^6}$

Final case base

$(ab)^1$	A^1B^1
$(ab)^2$	A^2B^2
$(ab)^3$	A^3B^3
$(ab)^4$	A^4B^4
$(ab)^5$	A^5B^5
$(ab)^6$	A^6B^6

Fig. 1. Trace for the translation from a regular language into a context-free language. The target problem is $(ab)^6$. It is correctly translated into A^6B^6 after enrichment of the case base. The problem space is on the left, the solution space in the middle. The case base and its enrichment are shown on the right. The distance to the target problem is denoted by δ. New source problems, solutions and cases are boldfaced; old ones are grayed out.

Size of the Case Base. For case-based reasoning, the case base here is quite large: 109,390 cases. However, in the field of machine translation, on the contrary, it is considered rather small.

Nature of the Ratios: they are defined as in Sect. 2.2 (analogies on strings) as we deal with strings. A ratio is a vector made of the difference between the Parikh vectors of the two strings considered, plus an extra dimension with the LCS distance between the two strings. Notice again that the equality between ratios does not imply the existence of an analogy, contrary to arithmetic or geometric analogies on numbers or tuples: $\mathtt{dist}(A, B) = \mathtt{dist}(C, D)$ does not imply $\mathtt{dist}(A, C) = \mathtt{dist}(B, D)$. Also, conformity is not transitive, so it is not an equivalence relation.

Nature of the Variations: Analogical Clusters. Because of the nature of the ratios and the nature of conformity, variations are defined as sets of ratios in which any pair of ratios is an analogy. We use the tools[10] described in Fam and Lepage [5] to extract all analogical clusters containing at least 2 ratios from the case base in the problem and the solution spaces. As a note, the use a more semantically justified distance, instead of the purely formal LCS distance, is of course worth exploring: a distance associated with the cosine similarity of distributional semantics can produce semantically justified analogies (Lepage [11]) and could hence allow to produce semantic variations.

Number of Variations: It is rather large: almost 8 million analogical clusters were extracted in French, more than half a million in English. The extraction of such variations from an actual corpus is time-consuming. For efficiency, we retain only the first 3,000 largest analogical clusters in number of ratios. Three examples of analogical clusters are given in Fig. 3. Typically, variations reflect grammatical oppositions. In the examples of Fig. 3, affirmative/negative, masculine/feminine and insertion of the adverb *just.*

As an example, the translation process of the tokenized sentence *vous êtes vraiment trop tatillon* . from French into English is given in Fig. 4. The seed problems are the 5 most similar French sentences in the case base. The process

French	English
regardez comment je le fais !	*watch how i do it .*
vous m' avez oublié , n' est-ce pas ?	*you 've forgotten me , haven 't you ?*
pensez-vous sérieusement à vendre cela sur internet ?	*are you seriously thinking about selling this online ?*
les travailleurs se plaignent de leurs conditions de travail .	*the workers are complaining about their working conditions .*

Fig. 2. French and English example sentences, i.e., problems and solutions, in the case base for case-based machine translation

---

[10] https://lepage-lab.ips.waseda.ac.jp/ > Kakenhi 15K00317 > Tools.

get up . : don 't get up .

tell me . : don' t tell me .

leave us . : don 't leave us .

leave me . : don 't leave me .

he runs . : she runs .

he is young . : she is young .

he helps us . : she helps us .

he loves singing . : she loves singing .

he just left . : she just left .

i can 't believe this . : i just can 't believe this .

i need a little more time . : i just need a little more time .

i want this to be over . : i just want this to be over .

i want someone to talk to . : i just want someone to talk to .

i felt a drop of rain . : i just felt a drop of rain .

Fig. 3. Three analogical clusters which stand for variations in the solution space

takes three steps corresponding to each block of sentences in Fig. 4. The distance to the target problem is given on the left. Notice the infelicitous enrichment of the case base with an invalid sentence: *you really not 're finicky* .

An example of a variation applied in the problem and solution spaces during the above translation process is shown in Fig. 5. It corresponds to a variation applied on the fourth seed case (boldfaced) in Fig. 4. As a result, the case base is enriched with the first case marked as a new case in the above table ($\delta = 4$).[11]

$\delta = 12$ *vous êtes vraiment très bons .*	*you are really very good .*	(in **CB**)
$\delta = 13$ *vous êtes venu trop tôt .*	*you 've come too early .*	"
$\delta = 14$ *vous êtes venu trop tard .*	*you came too late .*	"
$\delta = 14$ **vous êtes tatillon .**	**you 're finicky .**	"
$\delta = 14$ *vous êtes venue trop tôt .*	*you 've come too early .*	"
$\delta = 4$ *vous êtes vraiment très tatillon .*	*you are really very finicky .*	(new case)
$\delta = 18$ *ne vous vraiment pas êtes tatillon .*	*you really not 're finicky .*	"
$\delta = 6$ *vous êtes vraiment fort tatillon .*	*you are really very finicky .*	"
$\delta = 0$ *vous êtes vraiment trop tatillon .*	*you are really too finicky .*	"

Fig. 4. Translation process of a French sentence into English

The standard metric BLEU (Papineni et al. [15]) is used for the evaluation of the accuracy of a machine translation output against a given reference set. BLEU scores range from 0 to 1; the higher, the better. The system described

---

[11] Remember that LCS distance is used: dist(très, trop) = 4 (two deletions and two insertions), not 2 (two substitutions) as would be the case with Levenshtein distance.

$$\textit{vous êtes bons .} : \frac{\textit{vous êtes vrai-}}{\textit{ment très bons .}} :: \textbf{vous êtes tatillon .} : \frac{\textit{vous êtes vraiment}}{\textit{très tatillon .}}$$

$$\textit{you 're good .} : \frac{\textit{you are really}}{\textit{very good .}} :: \textbf{you 're finicky .} : \frac{\textit{you are really very}}{\textit{finicky .}}$$

Fig. 5. Variation in the problem and solution spaces resulting in the enrichment of the case base

above achieves a BLEU score of 0.51 in translating the 10,998 sentences in the test set. Although the differences in scores are statistically significant, this is a reasonable score when compared with the scores of two much more elaborated systems, a neural system (OpenNMT[12]), and a statistical system (GIZA++, Moses, KenLM, MERT[13]), which achieve 0.60 and 0.65 respectively, on exactly the same data.

7   Conclusion, Discussion and Related Work

In this paper, we proposed a new approach to case-based reasoning which consists in enriching the case base while performing reasoning. Enrichment results from the application of adaptation rules to seed cases, i.e., cases taken from the case base as starting points. New cases are created, which get closer and closer to the target problem, but they get penalties characteristic of the uncertainty brought by the application of the adaptation rules. Adaptation rules are given by variations in the problem space and variations in the solution space. The last variations should approach the target problem itself, so that corresponding variations in the solution space will produce (hypothetical) solutions to the target problem. We implemented such a new approach and illustrated it with two applications which shared the fact that the solution and problem spaces were spaces of strings of characters: formal and natural languages.

The notion of penalized case is similar to the notion of *ghost case* introduced by D. Leake and B. Schack [9], a ghost case being a hypothetical case generated by adaptation. The main difference between these notions is the use of penalty values when reasoning with penalized cases. Although the aim of both works are quite opposite (case base enrichment versus case base contraction), they use these similar notions in similar ways: in [9], it is argued that ghost cases are used to compensate for the lack of expressiveness of the case base and this argument can be reused in the current work.

The general framework can be adapted to various scenarios. Several points can be adapted to the specificity of the domains at hand. For instance, the selection of the seed cases can be performed in various ways suggested in Sect. 5.2, at random or according to some selection method specific to the domain.

Our approach to case-based reasoning can be seen as a variant of gradient descent or hill climbing. Similarly, our approach exhibits the risk of reaching

---

[12] http://opennmt.net.

[13] http://www.statmt.org.

local minima (or maxima) instead of global minima (or maxima). Here, the landscape is shaped by the variations observed between the cases present in the initial case base. This issue of local optimality can be partially addressed by considering several branches generated from each seed case: instead of a single path approaching the target problem, a tree can be generated rooted at this seed case, whose breadth should be controlled to avoid an explosion of the enriched case base size. This way, several new cases can be generated from a single seed case. The precise study of this idea remains to be done.

From a theoretical viewpoint, the approach is presented in a very constrained framework, in particular for the definition of variations. This makes the explanations simpler, but, in particular for the considered applications in machine translation, some of these constraints do not hold. Thus, a theoretical study on less constrained variation spaces must be carried on.

References

1. Aamodt, A., Plaza, E.: Case-based reasoning: foundational issues, methodological variations, and system approaches. AI Commun. **7**(1), 39–59 (1994)
2. Couceiro, M., Hug, N., Prade, H., Richard, G.: Analogy-preserving functions: a way to extend Boolean samples. In: IJCAI 2017, pp. 1575–1581 (2017)
3. Craw, S., Wiratunga, N., Rowe, R.C.: Learning adaptation knowledge to improve case-based reasoning. Artif. Intell. **170**(16–17), 1175–1192 (2006)
4. d'Aquin, M., Badra, F., Lafrogne, S., Lieber, J., Napoli, A., Szathmary, L.: Case base mining for adaptation knowledge acquisition. In: IJCAI 2007, pp. 750–755 (2007)
5. Fam, R., Lepage, Y.: Tools for the production of analogical grids and a resource of n-gram analogical grids in 11 languages. In: Proceedings of LREC 2018, pp. 1060–1066. ELRA, May 2018
6. Hanney, K., Keane, M.T.: Learning adaptation rules from a case-base. In: Smith, I., Faltings, B. (eds.) EWCBR 1996. LNCS, vol. 1168, pp. 179–192. Springer, Heidelberg (1996). https://doi.org/10.1007/BFb0020610
7. Hussain, Z., Gimenez, F., Yi, D., Rubin, D.: Differential data augmentation techniques for medical imaging classification tasks. In: Annual Symposium Proceedings, pp. 979–984 (2017)
8. Jalali, V., Leake, D., Forouzandehmehr, N.: Learning and applying adaptation rules for categorical features: an ensemble approach. AI Commun. **30**(3–4), 193–205 (2017)
9. Leake, D., Schack, B.: Exploration vs. Exploitation in case-base maintenance: leveraging competence-based deletion with ghost cases. In: Cox, M.T., Funk, P., Begum, S. (eds.) ICCBR 2018. LNCS (LNAI), vol. 11156, pp. 202–218. Springer, Cham (2018). https://doi.org/10.1007/978-3-030-01081-2_14
10. Lepage, Y.: Proportional analogy in written language data. In: Gala, N., Rapp, R., Bel-Enguix, G. (eds.) Language Production, Cognition, and the Lexicon. TSLT, vol. 48, pp. 151–173. Springer, Cham (2015). https://doi.org/10.1007/978-3-319-08043-7_10
11. Lepage, Y.: Semantico-formal resolution of analogies between sentences. In: Proceedings of LTC 2019, pp. 57–61 (2019)

12. Lepage, Y., Denoual, E.: Purest ever example-based machine translation: detailed presentation and assessment. Mach. Transl. **19**, 251–282 (2005)
13. Lepage, Y., Lieber, J.: Case-based translation: first steps from a knowledge-light approach based on analogy to a knowledge-intensive one. In: Cox, M.T., Funk, P., Begum, S. (eds.) ICCBR 2018. LNCS (LNAI), vol. 11156, pp. 563–579. Springer, Cham (2018). https://doi.org/10.1007/978-3-030-01081-2_37
14. Lieber, J., Nauer, E., Prade, H., Richard, G.: Making the best of cases by approximation, interpolation and extrapolation. In: Cox, M.T., Funk, P., Begum, S. (eds.) ICCBR 2018. LNCS (LNAI), vol. 11156, pp. 580–596. Springer, Cham (2018). https://doi.org/10.1007/978-3-030-01081-2_38
15. Papineni, K., Roukos, S., Ward, T., Zhu, W.J.: BLEU: a method for automatic evaluation of machine translation. Technical report, IBM (2001)
16. Richter, M.M., Weber, R.: Case-Based Reasoning: A Textbook. Springer, Heidelberg (2013). https://doi.org/10.1007/978-3-642-40167-1
17. Taylor, L., Nitschke, G.: Improving deep learning using generic data augmentation. CoRR (2017)

Improving Analogical Extrapolation
Using Case Pair Competence

Jean Lieber[1]([✉]), Emmanuel Nauer[1]([✉]), and Henri Prade[2]([✉])

[1] Université de Lorraine, CNRS, Inria, LORIA, 54000 Nancy, France
`Emmanuel.Nauer@loria.fr`
[2] IRIT, Université de Toulouse, Toulouse, France

Abstract. An analogical proportion is a quaternary relation that is to be read "a is to b as c is to d", verifying some symmetry and permutation properties. As can be seen, it involves a pair of pairs. Such a relation is at the basis of an approach to case-based reasoning called analogical extrapolation, which consists in retrieving three cases forming an analogical proportion with the target problem in the problem space and then in finding a solution to this problem by solving an analogical equation in the solution space. This paper studies how the notion of competence of pairs of source cases can be estimated and used in order to improve extrapolation. A preprocessing of the case base associates to each case pair a competence given by two scores: the support and the confidence of the case pair, computed on the basis of other case pairs forming an analogical proportion with it. An evaluation in a Boolean setting shows that using case pair competences improves significantly the result of the analogical extrapolation process.

Keywords: Analogical proportion · Analogical inference ·
Case-based reasoning · Competence · Extrapolation

1 Introduction

In a recent paper [16], the authors have advocated that reasoning about cases (or case-based reasoning, CBR [20,21]) may not be only based on similarity-based reasoning, looking for the nearest solved cases, but may also use analogical proportions for extrapolation purposes. Extrapolation is based on analogical inference, that uses triples of cases (a, b, c) for building the solution of a fourth (new) case d through an adaptation mechanism. An illustration of this is given in [2] where in three distinct situations (problems) the recommended actions (solutions) are respectively to (a) serve tea without milk without sugar, (b) serve tea with milk without sugar, (c) serve tea without milk with sugar, while in a fourth situation (d) that makes an analogical proportion with the three others, the action to do would be "serve tea with milk and with sugar".

Usually, several triples (a, b, c) in the case base can be used for predicting the solution of the fourth case d and predictions may diverge. In fact, it has been

K. Bach and C. Marling (Eds.): ICCBR 2019, LNAI 11680, pp. 251–265, 2019.
https://doi.org/10.1007/978-3-030-29249-2_17

established for Boolean features that such an inference makes no error (thus all the triples agree on the same prediction) if and only if the function that associates the solution to the description of a case is an affine Boolean function [6]. This is why, when the function is not assumed to be affine, a voting procedure is organized between the predicting triples.

Such a procedure is quite brute-force, and did not take really lesson from the case base. Indeed, it may happen that some triples in the case base fail to predict the correct answer of another case of the case base. In this paper, we propose to take into account this kind of information for restricting the number of triples used for making a prediction in a meaningful way.

The paper is organized as follows. The next section provides the necessary background on analogical proportions and the notations about CBR used throughout the paper. Section 3 discusses how to restrict the set of triples allowed to participate to a given prediction. Section 4 reports experimentations showing the gain in accuracy of the new inference procedure. Section 5 discusses related work, before concluding.

2   Preliminaries

This section presents first the formal framework of this study: the nominal representations and, more specifically, the representation by tuples of Boolean values. Then, it recalls some notions and gives some notations about analogical proportions and about case-based reasoning.

2.1   Nominal Representations and Boolean Setting

Feature-value representations are often used in CBR (see, e.g., [14]). A nominal representation is a feature-value representation where the range of each feature is finite (and, typically, small). More formally, let $\mathcal{U}_1$, $\mathcal{U}_2$, ..., $\mathcal{U}_p$ be p finite sets and $\mathcal{U} = \mathcal{U}_1 \times \mathcal{U}_2 \times \ldots \times \mathcal{U}_p$. A feature on $\mathcal{U}$ is one of the p projections $(x_1, x_2, \ldots, x_p) \in \mathcal{U} \mapsto x_i \in \mathcal{U}_i$ ($i \in \{1, 2, \ldots, p\}$).

A Boolean representation is a nominal representation where $\mathcal{U}_1 = \mathcal{U}_2 = \ldots = \mathcal{U}_p = \mathbb{B}$, where $\mathbb{B} = \{0, 1\}$ is the set of Boolean values: the value "false" is assimilated to the integer 0, and "true" is assimilated to 1. The Boolean operators $\neg$, $\wedge$ and $\vee$ are defined, for $a, b \in \mathbb{B}$, by $\neg a = 1 - a$, $a \wedge b = \begin{cases} 1 & \text{if } a = b = 1 \\ 0 & \text{otherwise} \end{cases}$, $a \vee b = \neg(\neg a \wedge \neg b)$, and $a \equiv b = (\neg a \vee b) \wedge (\neg b \vee a)$. An element of $\mathbb{B}^p$ is denoted without commas and parentheses, e.g., 01101 stands for $(0, 1, 1, 0, 1)$.

2.2   Analogical Proportions

Given a set $\mathcal{U}$, an analogical proportion on $\mathcal{U}$ is a quaternary relation on $\mathcal{U}$, denoted by $a : b :: c : d$ for $(a, b, c, d) \in \mathcal{U}^4$, and satisfying the following postulates (for $a, b, c, d \in \mathcal{U}$):

(**Reflexivity**) $a\!:\!b\!::\!a\!:\!b$.
(**Symmetry**) If $a\!:\!b\!::\!c\!:\!d$ then $c\!:\!d\!::\!a\!:\!b$.
(**Exchange of the means**) If $a\!:\!b\!::\!c\!:\!d$ then $a\!:\!c\!::\!b\!:\!d$.

In the Boolean case, an analogical proportion is a quaternary connective that can be defined as [19]:

$$a\!:\!b\!::\!c\!:\!d \stackrel{\text{def}}{=} (a \equiv b \wedge c \equiv d) \vee (a \equiv c \wedge b \equiv d)$$

This expression can be equivalently written [17,19]:

$$a\!:\!b\!::\!c\!:\!d = (\neg a \wedge b \equiv \neg c \wedge d) \wedge (\neg b \wedge a \equiv \neg d \wedge c)$$

This makes clear that "a is to b as c is to d" is understood as "a differs from b as c differs from d and b differs from a as d differs from c". Therefore, the set of $(\mathsf{a}, \mathsf{b}, \mathsf{c}, \mathsf{d}) \in \mathbb{B}^4$ such that $\mathsf{a}\!:\!\mathsf{b}\!::\!\mathsf{c}\!:\!\mathsf{d}$ is $\{0000, 0011, 0101, 1111, 1100, 1010\}$. In the nominal representation, the 4-tuples (a, b, c, d) in analogical proportion have one of the three following forms [4]: (s, s, s, s), (s, t, s, t) and (s, s, t, t) for $s, t \in \mathcal{U}_i$. Note that it amounts to make the following expression true: $((a = b) \wedge (c = d)) \vee ((a = c) \wedge (b = d))$, which generalizes the Boolean case.

Given a finite $\mathcal{U} = \mathcal{U}_1 \times \mathcal{U}_2 \times \ldots \times \mathcal{U}_p$ the following analogical proportion can be defined:

$$a\!:\!b\!::\!c\!:\!d \stackrel{\text{def}}{=} a_1\!:\!b_1\!::\!c_1\!:\!d_1 \quad \wedge \quad a_2\!:\!b_2\!::\!c_2\!:\!d_2 \quad \wedge \quad \ldots \quad \wedge \quad a_p\!:\!b_p\!::\!c_p\!:\!d_p$$

Given $a, b, c \in \mathcal{U}$, solving the *analogical equation* $a\!:\!b\!::\!c\!:\!y$ aims at finding the $y \in \mathcal{U}$ such that this relation holds. In a nominal representation, such an equation has 0 or 1 solution. More precisely:

- If $a = b$, the solution is $y = c$.
- If $a = c$, the solution is $y = b$.
- Otherwise, $a\!:\!b\!::\!c\!:\!y$ has no solution.

2.3 Notations and Assumptions on CBR

Let $\mathcal{P}$ and $\mathcal{S}$ be two sets. A *problem* x is by definition an element of $\mathcal{P}$ and a *solution* y, an element of $\mathcal{S}$. If $a \in \mathcal{P} \times \mathcal{S}$, then x^a and y^a denote its problem and solution parts: $a = (\mathsf{x}^a, \mathsf{y}^a)$. Let $\rightsquigarrow$ be a relation on $\mathcal{P} \times \mathcal{S}$. For $(\mathsf{x}, \mathsf{y}) \in \mathcal{P} \times \mathcal{S}$, $\mathsf{x} \rightsquigarrow \mathsf{y}$ is read "x has for solution y" or "y solves x". A *case* is a pair (x, y) such that $\mathsf{x} \rightsquigarrow \mathsf{y}$. The aim of a CBR system is to solve problems, i.e., it should approximate the relation $\rightsquigarrow$: given $\mathsf{x}^{\mathsf{tgt}} \in \mathcal{P}$ (the *target* problem), it aims at proposing $\mathsf{y}^{\mathsf{tgt}} \in \mathcal{S}$ such that it is plausible that $\mathsf{x}^{\mathsf{tgt}} \rightsquigarrow \mathsf{y}^{\mathsf{tgt}}$. For this purpose, a finite set of cases, called the case base and denoted by CB, is used. An element of CB is called a *source case*. Besides the case base, other knowledge containers are often used [20], but they are not considered in this paper.

The classical way of defining a CBR process consists in selecting a set of k source cases related to $\mathsf{x}^{\mathsf{tgt}}$ (retrieve phase) and solve $\mathsf{x}^{\mathsf{tgt}}$ with the help of the retrieved cases (reuse phase). Other steps are considered in the classical 4 Rs model [1], but not in this paper. In [16], three approaches are presented for $k \in \{1, 2, 3\}$. The approach for $k = 3$, called analogical extrapolation, is recalled in the next section.

3   Improving Extrapolation Thanks to Case Pair Competence

This section presents the proposed approach. First, it is shown how a notion of competence associated to case pairs can be used to improve extrapolation, an approach to CBR based on analogical proportions. Then, this notion of competence is formally defined. Finally, strategies for exploiting case pair competence are described.

3.1   Principles

The analogical proportion-based inference principle [23] can be stated as follows (using the notations on CBR introduced above; $a = (\mathbf{x}^a, \mathbf{y}^a)$, $b = (\mathbf{x}^b, \mathbf{y}^b)$, $c = (\mathbf{x}^c, \mathbf{y}^c)$ and $d = (\mathbf{x}^d, \mathbf{y}^d)$ are four cases):

$$\frac{\mathbf{x}^a : \mathbf{x}^b :: \mathbf{x}^c : \mathbf{x}^d \text{ holds}}{\mathbf{y}^a : \mathbf{y}^b :: \mathbf{y}^c : \mathbf{y}^d \text{ holds}}$$

In order to solve a new problem $\mathbf{x}^{\text{tgt}}$, this leads to look for all triples of source cases (a, b, c) such that $\mathbf{x}^a : \mathbf{x}^b :: \mathbf{x}^c : \mathbf{x}^{\text{tgt}}$ holds and such that the equation $\mathbf{y}^a : \mathbf{y}^b :: \mathbf{y}^c : \mathbf{y}$ is solvable. Let $\mathcal{T}$ be the set of all these triples. Then the implementation of this inference pattern uses a vote among all triples of $\mathcal{T}$ and chooses the solution $\mathbf{y}$ found for the largest number of triples. This is the principle called analogical extrapolation (or, simply, extrapolation) in [16].

In the following, we assume for simplicity that all the features are nominal (e.g., Boolean). When there is only one feature for the solutions, the problem-solving task is a classification task (finding the class $\mathbf{y}^{\text{tgt}} \in \mathcal{S}$ to be associated with $\mathbf{x}^{\text{tgt}}$). When there are several features, one can handle them one by one only if they are logically independent, otherwise the vote should be organized between the whole vectors describing the different solutions. In the following, we assume independence, and we consider one of the components $\mathbf{y}_i$ of a solution $\mathbf{y}$ (thus, the index i is useless: $\mathbf{y}_i$ is denoted by $\mathbf{y}$).

Still, one may wonder if all triples of $\mathcal{T}$ involved in a vote for making a particular prediction have the same legitimacy. Indeed, one may take lesson from $\mathcal{T}$ by observing that if one wants to predict a solution for one problem taken from $\mathcal{T}$ from the rest of the examples, there may exist triples that make a wrong prediction, as suggested in [18]. The situation may be better analyzed in terms of pairs, as shown now.

Indeed look at Table 1. It exhibits three Boolean pairs such that a: b:: c: d and a': b':: c: d hold in all columns, except the last one (column 'sol', as solution). Note that the 'D' columns (first two columns, 'D' as in disagreement) show the possible patterns expressing that a and b differ in the same way as c and d and as a' and b'.[1] The 'A' columns (as in agreement) show all the ways a and b agree,

---

[1] $D(0/1)$ indicates the disagreement between a and b (respectively between c and d and between a' and b') when the former is equal to 0 and the latter is equal to 1. $D(1/0)$ is the reverse disagreement.

Table 1. Double pairing of pairs (a, b), (c, d) and (a', b'): Analogy breaking on *sol*.

	D(0/1)	D(1/0)	A(0, 0, 0)	A(0, 0, 1)	A(0, 1, 0)	A(0, 1, 1)	A(1, 0, 0)	A(1, 0, 1)	A(1, 1, 0)	A(1, 1,1)	sol
a	0	1	0	0	0	0	1	1	1	1	0
b	1	0	0	0	0	0	1	1	1	1	1
c	0	1	0	0	1	1	0	0	1	1	0
d	1	0	0	0	1	1	0	0	1	1	1
a'	0	1	0	1	0	1	0	1	0	1	0
b'	1	0	0	1	0	1	0	1	0	1	0

while c and d agree, and a' and b' also agree, maybe in different manners.[2] If we take out the value of d in the column '*sol*' and we try to predict it from the other values from this column, the equation a: b:: c: y yields the good result (i.e., 0: 1:: 0: y gives $y = 1$, i.e., the value of d in the table, column '*sol*'), while the equation a': b':: c: y gives a wrong result (i.e., 0: 0:: 0: y gives $y = 0$, whereas d = 1 in the table, column '*sol*').

So, for each pair, like (a, b) or (a', b') in the table, one may count the numbers of times where the pairs leads to a correct and to a wrong prediction for an example taken from the case base. This provides a basis for favoring triples containing pairs leading often to good predictions, in the voting procedure.

The above idea of looking at pairs of cases can be related to the reading of a pair of cases (a, b) as a potential rule expressing either that the change from x^a to x^b explains the change from y^a to y^b, whatever the context (encoded by the features where the two examples agree), or that the change from x^a to x^b does not modify the solution (if $y^a = y^b$). This view of pairs as rules has already been proposed in CBR for finding adaptation rules [7,8,11] and later in an analogical proportion-based inference perspective in [3,5].

So, roughly speaking, we are interested in a preprocessing process, in order to discover *analogy breakings* in T. By an analogy breaking, we mean the existence of a quadruple of cases (a, b, c, d) such that (i) $x^a: x^b:: x^c: x^d$ holds, while (ii) $y^a: y^b:: y^c: y^d$ *does not* hold. If some analogy breaking(s) can be found in T, this means that the partially unknown Boolean function associating to a problem a solution (or a class) cannot be affine [6]. In such a situation, analogical inference cannot be blindly applied with any triple, and we should take into account the analogy breaking(s), by introducing some further restrictions in the choice of the suitable triples.

More precisely, the idea is to make a preliminary preprocessing of the pairs $(a, b) \in CB^2$, by associating with each of them a *competence*. The intuition behind this notion is that the more a case pair is competent for solving problems, the more it can play a role during the voting and selection process. To assess the competence of a pair $(a, b) \in CB^2$, it has to be compared to other pairs $(c, d) \in CB^2$ such that the triple (a, b, c) can be used to solve the problem x^d by extrapolation. When the outcome y of the extrapolation is equal to y^d, then it increases the

---

[2] In Table 1, $A(u, v, w)$ means that a = b = u, c = d = v and a' = b' = w.

competence of the case pair (a, b). Otherwise, it lowers it. The definition of competence is detailed in the next section.

The case pair competence can be used at problem-solving time according to different strategies. Section 3.3 presents some of these strategies that are experimentally evaluated in a Boolean setting in Sect. 4.

3.2   Case Pair Competence: Definition

Let (a, b) be a pair of source cases: $a = (\mathbf{x}^a, \mathbf{y}^a) \in$ CB and $b = (\mathbf{x}^b, \mathbf{y}^b) \in$ CB. The *competence* of the pair (a, b) is defined by two scores: the support and the confidence of (a, b), defined below following the principle presented above.

First, let $\texttt{SolvableBy}(a, b)$ be the set of source case pairs $(c, d) \neq (a, b)$ such that the triple (a, b, c) can be used to solve $\mathbf{x}^d$ by extrapolation: $\mathbf{x}^a : \mathbf{x}^b :: \mathbf{x}^c : \mathbf{x}^d$ and the equation $\mathbf{y}^a : \mathbf{y}^b :: \mathbf{y}^c : y$ is solvable (and so, its solution y is unique in a nominal representation). Formally:

$$\texttt{SolvableBy}(a, b) = \left\{ (c, d) \in \mathtt{CB}^2 \;\middle|\; \begin{array}{l} (c, d) \neq (a, b), \quad \mathbf{x}^a : \mathbf{x}^b :: \mathbf{x}^c : \mathbf{x}^d \\ \text{and the equation } \mathbf{y}^a : \mathbf{y}^b :: \mathbf{y}^c : y \text{ is solvable} \end{array} \right\}$$

In other words, $\texttt{SolvableBy}(a, b)$ is the set of source case pairs such that c can be adapted into a solution of $\mathbf{x}^d$ using (a, b) as an adaptation rule (without considering the trivial case when $(a, b) = (c, d)$). The support of (a, b), $\texttt{supp}(a, b)$, is simply the number of such pairs:

$$\texttt{supp}(a, b) = |\texttt{SolvableBy}(a, b)|$$

Among the $(c, d) \in \texttt{SolvableBy}(a, b)$ some leads to a correct solution $(y = \mathbf{y}^d)$ and some does not. The formers constitute the following set:

$$\texttt{CorrectlySolvableBy}(a, b) = \left\{ (c, d) \in \texttt{SolvableBy}(a, b) \mid \mathbf{y}^a : \mathbf{y}^b :: \mathbf{y}^c : \mathbf{y}^d \right\}$$

For example, if $\texttt{supp}(a, b) = 6$ and $|\texttt{CorrectlySolvableBy}(a, b)| = 4$, it means that (a, b), considered as a rule, has been tested 6 times on the case base and has given 4 correct answers. Thus, the proportion of correct answers is $4/6 = 2/3$. This proportion is called the confidence of (a, b), denote by $\texttt{conf}(a, b)$. A special case has to be considered when $\texttt{supp}(a, b) = 0$. This means that the "adaptation rule" (a, b) cannot be tested on the case base. In such a situation, the value of the confidence is set to 0.5 (better than a confidence of, say, 3/7 for which the rule fails more often then it succeeds and worse then a confidence of 4/7 for which it succeeds more often then in fails). To summarize, the confidence of a pair (a, b) is:

$$\texttt{conf}(a, b) = \begin{cases} \dfrac{|\texttt{CorrectlySolvableBy}(a, b)|}{\texttt{supp}(a, b)} & \text{if } \texttt{supp}(a, b) \neq 0 \\ 0.5 & \text{otherwise} \end{cases}$$

3.3   Using Case Pair Competence for Selection and Vote Strategies

Given a target problem $\mathrm{x}^{\mathrm{tgt}}$, extrapolation consists in retrieving triples $(a, b, c) \in \mathrm{CB}^3$ such that $\mathrm{x}^a : \mathrm{x}^b :: \mathrm{x}^c : \mathrm{x}^{\mathrm{tgt}}$ and in adapting this triple by solving the equation $\mathrm{y}^a : \mathrm{y}^b :: \mathrm{y}^c : y$ for each such triples (the triples (a, b, c) for which the equation has no solution are not considered). So, the result of extrapolation is the set $\mathcal{R}$ of $((a, b, c), y) \in \mathrm{CB}^3 \times \mathcal{S}$, y being the result of the extrapolation of (a, b, c) in order to solve $\mathrm{x}^{\mathrm{tgt}}$. Now, the question is how to consider all these solutions y to propose a sole solution $\mathrm{y}^{\mathrm{tgt}}$ of $\mathrm{x}^{\mathrm{tgt}}$. Four strategies for that purpose are detailed below.

The first one, called `withoutComp`, just makes a vote on all values of y, regardless of the competences. The proposed solution is thus

$$\mathrm{y}^{\mathrm{tgt}} = \underset{\widehat{y}}{\mathrm{argmax}} \, |\{((a, b, c), y) \in \mathcal{R} \mid y = \widehat{y}\}|$$

This is the strategy used in [16] and the baseline for the evaluation.

The second strategy, called `allConf`, considers all the $((a, b, c), y) \in \mathcal{R}$ and makes a vote weighted by the confidence:

$$\mathrm{y}^{\mathrm{tgt}} = \underset{\widehat{y}}{\mathrm{argmax}} \sum_{((a,b,c),y) \in \mathcal{R}, y = \widehat{y}} \mathtt{conf}(a, b)$$

The third strategy, called `topConf`, considers only the $((a, b, c), y) \in \mathcal{R}$ with the highest confidence, then makes a vote among them. Formally:

$$\text{with } \mathtt{conf}_{\max} = \max \{\mathtt{conf}(a, b) \mid ((a, b, c), y) \in \mathcal{R}\}$$
$$\text{and } \mathcal{R}^* = \{((a, b, c), y) \in \mathcal{R} \mid \mathtt{conf}(a, b) = \mathtt{conf}_{\max}\}$$
$$\mathrm{y}^{\mathrm{tgt}} = \underset{\widehat{y}}{\mathrm{argmax}} \, |\{((a, b, c), y) \in \mathcal{R}^* \mid y = \widehat{y}\}| \tag{1}$$

The fourth strategy, called `topConfSupp`, is similar to the previous one, except that it uses both the confidence and the support to make a preference. More precisely, it is based on the preference relation $\succcurlyeq$ on case pairs defined below (for $(a, b), (a', b') \in \mathrm{CB}^2$):

$$(a, b) \succcurlyeq (a', b') \quad \text{if} \quad \begin{vmatrix} \mathtt{conf}(a, b) > \mathtt{conf}(a', b') \textbf{ or} \\ (\mathtt{conf}(a, b) = \mathtt{conf}(a', b') \textbf{ and } \mathtt{supp}(a, b) \geq \mathtt{supp}(a', b')) \end{vmatrix}$$

In other words, confidence is the primary criterion, but in case of equality, the higher the support is, the more competent the case pair (a, b) is considered. For instance, if $\mathtt{conf}(a, b) = \mathtt{conf}(a', b') = 0.75$, $\mathtt{supp}(a, b) = 8$ and $\mathtt{supp}(a', b') = 4$, then (a, b) gives the good answer in 6 situations over 8, whereas (a', b') gives the good answer in 3 situations over 4. In this example, (a, b) is strictly preferred to (a', b') —$(a, b) \succ (a', b')$. Now, let $\mathcal{R}^*$ be the set of $((a, b, c), y) \in \mathcal{R}$ such that (a, b) is maximal for $\succcurlyeq$. Then, $\mathrm{y}^{\mathrm{tgt}}$ results from a vote, as described above in Eq. (1).

The interest of considering a triple (a, b, c) in the voting procedure at the end of the inference process is evaluated in terms of the competence of the pair (a, b).

Since analogical proportions are stable under central permutation, one might think of considering the pair (a, c) as well. Preliminary investigations using different combinations (minimum, maximum, sum or product of the confidences of (a, b) and (a, c)) have not shown any clear improvement with respect to the simple use of the competence of (a, b); it is why we have restricted ourselves to this latter type of competence assessment. However, these preliminary investigations were only based on the allConf strategy, so it deserves to be reconsidered: this constitutes a future work.

4   Evaluation

The objective of the evaluation is to study the impact of the strategies for case pair selection and vote presented before on various types of Boolean functions.

4.1   Experiment Setting

In the experiment, $\mathcal{P} = \mathbb{B}^8$ and $\mathcal{S} = \mathbb{B}$. $\rightsquigarrow$ is assumed to be functional: $\rightsquigarrow = \mathtt{f}$, meaning that y is a solution to x if $y = \mathtt{f}(x)$.

The function f is randomly generated using the following generators that are based on two normal forms, with the purpose of having various types of functions:

DNF f is generated in a disjunctive normal form, i.e., $\mathtt{f}(x)$ is a disjunction of n_{disj} conjunctions of literals, for example $\mathtt{f}(x) = (x_1 \wedge \neg x_7) \vee (\neg x_3 \wedge x_7 \wedge x_8) \vee x_4$. The value of n_{disj} is randomly chosen uniformly in $\{3, 4, 5\}$. Each conjunction is generated on the basis of two parameters, $p^+ > 0$ and $p^- > 0$, with $p^+ + p^- < 1$: each variable x_i occurs in the disjunct in a positive (resp. negative) literal with a probability p^+ (resp., p^-). In the experiment, the values $p^+ = p^- = 0.1$ were chosen.[3]

Pol f is generated in polynomial normal form: it is the same as DNF, except that the disjunctions $(\vee)$ are replaced with exclusive or's $(\oplus)$. As only positive literals occur in the polynomial normal form, the parameter $p^- = 0$.

The case base CB is generated randomly, with the values for its size: $|\mathtt{CB}| \in \{32, 64, 96, 128\}$, i.e., $|\mathtt{CB}|$ is between $\frac{1}{8}$ and $\frac{1}{2}$ of $|\mathcal{P}| = 2^8 = 256$. Each source case (x, y) is generated as follows: x is randomly chosen in $\mathcal{P}$ with a uniform distribution and $y = \mathtt{f}(x)$.

---

[3] A generator CNF, generating formulas in CNF (conjunctive normal form: conjunction of disjunctions of literals) could also have been considered. However, this does not add anything new since it is dual with the DNF generator for two reasons. First, the drawn inferences are code-independent, meaning that replacing the attributes by their negations does not change the result of the inference, in particular, for $a, b, c, d \in \mathbb{B}$, $a{:}b{::}c{:}d$ iff $\neg a{:}\neg b{::}\neg c{:}\neg d$. Second, if f is obtained from the DNF generator then $\neg \mathtt{f}$ can be put easily in a function g written in CNF using De Morgan laws, and the distribution of g obtained this way would be the same as the distribution from a CNF generator with the same parameters.

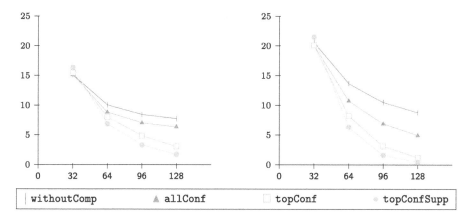

Fig. 1. Error rate function of |CB|, for each generator (DNF at the left, Pol at the right).

Let #tgt_pb be the number of target problems posed to the system, #ans be the number of (correct or incorrect) answers (#tgt_pb − #ans is the number of target problems for which the system fails to propose a solution), and #corr_ans be the number of correct answers. For each selection and vote strategy, the following scores are computed:

The error rate %err is the average of $\left(1 - \dfrac{\#\texttt{corr\_ans}}{\#\texttt{ans}}\right) \times 100 \in [0, 100]$.

The answer rate %ans is the average of the ratios $\dfrac{\#\texttt{ans}}{\#\texttt{tgt\_pb}} \times 100 \in [0, 100]$.

If the system always gives an answer (correct or not) then %ans = 100.

The average is computed on 1 million problem solving for each function generator, requiring the generation of 1420 f for each of them. The average computing time of a CBR session (retrieval and adaptation for solving one problem) is about 2 ms on a current standard laptop. From a complexity point of view, using a hashtable representing the differences within case pairs in CB allows to reduce the complexity of the retrieval step to $O(|CB|^2)$ in the worst case (and frequently closer to $O(|CB|)$ in practice), in addition to an offline part in $O(|CB|^2)$ to generate the hashtable [16].

For the sake of reproducibility, the code for this experiment is available at https://tinyurl.com/analogyCBRTests.

4.2   Results

Table 2 presents the error rate and the answer rate for the different case selection and vote strategies for the two different generators with an application on the different case base sizes. Error rate curves are given in Fig. 1.

Given a function generator and a case base size, the answer rate is the same for the four strategies because all case pair selection strategies provide results

Table 2. %err and %ans for the different selection and vote strategies for the different generators.

		$|CB| = 32$		$|CB| = 64$		$|CB| = 96$		$|CB| = 128$	
		%err	%ans	%err	%ans	%err	%ans	%err	%ans
DNF	withoutComp	15.1	97.5	10.1	100.0	8.4	100.0	7.7	100.0
	allConf	15.1		8.8		7.0		6.3	
	topConf	15.5		8.0		4.9		3.1	
	topConfSupp	16.4		**6.9**		**3.3**		**1.7**	
POL	withoutComp	20.6	95.8	13.7	100.0	10.5	100.0	8.8	100.0
	allConf	**20.1**		10.8		6.9		4.9	
	topConf	20.1		8.2		3.1		1.2	
	topConfSupp	21.5		**6.3**		**1.6**		**0.5**	

for a problem that could be solved, without using competences, by withoutComp (i.e. if a triple was found to solve a case x^{tgt} by the withoutComp strategy, this triple is considered by the three case pair selection strategies and either it will participate in solving x^{tgt} or it exists another "better" triple according to the selection procedure. The answer rate is high for all the methods: over 96% for $|CB| = 32$ and 100% for $|CB| \geq 64$.

Except for $|CB| = 32$, which seems to be a too small training data set for computing competences, the error rate shows that the hypothesis of pair selection improves the precision. For both generators, all pair selection strategies give better results than the baseline (withoutComp). However, the improvement is rather different depending of the selection strategy: the more the selection of pairs is constrained, the more the error rate decreases. allConf decreases the error rate a little bit, topConf decreases the error rate a little bit more, and the best results are given by topConfSupp.

The benefit of all strategies is related to the case base size: the more the case base contains cases for competence acquisition, the better the results are. Comparing to the baseline, the benefit of the best selection strategy topConfSupp is noteworthy. Even if the error rate is already rather good with the baseline, and especially with a 100% answer rate, topConfSupp improves it, making it close to a 100% of correct answers. For DNF, according to the size of the case base (64, 96 and 128), the error rate %err decreases from 10.1 to 6.9 (decreasing of 32%), from 8.4 to 3.3 (decreasing of 61%) and from 7.7 to 1.7 (decreasing of 78%). For Pol, the results are even more impressive: according to the size of the case base (64, 96 and 128), the error rate %err decreases from 13.7 to 6.3 (decreasing of 54%), from 10.5 to 1.6 (decreasing of 85%), and from 8.8 to 0.5 (decreasing of 94%).

So, these first experimental results show that from a given case base size, the topConfSupp strategy overcomes all others and decreases drastically the error rate, while using less triples.

5   Discussion and Related Work

In this section, the approach presented in this paper is compared to related work in CBR according to two viewpoints: the notion of competence in CBR and the adaptation knowledge learning approaches.

Competence in CBR. In [16], three types of CBR processes are distinguished, in particular extrapolation, that retrieves and reuses cases by triples and approximation, that retrieves and reuses cases by singletons. It is argued here that previous researches on competence are related to approximation, whereas the work presented in this paper considers a notion of competence related to extrapolation.

The notion of competence in CBR is used in general for the purpose of case base maintenance, either for deleting the least competent cases [22] or adding the most competent ones [24]. In these previous studies, competence is assessed to individual source cases, in relation to other cases from the case base. In particular, in the seminal paper [22], the competence of cases is assessed by putting source cases into categories (from pivotal cases who are the most competent ones to auxiliary cases), these categories being defined with the help of the binary relation of adaptability between a case and a problem. Thus, this notion of competence is linked with the approximation process (considering individually source cases).

By contrast, the current paper is concerned by competence related to the extrapolation process: cases are retrieved by triples. The competence of a triple $(a, b, c) \in \mathsf{CB}^3$ is reduced to the competence of a pair $(a, b) \in \mathsf{CB}^2$, which is related to the set of the other pairs $(c, d) \in \mathsf{CB}^2$. A common point of these two notions of competence is that the competence of an object (an object being a case for approximation and a case pair for extrapolation) is not an intrinsic property of the object, but is related to other objects (from CB or CB^2).

A minor difference between previous studies on competence and the one defined in this paper is related to the use of competence: case base maintenance for the formers and problem-solving for the latter.

Relations with adaptation knowledge learning. The work presented in this paper has strong links with the issue of adaptation knowledge learning (AKL). The adaptation considered here is the one that follows the retrieval of a sole case (i.e., it is a single case adaptation). Such an adaptation has profit of the adaptation knowledge AK, that can be informally defined by:

AK = "How does the solution changes when the problem changes."

The approach generally applied for AKL is modelled in the seminal work of Hanney and Keane [11]. It uses the case base for learning adaptation knowledge according to the following principle. A set TS of source case pairs (a, b), with $a \neq b$, is built, either by considering all the distinct pairs from CB or by considering only the pairs (a, b) where a and b are judged as enough similar, according to some criterion. Then, TS is used as training set of a supervised learning process: for each pair (a, b) the input of an example is the pair $(\mathbf{x}^a, \mathbf{x}^b)$ and its output is the pair $(\mathbf{y}^a, \mathbf{y}^b)$. The supervised learning process provides a model of this knowledge AK, used by the adaptation process.

Several works are based on this general scheme. In [13], AK consists in the representation of "adaptation cases". In [7], different techniques are used, in particular, decision tree induction and ensemble learning techniques. In [8] the frequent closed itemset extraction is used. The expert interpretation following this extraction produces adaptation rules to be added to AK. In [10], similar techniques as in [8] are used (formal concept analysis and frequent closed itemset extraction are similar data-mining techniques), but, in this work, negative cases (i.e., pairs $(\mathbf{x}, \mathbf{y}) \in \mathcal{P} \times \mathcal{S}$ such that $\mathbf{y}$ is not a correct solution of $\mathbf{x}$) are used, which improves significantly the results of the learning process. In [12], an ensemble approach provides adaptation rules with categorical features.

The work presented in this paper could also be considered as an AKL approach. In fact, in this paper, the term of adaptation rule for considering a case pair (a, b) has been used. Let us make this idea more accurate. Let $\sim$ be the relation defined, for (a, b) and (a', b'), two case pairs, by:

$$(a, b) \sim (a', b') \quad \text{if} \quad \mathbf{x}^a : \mathbf{x}^b :: \mathbf{x}^{a'} : \mathbf{x}^{b'} \text{ and } \mathbf{y}^a : \mathbf{y}^b :: \mathbf{y}^{a'} : \mathbf{y}^{b'}$$

For analogical proportions on nominal representations defined in Sect. 2.2, $\sim$ is an equivalence relation.[4] Thus, solving a problem $\mathbf{x}^{\mathtt{tgt}}$ by extrapolation from a triple $(a, b, c) \in \mathtt{CB}^3$ or from a triple $(a', b', c) \in \mathtt{CB}$ (with the same c) such that $(a, b) \sim (a', b')$ will give the same result: extrapolation is independent from the choice of a representative of the equivalent class of (a, b) for $\sim$. Such an equivalence class $C\ell$ can be used as an adaptation rule (where c is the retrieved case and $\mathbf{x}^{\mathtt{tgt}}$ is the problem to be solved):

> **with** (a, b) arbitrarily chosen in $C\ell$
>
> **if** $\mathbf{x}^a : \mathbf{x}^b :: \mathbf{x}^c : \mathbf{x}^{\mathtt{tgt}}$ and $\mathbf{y}^a : \mathbf{y}^b :: \mathbf{y}^c : y$ has a solution
>
> **then** this solution is a plausible solution to $\mathbf{x}^{\mathtt{tgt}}$

Thus, the set of equivalent classes of the restriction of $\sim$ to $\mathtt{CB}^2$ gives a set of candidate adaptation rules, but all these rules are not equivalently interesting: some gives more plausible results than the other ones. So, a criterion has to be defined for making a preference between these rules and, if it is decided to apply all of them, to do so by making a weighted vote (the more an adaptation rule is preferred, the higher its weight in the vote should be).

A simple way of doing this (used, e.g., in [8]) consists in using the cardinality of $C\ell$. This can be related to the notion of competence of a case pair: if $(a, b) \in C\ell$ then $|C\ell| = \mathtt{supp}(a, b) \times \mathtt{conf}(a, b)$. One limitation of this approach is

---

[4] Reflexivity and symmetry are direct consequences of the postulates with the same names. By contrast, there exist analogical proportions for which transitivity does not hold [15].

that it counts only the examples (supporting the rule), not the counterexamples (penalizing the rule). By contrast, the approach presented in this paper takes into account counterexamples. For example, if $\mathtt{conf}(a, b) = 1/3$, then, for each example of the rule, there are two counterexamples, so, even if $\mathtt{supp}(a, b)$ is large, the rule associated to (a, b) is, at best, questionable.

Another difference with the work of [8] is that, in [8], when several case pairs have the same variations only on a subset of the features, they are still used to build an adaptation rule. For example, if (a, b) and (a', b') are two source case pairs such that for *most* attributes i, $\mathbf{x}_i^a : \mathbf{x}_i^b :: \mathbf{x}_i^{a'} : \mathbf{x}_i^{b'}$, the rules built on these common attributes are considered, neglecting the other attributes. In a formal framework in which analogies are rare (for example, when there are features with real number values), it could be justified to replace the exact analogical proportion with a gradual analogical proportion [9] in the approach described in this paper. Studying it constitutes a potential future work.

This discussion shows how some ideas related to AKL from the case base can be easily reformulated in the framework of analogical proportions: the links so established between these two fields is therefore potentially fruitful.

6   Conclusion

Classical case-based reasoning relies on the individual similarities of the problem at hand with each already solved problem that is known. We have shown that it may be also of interest to consider triples of cases (a, b, c) in order to equalize the change from a to b with the change from c to the problem to be solved with its tentative solution. This is the basis of analogical extrapolation based on analogical proportions. Still it has been observed that some triples may lead to wrong inferences.

In this paper, we have proposed to discriminate triples according to an evaluation of the "competence" of the pairs involved in the triples. Indeed an analogical proportion "a is to b as c is to d" can be viewed as establishing a parallel between two pairs. The differences between the components of a pair of problems are naturally related to the differences between solutions, but this relation may depend on the context expressed by the component values that do not change. We have shown that it was possible, at least to some extent, to evaluate the competence of pairs for selecting "good" triples and improving analogical inference results. This contributes to confirm the interest of analogical extrapolation for case-based reasoning.

Several future works follow these studies.

The first one has been mentioned at the end of Sect. 3. It consists, when choosing a triple (a, b, c), in considering not only the competence of the pair (a, b) but also the competence of the pair (a, c). Preliminary studies with the strategy $\mathtt{allConf}$ where carried out that does not give significant changes in the result. However, this may be different for the other strategies, and this remains to be studied.

Another future work will be to transfer contributions from the adaptation knowledge learning field to improve furthermore the performance of analogical extrapolation (cf. Section 5). In particular, a promising direction of work is the use of a base of negative cases, as it has been used in [10].

The representation framework of this research is the one of nominal representations, especially the Boolean setting. Another research direction would be to study how the idea of case pair competence introduced here can be handled in practice when numerical representations are used.

A related future work concerns the evaluation in various settings: this could use benchmarks as the ones developed in the machine learning community. Actually, in [4], a work similar to analogical proportion in CBR has been evaluated with such benchmarks with good results, and the idea could be to extend this work by using the notion of case pair competence.

Finally, the competence of case pairs can be used in order to associate to a solution proposed by extrapolation an indication of its plausibility, according to the following idea: the higher are the competences of the case pairs used for giving a solution, the more plausible the proposed solution is. This can be used in order to combine analogical extrapolation with other approaches to CBR that also provides an indication of plausibility.

References

1. Aamodt, A., Plaza, E.: Case-based reasoning: foundational issues, methodological variations, and system approaches. AI Commun. **7**(1), 39–59 (1994)
2. Billingsley, R., Prade, H., Richard, G., Williams, M.-A.: Towards analogy-based decision - a proposal. In: Christiansen, H., Jaudoin, H., Chountas, P., Andreasen, T., Legind Larsen, H. (eds.) FQAS 2017. LNCS (LNAI), vol. 10333, pp. 28–35. Springer, Cham (2017). https://doi.org/10.1007/978-3-319-59692-1_3
3. Bounhas, M., Prade, H., Richard, G.: Analogical classification: a rule-based view. In: Laurent, A., Strauss, O., Bouchon-Meunier, B., Yager, R.R. (eds.) IPMU 2014. CCIS, vol. 443, pp. 485–495. Springer, Cham (2014). https://doi.org/10.1007/978-3-319-08855-6_49
4. Bounhas, M., Prade, H., Richard, G.: Analogy-based classifiers for nominal or numerical data. Int. J. Approx. Reason. **91**, 36–55 (2017)
5. Correa, W.F., Prade, H., Richard, G.: Trying to understand how analogical classifiers work. In: Hüllermeier, E., Link, S., Fober, T., Seeger, B. (eds.) SUM 2012. LNCS (LNAI), vol. 7520, pp. 582–589. Springer, Heidelberg (2012). https://doi.org/10.1007/978-3-642-33362-0_46
6. Couceiro, M., Hug, N., Prade, H., Richard, G.: Analogy-preserving functions: a way to extend Boolean samples. In: Proceedings of the 26th International Joint Conference on Artificial Intelligence (IJCAI 2017), pp. 1575–1581. Morgan Kaufmann, Inc. (2017)
7. Craw, S., Wiratunga, N., Rowe, R.C.: Learning adaptation knowledge to improve case-based reasoning. Artif. Intell. **170**(16–17), 1175–1192 (2006)
8. d'Aquin, M., Badra, F., Lafrogne, S., Lieber, J., Napoli, A., Szathmary, L.: Case base mining for adaptation knowledge acquisition. In Veloso, M.M. (ed.) Proceedings of the 20th International Joint Conference on Artificial Intelligence (IJCAI 2007), pp. 750–755. Morgan Kaufmann, Inc. (2007)
9. Dubois, D., Prade, H., Richard, G.: Multiple-valued extensions of analogical proportions. Fuzzy Sets Syst. **292**, 193–202 (2016)

10. Gillard, T., Lieber, J., Nauer, E.: Improving adaptation knowledge discovery by exploiting negative cases: first experiment in a Boolean setting. In: Proceedings of ICCBR 2018–26th International Conference on Case-Based Reasoning, Stockholm, Sweden, July 2018

11. Hanney, K., Keane, M.T.: Learning adaptation rules from a case-base. In: Smith, I., Faltings, B. (eds.) EWCBR 1996. LNCS, vol. 1168, pp. 179–192. Springer, Heidelberg (1996). https://doi.org/10.1007/BFb0020610

12. Jalali, V., Leake, D., Forouzandehmehr, N.: Learning and applying adaptation rules for categorical features: an ensemble approach. AI Commun. **30**(3–4), 193–205 (2017)

13. Jarmulak, J., Craw, S., Rowe, R.: Using case-base data to learn adaptation knowledge for design. In: Proceedings of the 17th International Joint Conference on Artificial Intelligence (IJCAI 2001), pp. 1011–1016. Morgan Kaufmann, Inc. (2001)

14. Kolodner, J.: Case-Based Reasoning. Morgan Kaufmann Inc., Burlington (1993)

15. Lepage, Y.: Proportional analogy in written language data. In: Gala, N., Rapp, R., Bel-Enguix, G. (eds.) Language Production, Cognition, and the Lexicon. TSLT, vol. 48, pp. 151–173. Springer, Cham (2015). https://doi.org/10.1007/978-3-319-08043-7_10

16. Lieber, J., Nauer, E., Prade, H., Richard, G.: Making the best of cases by approximation, interpolation and extrapolation. In: Cox, M.T., Funk, P., Begum, S. (eds.) ICCBR 2018. LNCS (LNAI), vol. 11156, pp. 580–596. Springer, Cham (2018). https://doi.org/10.1007/978-3-030-01081-2_38

17. Miclet, L., Prade, H.: Handling analogical proportions in classical logic and fuzzy logics settings. In: Sossai, C., Chemello, G. (eds.) ECSQARU 2009. LNCS (LNAI), vol. 5590, pp. 638–650. Springer, Heidelberg (2009). https://doi.org/10.1007/978-3-642-02906-6_55

18. Prade, H., Richard, G.: A discussion of analogical-proportion based inference. In Sánchez-Ruiz, A.A., Kofod-Petersen, A. (eds.) Proceedings of ICCBR 2017 Workshops (CAW, CBRDL, PO-CBR), Doctoral Consortium, and Competitions Co-located with the 25th International Conference on Case-Based Reasoning (ICCBR 2017), Trondheim, 26–28 June, vol. 2028 of CEUR Workshop Proceedings, pp. 73–82 (2017)

19. Prade, H., Richard, G.: Analogical proportions: from equality to inequality. Int. J. Approx. Reason. **101**, 234–254 (2018)

20. Richter, M.M., Weber, R.O.: Case-Based Reasoning, A Textbook. Springer, Heidelberg (2013). https://doi.org/10.1007/978-3-642-40167-1

21. Riesbeck, C.K., Schank, R.C.: Inside Case-Based Reasoning. Lawrence Erlbaum Associates Inc., Hillsdale (1989). Available on line

22. Smyth, B., Keane, M.T.: Remembering to forget. In: Proceedings of the 14th International Joint Conference on Artificial Intelligence (IJCAI 1995), Montréal (1995)

23. Stroppa, N., Yvon, F.: Analogical learning and formal proportions: definitions and methodological issues. Technical Report D004, ENST-Paris (2005)

24. Zhu, J., Yang, Q.: Remembering to add: competence-preserving case-addition policies for case base maintenance. In: Proceedings of the 16th International Joint Conference on Artificial Intelligence (IJCAI 1999), pp. 234–241 (1999)

Towards Finding Flow in Tetris

Diana Sofía Lora Ariza$^{(\boxtimes)}$, Antonio A. Sánchez-Ruiz,
and Pedro A. González-Calero

Dep. Ingeniería del Software e Inteligencia Artificial,
Universidad Complutense de Madrid, Madrid, Spain
{dlora,antsanch,pagoncal}@ucm.es

Abstract. One of the most challenging goals in the game industry is to design games which are difficult enough to be a fun challenge but not so hard to provoke frustration among a wide range of different types of players. Dynamic difficulty adjustment (DDA) is a set of techniques used to customize the difficulty of a game according to the skill level of the player so that the game can keep the player "flowing".

In this paper, we present a novel DDA architecture that we implement using case-based reasoning and we integrate into a Tetris game. In particular, we dynamically change the difficulty of the game by selecting the next piece the player has to place on the board to make the current game more similar to one of the "good" games in our case base. Games are modeled using time series representing the evolution of different game features and evaluated by the players according to their level of entertainment. This way, we alter the difficulty of the game so that it evolves similarly to other previous good games and we expect the current player also experience the same *flow*.

Keywords: Dynamic difficulty adjustment · Flow ·
Case-based reasoning · Video games · Tetris

1 Introduction

Video games are part of daily life for many of us. Although the main goal of a video game is entertainment, they have also been used successfully with other goals such as learning or even health but, regardless of the goals, games have to entertain to keep the players engaged. Entertainment can be measured by the level of immersion of the player within the game [23,24], where immersion is a state of consciousness in which awareness of physical self is lost by being in an artificial environment [19]. Csikszentmihalyi [8] used the term *flow* to describe this state of optimal or "peak" experience. To achieve it, players should perceive challenges that enhance their skills, clear goals to overcome and receive immediate feedback about their actions. Moreover, the challenges in the game

This work has been partially supported by the Spanish Committee of Economy and Competitiveness (TIN2017-87330-R) and the UCM (Group 921330).

K. Bach and C. Marling (Eds.): ICCBR 2019, LNAI 11680, pp. 266–280, 2019.
https://doi.org/10.1007/978-3-030-29249-2_18

should be at the "right" difficulty level so that the players feel challenged but not overwhelmed [3,5].

Nowadays, most of the game content is static and created during the development phase of video game production. Game designers try to anticipate all the possible ways to play the game, so the players never get into scenarios with obstacles that are too difficult or too easy for them. Due to the complexity of this task, game designers value the feedback provided by the players, because it helps them to distinguish excessively difficult tasks from good challenges [20]. However, computer games can also benefit from a certain level of "intelligence" to decide what to do in those scenarios the game designers could not anticipate, and provide appropriate responses to changing circumstances in realistic environments [6].

Dynamic difficulty adjustment (DDA) can be used to automatically alter the difficulty of the game based on the performance of the player [14,17,25]. There exist different player modeling techniques to recognize typical user behaviors and personalize the game experience for different player profiles. In particular, several researchers have successfully applied various machine learning techniques to automatically model different types of players [10,11,16,21].

In our previous work regarding DDA and Tetris [1], we used case-based reasoning to predict the skill level of the player dynamically, and then we made the game easier for newbie and average players by providing easy-to-place next pieces now and then. Our experiments showed that players perceived these games as more entertaining as long as they did not realize we were modifying the difficulty of the game.

In this paper we propose a different approach, instead of helping the players depending on their skill level, we try to improve the player experience by altering the game in such a way that it evolves similarly to some of the "good" games in our case base. The characterization of a game is based on time series representing the evolution of game features such as the score, the height of the board or the number of "holes", and evaluated by the players according to their level of entertainment. By making the current game similar to one of the previous good games, we expect the current player to experience a similar state of flow. The main contributions of this work are: (1) a general DDA architecture that can be used in different games, (2) the implementation of this architecture using case-based reasoning (CBR) on game features collected in a time window, and (3) a mechanism to alter the difficulty of Tetris taking into account both the current game state and the player profile.

The paper continues as follows. Section 2 provides an overview of The Flow Theory, particularly in the context of video games. Section 3 describes the specific version of Tetris used in our work. Next, Sect. 4 explains the features we use to characterize Tetris games and how we transform that data into cases. Section 5 explains our DDA architecture and how its different components work to adjust the difficulty of a game dynamically. After that, Sect. 6 describes some experiments performed with new players and our results. Finally, the paper closes with related work, conclusions, and future work.

2   The Flow Theory

According to Nakamura and Csikszentmihalyi [18], entering flow depends on creating the perfect balance between the perceived challenges of the task at hand and their own perceived skills. More concretely, they stated that *"It is the subjective challenges and subjective skills, not objective ones, that influence the quality of a person's experience."*. So to enter flow, one must have confidence in his ability to complete the task at hand. Attention also plays an essential role in achieving a long-lasting flow feeling. That is why it is crucial to have clear goals (they add direction and structure to the task) and immediate feedback (to be able to adjust performance and maintain flow).

There are eight emotions a person might experience while doing a task (Fig. 1). The state of flow happens when the task at hand is exciting and challenging, but achievable, between the emotions of control and arousal. These feelings intensify when challenges and skills move away from the player's average levels. Moreover, the challenges present in an activity need to grow in difficulty as the person masters his skills; otherwise, the task stops being entertaining. That is why it is so difficult to achieve and maintain that state of equilibrium over time.

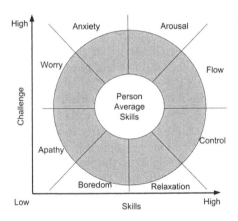

Fig. 1. The emotions a person might experience while doing a task [18].

In the context of video games, dynamic difficulty adjustment techniques can be used to maintain the player in that state of equilibrium as the player progresses in the game and masters his abilities, so that the challenges in the game (enemies or puzzles) evolve at the same rate. In this work, we propose to subtly modify the difficulty of the game so that different game parameters change similarly as they did in previous games that were rated as good by the players.

3   Tetris Analytics

Tetris is a very popular video game in which the player must place different tetromino pieces that fall from the top of the screen in a rectangular game

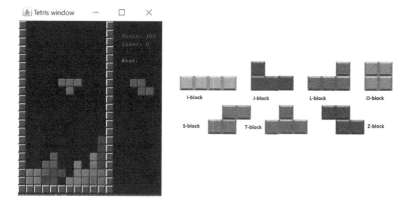

Fig. 2. Tetris analytics screen capture and Tetris pieces names.

board. When a row of the game board is complete, i.e. it has no holes, the row disappears, all pieces above drop one row, and the player is rewarded with some points. As the game progresses, the new pieces fall faster and faster, gradually increasing the difficulty, until a piece is placed such that it exceeds the top of the board and the game ends. The goal of the game is to complete as many lines as possible to increase the score and make room for the next pieces. Although the game is quite simple to play, it is also very difficult to master and hard to solve from a computational point of view [2].

In our experiments, we use *Tetris Analytics* (Fig. 2), a version of the game implemented in Java that looks like ordinary Tetris game from the point of view of the player, but internally provides extra functionality to extract, store and reproduce game traces. From these traces, we can select the most significant features to characterize the playing style of each player, determine her skill level and dynamically adjust the difficulty of the game.

Each time a new piece appears on the top of the board, players have to make two different decisions. The first one, that we call *tactical*, is to decide where to settle the piece, that is, its final location. The second decision involves all the specific *moves* (side-to-side and rotations) required to lead the piece to that final location.

Currently, we only consider the tactical decisions to define the skill level of the player in this paper, but we are aware that we could also extract valuable information from the particular moves (looking at parameters like speed, cadence and moves undone) we plan to extend our work to consider them in the future.

4   Game Features and Cases

Tetris Analytics provides the infrastructure to extract different game features both on-line during the game and off-line from the game traces. Since we only consider the tactical decisions of the player, we only analyze the game state when a new piece is settled in its final location. We extract the following features:

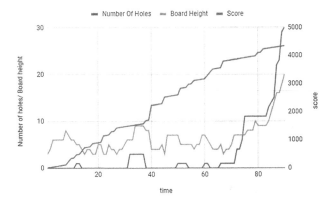

Fig. 3. Feature evolution of an average player.

- *Number of piece* from the beginning of the game. It represents the current time in the game.
- *Current score* obtained by the player after placing the current piece.
- *Number of holes* or empty spaces under other pieces in the same column. Good players tend to compact the pieces to ease the completion of new lines.
- *Board height* or the highest row occupied by a piece in the board. Good players tend to play most of the game in the lowest half of the board because each time they complete a line, the height of the board decreases by one unit.
- *Piece type* among the 7 different pieces available in the game (Fig. 2).

For example, Fig. 3 shows the evolution of these features in a particular game. This player was able to settle 90 pieces before the end of the game (x-axis). The yellow line describes the evolution of the score that grows slightly with each new piece and more abruptly when a new line is completed. The red line represents the height of the board (1 to 20). During most of the game the player plays in the lower half of the board, but at the end the height grows really fast probably because the pieces fall too fast. Similarly, the blue line represents the number of holes in the board and during the last quarter of the game increases very fast.

Once the game ends, we split the time series into non-overlapping chunks of 10 pieces to create cases (see [1] for other case representations). Therefore, a case represents the evolution of a game during a small time window. Each case contains the following features:

- Number of the piece counting from the beginning of the game.
- Time series with the evolution of the score.
- Time series with the evolution of the number of holes.
- Time series with the evolution of the board height.
- Skill level of the player.
- How good the game is.

The last 2 features can only be computed when the game ends. Regarding the skill level, we differentiate 3 different skill levels depending on the final game score: *newbie* (0–2999), *average* (3000–5499) and *expert* (5500 or more). Although the score frontier between categories is somewhat arbitrary, the separation of players into three groups according to their skill level is quite common in video games.

When the game ends, we ask the player to evaluate the sentence *It was a good game* with a Likert scale of 5 values where 1 is *strongly disagree* and 5 is *strongly agree*. We use a simple and open question to evaluate the level of satisfaction and general positive or negative feeling about the game. We consider good game as only those with a score of 4 or 5. Then we annotate that score in all the cases extracted from that game.

In order to retrieve similar cases from the case base we use a linear combination of the similarities between the time series, that are compared using an Euclidean distance:

$$sim_c(c_1, c_2) = \alpha_1 \times sim_{ts}(c_1.score, c_2.score) +$$
$$\alpha_2 \times sim_{ts}(c_1.holes, c_2.holes) +$$
$$\alpha_3 \times sim_{ts}(c_1.height, c_2.height)$$

$$sim_{ts}(r, s) = 1 - \sqrt{\sum_{i=1}^{n}(r_i - s_i)^2}$$

where the weights (α_1, α_2, α_3) can be tuned to give more or less importance to each feature and n is the size of the time series containing the evolution of the player in a time range. The most significant feature in the similarity measure is the game score ($\alpha_1 = 0.70$), followed by the number of holes ($\alpha_2 = 0.25$) and finally the board height ($\alpha_3 = 0.05$). These weights were computed using an exhaustive grid search with increments of 0.05.

The case base has 266 games with 22497 tactical moves; each one classified adequately with the player profile based on the total score of each match. Since each case contains the evolution of the player in 10 consecutive tactical moves, we have a total of 2249 cases in our case base. From all the games, 38% are newbies, 47% are averages, and 15% are experts. The average duration time of a newbie game is 53.5 tactical moves with a standard deviation of 13.3. For average games is 93.9 tactical moves with a standard deviation of 11.6. Experts have a game-time duration of 132.0 tactical moves with 8.1 of standard deviation.

Additionally, players with better skill set give higher flow score to the games they play. The average flow score for newbies is 3.19, for averages 3.61, and experts 4.00.

As we will explain in the following Section, the case base is used to solve two different problems: to predict the player profile or skill level, and to decide whether or not to alter the difficulty of the game. In both cases, we only construct cases every ten pieces, and we only consider cases from the same instant in the game. Note that the same values of the score, the board height, and the number of holes have very different meanings depending on the instant of the game when they are measured.

5   DDA System Architecture

Our DDA architecture consists of three main modules (Algorithm 1). The first one decides whether or not to alter the difficulty based on the evolution of the game during the last time window. The second module analyzes the performance of players to predict their profile dynamically. The third module takes into account both the evolution of the game and the profile of the player, to decide how to alter the difficulty of the game. When the DDA system is not active, or it decides not to change the difficulty, the game goes on as it is designed.

```
while game is active do
    game step;
    if DDA is active and is time to intervene then
        get latest performance of the player;
        predict player profile;
        if player has low-average skill level then
            compare if player is doing better or worse against past games with
            same skill level;
            if player is doing worse then
                | return helpful behavior;
            else
                | return random behavior;
            end
        else
            | return random behavior;
        end
    else
        | return random behavior;
    end
end
```

Algorithm 1: DDA system architecture flow.

In the following sections we explain our current implementation of these modules in the context of Tetris.

5.1   To Adjust or Not to Adjust the Difficulty

The DDA system decides to weather help the player or not every 10 pieces. This time threshold helps to provide some inertia to the decisions, so the system is not activating and deactivating too frequently. The 10 pieces time window it is also essential to measure the evolution of the game and compare it with other game fragments from our case base. Moreover, our current implementation of DDA only changes the difficulty of the game in one direction: to make it easier. We plan to extend it soon also to increase the difficulty of the game, so it becomes more challenging for expert players.

In order to decide whether or not to help the player, we create a *query* with the evolution of the score, board height and number of holes during the last 10 pieces or tactical decisions. Then, we retrieve the most similar case from the case based using the similarity measure described in Sect. 4 and considering only the cases extracted from "good" games (those evaluated by players with a score of 4 or 5).

The module will decide to help the player if she is performing worse than the situation described in the retrieved case. In particular, the module will make the game easier for the next 10 pieces if at least 2 of the following conditions are met:

– The average score is less than the average scores in the retrieved case
– The average number of holes is greater than the average number of holes in the retrieved case
– The average board height is greater than the average board height in the retrieved case

Note that our approach could be easily extended to make the game more difficult if the player is performing better than the situation described in the retrieved case.

5.2 Predicting the Player Profile

As we explained in Sect. 4, we distinguish 3 different player profiles or skill levels (newbie, average and expert) depending on the final score of the game. Being able to predict the skill level of the player during gameplay is a challenge though. In our previous work [1], we presented a CBR approach based on comparing the evolution of the 3 game features during the last 10 tactical decisions with a case base created from previous games. The skill level is predicted using k-Nearest Neighbor and a majority vote. We also use an inertia function that forbids extreme prediction changes from newbie to expert or vice versa.

5.3 Changing the Game Difficulty

The mechanisms available to change the difficulty of the game are different in each game. In Tetris there are two obvious ways to adjust the difficulty of the game: either to alter the speed of the falling pieces or to change the type of piece to appear next. We use the second approach because it is more difficult to detect and, from previous experiences, we know some players are not happy when they realize we modify the difficulty of the game for each player.

In our previous work, when we wanted to make the game easier we provided easy-to-place next pieces for the current board configuration. We computed all the possible ways to place each type of piece in the board and then ranked each piece according to a measure that takes into account the game score, the number of holes and the board height after settling that piece in the best possible

Table 1. Rank of easiest piece to settle by profile. See Fig. 2.

Profile	1	2	3	4	5	6	7
Expert	T	J	L	Z	S	I	O
Average	T	L	J	I	S	Z	O
Newbie	I	J	O	T	L	Z	S

position, according to that measure. Then, we randomly selected one of the three best pieces as the next piece for the player.

In this work, we consider not only the current board configuration but also the player profile to select the next piece. We have realized that different players play better with some types of pieces. For example, Table 1 shows the Tetris pieces ordered by how easy it is to settle in a good place for our 3 player profiles. If a piece is at the end of the rank, it does not mean that the profile never settles it well or that the game is never going to give that piece to the player. For the reader that has previously played Tetris it may seem counter-intuitive that pieces like *O-block* or *I-block* are harder to place than a *T-block* or a *L-block*, for average and expert players. The rationale is that the more curves a piece has, the easier it is for the player to identify where to optimally place it, in terms of the number of holes and the board height. For example, in Fig. 2 you may see 3 possible locations where you would put the piece *T-block*. But if instead of a *T-block*, the piece falling is a *I-block*, you may think of more than three places to settle the piece, and that would make it harder for the player to identify the best place.

In order to rank the types of pieces for each profile, we analyzed 11600 tactical decisions from previous games. We considered a tactical decision to be correct if the piece was settled in one of the best 3 final positions according to our heuristic. Then we computed the percentage of pieces of each type that each player profile placed correctly (see Table 2). At this point, we have two sets of good pieces: (1) the 3 best pieces for the current board, according to the value of the resulting board after settling that piece in the optimum place, and (2) the 3 best pieces for current player profile. If the intersection of the sets is not empty, we randomly select one piece from the intersection. If there is no intersection, we randomly select one of the pieces from the first set. In any case, it is always important to add some randomness to the choice of the next piece so it is not easy for the player to detect strange patterns or pieces that appear much more often than others.

Table 2. Percentage of times a profile places a piece in a good location.

Piece name	Optimal Moves % Correctness			
	Expert	Average	Newbie	General
I	78.7	79.1	**83.3**	79.1
T	**86.3**	**83.3**	75.0	**84.4**
O	75.1	74.9	76.7	75.1
Z	83.2	75.4	69.5	78.9
S	83.1	77.6	67.5	80.0
J	84.3	82.1	77.6	83.0
L	84.0	82.5	73.2	82.9

6   Experiments and Results

We experimented to test whether our approach raises the satisfaction of the player compared to the original game. The level of flow during a game is a very complex and subjective feeling that depends on several factors, some of them external to the game itself. For these reasons, we need additional tests to verify our findings.

We asked 26 people with different levels of experience to play Tetris 6 times. Half of the games were the original version of Tetris and the other half games with our DDA systems active. Although Tetris is a very famous game and most of the people already knew how to play, we decided to rule out the first two games of each player to remove the effects due to the warm-up phase. After discarding the first two games of each player, we classified 104 games according to their final score. From these data we found 42 games correspond to newbie players, 60 games to average players and only two games to an expert player.

Players did not know the specific goal of our experiment; they only knew that we were interested in measuring their flow level in different games. After each game ended, the players had to evaluate the sentence *It was a good game* with a Likert scale of 5 values were one means *strongly disagree*, and five means *strongly agree*. We use a simple and open statement to evaluate the level of satisfaction of the player to collect their general impression if they have had a positive or negative feeling.

Figure 4 shows the distribution of flow levels in games with and without DDA. In other words, the chart shows the number of games vs. each possible flow level score (1–5). Most of the games with DDA (green bars) have a score of 4 or 5 while most of the games without DDA (red bars) obtain a score of 3 and 4. In particular, 70% of the games with DDA were evaluated as good (rating 4 or 5) while this percentage goes down to 50% for games without DDA.

As a collateral effect, DDA also influenced the game scores. The average final rate for regular games was 2775 and for games with DDA 3413 (22.9% higher). However, this is an expected result because we are making the games easier.

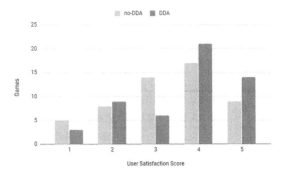

Fig. 4. Distribution of player's flow levels with and without DDA. (Color figure online)

When we implement DDA to make the game behavior also more difficult and not only easier, we expect this difference in the final scores will decrease.

We computed the average and standard deviation of the flow level scores for games with and without DDA. Without DDA the average rate is 3.27, while this value goes up to 3.63 for games with DDA active. Even though, we know we have a positive impact on players experience, the difference it is not statistically significant under a student's t-test. Seeing that, we decided to analyze both sub-groups more in-depth.

We characterized the player's behavior in good and bad games. We consider a game is a good one if the player gives it a rating of 4 or 5 in the user experience questionnaire, and a bad game gets a score of 1 or 2. We grouped the data per profile, good/bad games, and non-DDA/DDA, and then applied k-means clustering to the resulting subset. The idea is to analyze the average behavior of each player profile in a specific scenario through the centroids of the resulting clusters.

Figure 5 shows newbie's behavior when DDA occurs or not and how they rate their game experience. The red dotted line represents the bad games and the blue line the good games. The left figure has non-DDA games. Here we can observe that the bad games board height rise dramatically for this type of players, which might result in a high stress or frustration level. In the right side, we can see the good and bad DDA games. The bad games rise quite rapidly, but these do not increase as fast as non-DDA ones. Regarding the good DDA games, we can see time ranges were the variable rises, then lowers down, and then rises again; for example, the first 30 tactical moves.

Figure 6 shows the behavior of average players for games with and without DDA. The dotted red line represents the evolution of the board height variable for the games classified as bad, whereas the blue line is for the good games. In the left figure, we have the games without DDA, where we observe that the variable in the bad games is quite fluctuating. Between movement 30 and 50, the variable drops by approximately half the value it reached in the tactical movement 20. Players with a certain level of ability tend to like games that challenge them and does not have moments where it is too easy since it becomes boring. In

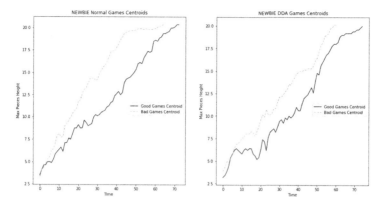

Fig. 5. Board height of Newbies player's games filtered by good/bad games and DDA/non-DDA. The left image shows the regular/non-DDA games filtered by good and bad games. The right image shows the DDA games filtered by good and bad games.

the right figure, we see the opposite behavior. The board height variable of the bad games with DDA goes up approximately half of the board and does not go back down for quite some time. This means that the game is not able to identify that although the player had a certain skill level, he still needs help at certain times. Having a game that creates a constant feeling of anxiety because it is too complicated is not optimal for the player's experience either. In contrast, for both cases of games with and without DDA, it is observed that the blue line rises a bit at the beginning but tends to fluctuate between tactical movements 20–80. In this interval of time, we see that the player tries to decrease the height of the board, and in some moments, he achieves it.

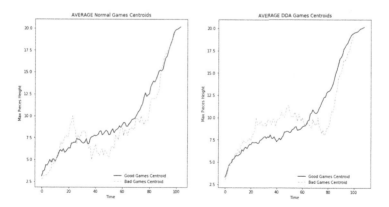

Fig. 6. Board height of Average player's games filtered by good/bad games and DDA/non-DDA. The left image shows the normal/non-DDA games filtered by good and bad games. The right image shows the DDA games filtered by good and bad games.

A key conclusion of this clustering analysis is that both in newbies and average players the evolution of good games, with and without DDA, are quite similar, while the evolution of bad games is more dissimilar when comparing bad games with and without DDA. This not only reminds us of famous Leo Tolstoy quote "All happy families are alike; each unhappy family is unhappy in its own way", but also shows us that our DDA technique does not fully capture the desired evolution represented in good cases, and introduces new types of bad cases. As future work we plan to use the results of this analysis to find new ways of reproducing the flow of good games.

7   Related Work

There is much interest in how dynamic difficulty adjustment improves the user experience. Mainly, machine learning techniques are used to adjust the game based on players performance creating complex behaviors [4,9,12,13,22]. Lopes and Bidarra [15] surveyed the state of art of adaptivity in games and simulations, from both academia and industry. They concluded that games and simulations adaptivity is establishing itself as a rapidly maturing field and that the advances show good results in adapting to an optimal challenge level and active states like fun, frustration, predictability, anxiety or boredom.

Missura and Gärtner [17] used a simple game where the player shoots down alien spaceships while those shoot back. They aimed to employ dynamic difficulty adjustments by grouping players into different profiles and supervised prediction from short traces of gameplay. Each game had a limit of 100 s. The first 30 s were used to acquire data, and the rest of the game, they adjust the aliens' spaceships speed based on the player's performance. We used this paper as inspiration. From the beginning of each game in Tetris Analytics, we wait for at least ten tactical moves, and then we start evaluating the player's performance. This interval of time to gather data about the player's behavior gives us the opportunity to assess the evolution of the player in a time range and increase the accuracy of the player profiling model. Once we determine the skill level of the player, the game decides whether to help him or not.

Furthermore, Arzate and Ramirez [7] describe a theoretical analysis on approaches of the Theory of Flow [18] in the video games domain and categorize them into four groups. One of them is mapping antecedents of flow to features of the game. In our work, we try to identify how the feature's behavior we monitored in Tetris Analytics can affect flow in games, and try to recreate this behavior in the current one.

8   Conclusions and Future Work

In this paper, we present a DDA approach to adjust a game based on a player's performance during gameplay. We use a game called *Tetris Analytics* for our experiments, which is like a simple Tetris game from the players' point of view, but it allows us to monitor and extract variables related to player performance and game state. The primary objective of the player during gameplay is to avoid the pieces reaching the top of the board by doing lines. Eventually, the pieces go down so fast that the player is not able to position them in the right spot, the pieces will reach the top of the board, and the player will lose. With the data extracted from player behavior and game state, we create a case base and use it with a DDA system that helps us improve player satisfaction.

Our experiments showed that players enjoy more games where the DDA system is active than when it is not. The players scored *good* in the user satisfaction questionnaire in most of the DDA games. Plus, the average total score of DDA games is higher than in regular games. Unfortunately, this improvement is not statistically significant between non-DDA and DDA games, which is our goal. However, we discovered several scenarios were players do not feel the challenge is at their level of expertise, e.g., when the game is too hard all the time, or when the game it is hard in a time range, and next is too easy. In the latter scenario, the player might lose interest when the game suddenly becomes too easy.

As future work, we would like to explore other options on how to recreate an ideal flow experience using previous games. At the moment, we only make the game easier for newbies and averages players, but we also want to make it harder if the game is starting to be too easy. Moreover, we would like to confirm the flow feeling of the player with standardized flow questionnaires and biometric data gathering. And create a general DDA architecture that other games can use improve users experience.

References

1. Lora Ariza, D.S., Sánchez-Ruiz, A.A., González-Calero, P.A.: Time series and case-based reasoning for an intelligent tetris game. In: Aha, D.W., Lieber, J. (eds.) ICCBR 2017. LNCS (LNAI), vol. 10339, pp. 185–199. Springer, Cham (2017). https://doi.org/10.1007/978-3-319-61030-6_13
2. Breukelaar, R., Demaine, E.D., Hohenberger, S., Hoogeboom, H.J., Kosters, W.A., Liben-Nowell, D.: Tetris is hard, even to approximate. Int. J. Comput. Geom. Appl. **14**(1–2), 41–68 (2004)
3. Buro, M., Furtak, T.: RTS games as test-bed for real-time AI research. In: Proceedings of the 7th Joint Conference on Information Science (JCIS 2003), vol. 2003, pp. 481–484 (2003)
4. Charles, D., Black, M.: Dynamic player modelling: a framework for player-centred digital games. In: Proceedings of 5th International Conference on Computer Games: Artificial Intelligence, Design and Education (CGAIDE 2004), pp. 29–35 (2004)
5. Charles, D., Kerr, A., McNeill, M.: Player-centred game design: player modelling and adaptive digital games. In: Proceedings of the Digital Games Research Conference, vol. 285, pp. 285–298 (2005)
6. Charles, D., et al.: Player-centred game design: player modelling and adaptive digital games. In: Proceedings of the Digital Games Research Conference, vol. 285, p. 00100 (2005)
7. Cruz, C.A., Uresti, J.A.R.: Player-centered game AI from a flow perspective: towards a better understanding of past trends and future directions. Entertain. Comput. **20**, 11–24 (2017). http://www.sciencedirect.com/science/article/pii/S1875952117300095

8. Csikszentmihalyi, M., Csikszentmihalyi, I.S.: Optimal experience: Psychological Studies of Flow in Consciousness. Cambridge University Press, Cambridge (1992)
9. Denisova, A., Cairns, P.: Adaptation in digital games: the effect of challenge adjustment on player performance and experience. In: Proceedings of the 2015 Annual Symposium on Computer-Human Interaction in Play, pp. 97–101. ACM (2015)
10. Drachen, A., Thurau, C., Sifa, R., Bauckhage, C.: A comparison of methods for player clustering via behavioral telemetry. arXiv preprint arXiv:1407.3950 (2014)
11. Drachen, A., Sifa, R., Bauckhage, C., Thurau, C.: Guns, swords and data: clustering of player behavior in computer games in the wild. In: 2012 IEEE Conference on Computational Intelligence and Games, CIG 2012, pp. 163–170 (2012)
12. Fagan, M., Cunningham, P.: Case-based plan recognition in computer games. In: Ashley, K.D., Bridge, D.G. (eds.) ICCBR 2003. LNCS (LNAI), vol. 2689, pp. 161–170. Springer, Heidelberg (2003). https://doi.org/10.1007/3-540-45006-8_15
13. Hintze, A., Olson, R.S., Lehman, J.: Orthogonally evolved AI to improve difficulty adjustment in video games. In: Squillero, G., Burelli, P. (eds.) EvoApplications 2016. LNCS, vol. 9597, pp. 525–540. Springer, Cham (2016). https://doi.org/10.1007/978-3-319-31204-0_34
14. Jennings-Teats, M., Smith, G., Wardrip-Fruin, N.: Polymorph: dynamic difficulty adjustment through level generation. In: Proceedings of the 2010 Workshop on Procedural Content Generation in Games, p. 11. ACM (2010)
15. Lopes, R., Bidarra, R.: Adaptivity challenges in games and simulations: a survey. IEEE Trans. Comput. Intell. AI Games 3(2), 85–99 (2011)
16. Lora, D., Sánchez-Ruiz, A.A., González-Calero, P.A., Gómez-Martín, M.A.: Dynamic difficulty adjustment in tetris. In: The Twenty-Ninth International Flairs Conference (2016)
17. Missura, O., Gärtner, T.: Player modeling for intelligent difficulty adjustment. In: Gama, J., Costa, V.S., Jorge, A.M., Brazdil, P.B. (eds.) DS 2009. LNCS (LNAI), vol. 5808, pp. 197–211. Springer, Heidelberg (2009). https://doi.org/10.1007/978-3-642-04747-3_17
18. Nakamura, J., Csikszentmihalyi, M.: The concept of flow. Flow and the Foundations of Positive Psychology, pp. 239–263. Springer, Dordrecht (2014). https://doi.org/10.1007/978-94-017-9088-8_16
19. Nechvatal, J.: Immersive ideals/Critical distances. LAP Lambert Academic Publishing, p. 14 (2009)
20. Pagulayan, R.J., Keeker, K., Wixon, D., Romero, R.L., Fuller, T.: User-Centered Design in Games. CRC Press, Boca Raton (2002)
21. Sharma, M., Mehta, M., Ontanón, S., Ram, A.: Player modeling evaluation for interactive fiction. In: Proceedings of the AIIDE 2007 Workshop on Optimizing Player Satisfaction, pp. 19–24 (2007)
22. Sutoyo, R., Winata, D., Oliviani, K., Supriyadi, D.M.: Dynamic difficulty adjustment in tower defence. Procedia Comput. Sci. 59, 435–444 (2015)
23. Sweetser, P., Wyeth, P.: Gameflow: a model for evaluating player enjoyment in games. Comput. Entertain. (CIE) 3(3), 3 (2005)
24. Taylor, L.N.: Video Games: Perspective, Point-of-view, and Immersion. Ph.D. thesis, University of Florida (2002)
25. Yannakakis, G.N., Hallam, J.: Real-time game adaptation for optimizing player satisfaction. IEEE Trans. Comput. Intell. AI Games 1(2), 121–133 (2009)

Scoring Performance
on the Y-Balance Test

Vivek Mahato$^{(\boxtimes)}$, William Johnston, and Pádraig Cunningham

School of Computer Science, University College Dublin, Dublin 4, Ireland
vivek.mahato@ucdconnect.ie, padraig.cunningham@ucd.ie

Abstract. The Y-Balance Test (YBT) is a dynamic balance assessment commonly used in sports medicine. In this research we explore how data from a wearable sensor can provide further insights from YBT performance. We do this in a Case-Based Reasoning (CBR) framework where the assessment of similarity on the wearable sensor data is the key challenge. The assessment of similarity on time-series data is not a new topic in CBR research; however the focus here is on working as close to the raw time-series as possible so that no information is lost. We report results on two aspects, the assessment of YBT performance and the insights that can be drawn from comparisons between pre- and post- injury performance.

Keywords: Wearable sensors · Y Balance Test · Time series data

1 Introduction

This research addresses the challenge of assessing human physical activity as measured using wearable sensors. We focus on the Y Balance Test (YBT) a test for assessing dynamic balance used in clinical and research settings [9]. We address two tasks, the task of scoring performance based on the sensor data (a regression task) and a classification task that identifies abnormal performance. We also seek to provide insight into *how* performance on a test is abnormal.

The YBT produces a normalised reach score which quantifies performance (Sect. 3). Our first objective is to see if we can estimate this directly from the sensor data. We report on what data streams from the sensor are most effective for this. This regression task is performed using k-Nearest Neighbour (k-NN) and we analyse a number of similarity mechanisms for identifying neighbours. The motivation for our first objective is to eliminate the manual task of measuring the reach distance.

In the second part of our evaluation, we examine data from six athletes recovering from a concussion. We explore the hypothesis that an in-depth analysis of the YBT sensor data provides insight into the extent to which the individual has recovered from the concussion. While the results we report are preliminary, this seems a promising strategy.

© Springer Nature Switzerland AG 2019
K. Bach and C. Marling (Eds.): ICCBR 2019, LNAI 11680, pp. 281–296, 2019.
https://doi.org/10.1007/978-3-030-29249-2_19

This paper is structured as follows. The next section provides an overview of relevant research on similarity measures for time-series. The Y Balance Test is described in Sect. 3. Our evaluation is presented in Sect. 4 and conclusions and directions for future work are presented in Sect. 5.

2   Similarity Measures for Time-Series

In reviewing relevant research on similarity measures it is worth separating research in a CBR context from the wider research in this area. In the next subsections we review the dominant methods for measuring similarity on time-series before focusing on CBR research on time series at the end of this section.

2.1   Dynamic Time Warping

To find the distance between two time series, the Euclidean distance formula is an obvious choice. But when dealing with time series data where the series may be displaced in time, the Euclidean distance may be large when the two series are similar, just off slightly on the time line (see Fig. 1(a)). To tackle this situation Dynamic Time Warping (DTW) offers us the flexibility of mapping the two data series in a non-linear fashion by warping the time axis [15]. It creates a cost matrix where the cells contain the distance value of the corresponding data-points and then finds the shortest path through the grid, which minimizes the total distance between them. Sakoe-Chiba [23] global constraint is introduced to the model to increase its performance and reduce time complexity.

The following are the steps DTW executes to find the optimum mapping path with forward Dynamic Programming (DP), which provides us with the minimum distance:

- Let t and r be two time-series vectors; then define $D(i,j)$ as the DTW distance between $t(1 : i)$ and $r(1 : j)$, with the mapping path starting from $(1, 1)$ to (i, j).
- With initial condition as $D(m, n) = |t(m) - r(n)|$, recursively calculate:

$$D(i,j) = |t(i) - r(j)| + min \left\{ \begin{array}{c} D(i-1,j) \\ D(i-1,j-1) \\ D(i,j-1) \end{array} \right\} \tag{1}$$

- The minimum distance then is $D(1, 1)$.

Simply said, we will construct a matrix D of dimensions $m \times n$ (where m and n are the sizes of time-series vectors t and r), and then insert the value of $D(1, 1)$ by using the initial condition. Using the recursive formula the whole matrix gets filled one element at a time, either following a column-by-column or row-by-row order. When completed, the minimum cost or distance between t and r will be available at $D(m, n)$. Thus, the computational complexity of DTW is $O(mn)$ when the time-series are unidimensional.

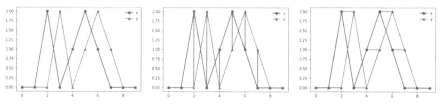

(a) Two similar time-series displaced in time. (b) Mapping of data-points without warping. (c) Mapping of data-points with warping.

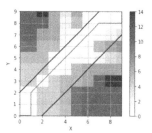

(d) DTW cost-matrix with global constraint.

Fig. 1. An example of DTW non-linearly mapping two time-series displaced in time, with Sakoe-Chiba global constraint.

In many cases, DTW may not provide the best mapping as required as it strives to find the minimum distance which can result in forming an unwanted path, which does not assist in discriminating two time-series belonging to different classes. Fixing this issue requires limiting the possible warping paths utilizing a global constraint. Sakoe-Chiba band (Fig. 1(d)), is one of the simplest and most popular global constraints applied to DTW. The warping path then is limited to the zone that falls under the band indices. Initially, when we restrict our algorithm with no warping allowed, the data points are linearly mapped between the two data series based on the common time axis value. As seen in Fig. 1(b), the algorithm does a poor job of matching the time series. But when we grant the DTW algorithm the flexibility of considering a warping window, the algorithm performs remarkably well when mapping the data-points following the trend of the time series data, which can be visualized in Fig. 1(c).

2.2 Symbolic Aggregate approXimation

Several symbolic representations of a time series data have been developed in recent decades with the objective of bringing the power of text processing algorithms to bear on time series problems. Keogh et al. provide an overview of these methods in their 2003 paper [16].

Symbolic Aggregate Approximation (SAX) is one such algorithm that achieves dimensionality and numerosity reduction and provides a distance measure that is a lower bound on the distance measures on the original series [16]. In this case numerosity reduction refers to a more compact representation of the data.

Piecewise Aggregate Approximation. SAX uses Piecewise Aggregate Approximation (PAA) in its algorithm for dimensionality reduction. The fundamental idea behind the algorithm is to reduce the dimensionality of a time series by slicing it into equal-sized fragments which are then represented by the average of the values in the fragment.

PAA approximates a time series X of length n into vector $\bar{X} = (\bar{x}_1, \bar{x}_2, ..., \bar{x}_m)$ of any arbitrary length $m \leqslant n$, where each of x_i is computed as follows:

$$\bar{x}_i = \frac{m}{n} \sum_{j=\frac{n}{m}(i-1)+1}^{\frac{n}{m}i} x_j \tag{2}$$

This simply means that in order to reduce the size from n to m, the original time series is first divided into m fragments of equal size and then the mean values for each of these fragments are computed. The series constructed from these mean values is the PAA approximation of the original time series. There are two cases worth noting when using PAA. When $m = n$ the transformed representation is alike to the original input, and when $m = 1$ the transformed representation is just the mean of the original series [14]. Before the transformation of original data into the PAA representation, SAX also normalizes each of the time series to have a mean of zero and a standard deviation of one, given the difficulty of comparing time series of different scales [13,16] (Fig. 2).

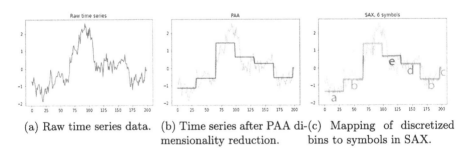

(a) Raw time series data. (b) Time series after PAA di-(c) Mapping of discretized
mensionality reduction.   bins to symbols in SAX.

Fig. 2. Symbolic Aggregate Approximation; The raw time-series in (a) will be represented by the sequence 'abfedbc' in (c) [17].

After the PAA transformation of the time series data, the output goes through another discretization procedure to obtain a discrete representation of the series. The objective is to discretize these levels into a bins of roughly equal size. These levels will typically follow a Gaussian distribution so these bins will get larger away from the mean. The breakpoints separating these discretized bins form a sorted list $B = \beta_1, ..., \beta_{a-1}$, such that the area under a $N(0,1)$ Gaussian curve from β_i to $\beta_{i+1} = \frac{1}{a}$. β_0 and β_a are defined as $-\infty$ and ∞ respectively [16].

When all the breakpoints are computed, the original time series is discretized as follows. First, the PAA transformation of the time series is performed. Then each of the PAA coefficients less than the smallest breakpoint β_1 is mapped to the symbol s_1, and all coefficients between breakpoints β_1 and β_2 (second smallest breakpoint) are mapped to the symbol s_2, and so on, until the last PAA coefficient gets mapped. Here, s_1 and s_2 belongs to a set of symbols $S = (s_1, s_2, ..., s_m)$ to which the time series is mapped by SAX, where m is the size of symbol pool.

SAX also has a sliding window implemented in its algorithm, the size of which can be adjusted. It extracts the symbols present in that window frame and creates a word, which is just the concatenated sequence of symbols in that frame. This sliding window is then shifted to the right and another word is extracted corresponding to the new frame. This goes on until the window hits the end of the time series, yielding a "bag-of-words" representing the series.

Once the data is converted to this symbolic representation, one can use this bag-of-words representation for calculating the distance between two time series using a string distance metric such as Levenshtein distance [27].

2.3 Symbolic Fourier Approximation

SFA was introduced by Schäfer et al. in 2012 as an alternative method to SAX built upon the idea of dimensionality reduction by symbolic representation. Unlike SAX which works on the time domain, SFA works on the frequency domain. In the frequency domain, each dimension contains approximate information about the whole time series. By increasing the dimensionality one can add detail, thus improving the overall quality of the approximation. In the time domain, we have to decide on a length of the approximation in advance and a prefix of this length only represents a subset of the time series [24].

Discrete Fourier Transform. In contrast to SAX which uses PAA as its dimensionality reduction technique, SFA, focusing on the frequency domain, uses the Discrete Fourier Transform (DFT). DFT is the equivalent of the continuous Fourier Transform for signals known only at N instants by sample times T, which is a finite series of data.

Let $X(t)$ be the continuous signal which is the source of the data. Let N samples be denoted $x[0], x[1], ..., x[N-1]$. The Fourier Transform of the original signal, $X(t)$, would be:

$$F(\omega_k) \triangleq \sum_{n=0}^{N-1} x(t_n)e^{-j\omega_k t_n}, k = 0, 1, 2, ..., N-1 \qquad (3)$$

Simply stated, DFT analyzes a time domain signal $x(n)$ to determine the signal's frequency content $X[k]$. This is achieved by comparing $x[n]$ against signals known as sinusoidal basis functions, using correlation. The first few basis functions correspond to gradually changing regions and describe the coarse distribution, while later basis functions describe rapid changes like gaps or noise. Thus employing only the first few basis functions yields a good approximation of the time series [24].

The DFT Approximation is a part of the preprocessing step of SFA algorithm, where all time series data are approximated by computing DFT coefficients. When all these DFT coefficients are calculated, multiple discretisations are determined from all these DFT approximations using Multiple Coefficient Binning (MCB) as shown in Fig. 3. MCB helps in mapping the coefficients to their symbols, and concatenates it to form an SFA word. Thus, this converts the time series into its symbolic representation.

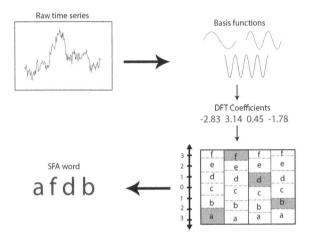

Fig. 3. Symbolic Fourier Approximation; The raw time-series will be represented by the sequence 'afdb' [17].

As in SAX, there is a sliding window present here which serves the same purpose of extracting a word representing the data in that frame. Thus, the output of SFA for a given source time series is a bag-of-words symbolically representing the entire series in lower dimension.

2.4   Time-Series Similarity in CBR

Research on temporal analysis within the CBR community has been strongly influenced by the Temporal Abstractions (TA) methodology. The idea with TA is to map low-level temporal data into higher level concepts that are meaningful for the domain in question [25]. This idea has its roots in the *The Knowledge Level* [21] view of Artificial Intelligence which fits well with the CBR paradigm. There has been significant research on TA in CBR with a particular focus on applications in medical decision support [19,20].

The objective with TA is to produce a high-level symbolic representation of the time-series that will reduce the dimension of the data by providing a high-level abstraction that fits with a *feature-value* case representation [19]. The attractiveness of TA for early CBR research was two fold, it delivered a feature value representation and it avoided the computational problem of dealing with the raw data. It is worth saying that while SAX and SFA also produce symbolic representations the motivations are different as the objective is not to produce *knowledge level* representations.

On the question of computational tractability, early work by Penta *et al.* [22] recognised the benefit of using DTW to quantify similarity on time-series for CBR but dismissed it as an option because of the computational cost.

More recently Elsayed *et al.* [5] did use DTW in CBR to classify *pseudo-times-series* in medical image analysis. This shows that the computational cost of DTW is no longer an issue. Bregón *et al.* [2] also report good results using DTW with CBR on a fault classification problem.

3   Y Balance Test

The Y Balance Test (YBT) is the most common dynamic balance assessment used within the sports medicine clinical context [6]. It requires an individual to transition from a position of bilateral to unilateral stance, perform a maximal reach excursion with the non-stance limb in three standardised directions (anterior; posteromedial; posterolateral), while maintaining controlled balance [6] (see Fig. 4). The individual is then required to return to the starting position in a controlled manner. A trial is deemed a fail if they remove their hands from their hips, make contact with the ground, weight bear through the slider, raise the stance leg heel or kick the slider forward for extra distance. Participants typically complete four practice trials prior to completion of three recorded trials in each direction (randomized order), bilaterally [6].

The traditional balance 'score' is obtained by manually measuring the distance the individual reaches outside of their base of support and normalising it to their leg length, allowing for appropriate comparison between individuals. Previous research has demonstrated the ability of this protocol to identify differences in dynamic performance between control and pathological groups, in conditions such as acute lateral ankle sprain [4] and anterior cruciate ligament injuries [7].

It has also been suggested that the YBT may have a role in evaluating concussed athletes. It can provide a means to challenge the sensorimotor subsystems of injured athletes, highlighting deficits that may increase their risk of sustaining further injury [10].

Johnston *et al.* have shown that a very good assessment of YBT performance can be obtained from a single wearable sensor [9]. "Normal" and "abnormal" balance performance can be assessed with a moderate level of accuracy.

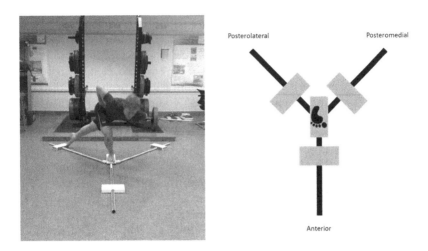

Fig. 4. A demonstration of the YBT test in operation.

3.1 The Datasets

The data set consists of data collected from two cohorts:

1. 29 young healthy adults (aged 23.3 ± 2.1 years; height 174.7 ± 9.2 cm; weight 71.6 ± 13.3 kg; left leg length 95.4 ± 4.8 cm; right leg length 95.5 ± 5.1 cm) were tested on one occasion in a university biomechanics laboratory.[1] In our evaluation we use 21 subjects for training and model/parameter selection and 8 for testing. For each subject there are 18 samples (3 trials, 3 directions, and 2 stances). So we have 378 training samples and 143 test samples (one sample is missing). The length of the time-series ranges from 114 to 645 data-points, having an average of ≈ 298 ticks.
2. Six elite rugby union players (aged 21 ± 1.5 years; height 182 ± 6.3 cm; weight 91 ± 15.4 kg; right leg length 95 ± 4.2 cm; left leg length 95 ± 4.22 cm) were baseline tested as part of a wider study protocol, as described in Johnston *et al.* [12]. These six athletes later went on to sustain a concussive injury, and were follow-up tested using the inertial sensor quantified YBT 48-hours post-injury and at the point of medical clearance to return to full contact training (RtP). The length of the time-series here ranges from 158 to 648 data-points, having an average of ≈ 345 ticks. Reliability control data was also obtained from two healthy young adults who were repeat tested on two occasions, separate by 7–10 days, as described in [11].

Ethical approval was sought and obtained from the university research ethics board, and all participants provided informed consent prior to completion of the testing protocol. Additional consent was provided from the young healthy participants (dataset 1) to allow open-access publication of the dataset.

---

[1] This dataset is available at http://mlg.ucd.ie/ybt.

3.2   Sensor Methods

The sensor used was a Shimmer3 sensor[2] that returns 10 data streams; accelerometer, gyroscope and magnetometer in three dimensions and an altimeter to provide the 10th data stream. The altimeter data was not used in our study; however pitch, roll and yaw were derived from the other data streams to provide 12 streams in all. The sensor was mounted at the level of the 4th lumbar vertebra, in line with the top of the iliac crests using a custom-made elastic belt. The sensor was configured to collect tri-axial accelerometer data ($\pm 2g$) and tri-axial gyroscope data ($\pm 500°/s$) at a sampling frequency of 51.2 Hz during each YBT reach excursion. The data collection procedure was consistent with previously describe methods [11,12].

4   Evaluation

In this evaluation we consider two questions:

1. Can we score YBT performance without actually measuring the reach?
2. Does a visual inspection of the sensor plots offer insights into performance?

For the first question we need to determine which data streams from the sensor are predictive of performance (Sect. 4.1) and identify which similarity measures are best for this task (Sect. 4.2). We take the tasks in this order, first we identify the best data streams then we tackle the similarity measures.

4.1   Feature Selection

The first task was to identify which subset of the 12 data streams would be effective for the regression task. A meta-analysis by Mitsa [18] states that when it comes to time-series classification, 1-NN-DTW is challenging to beat. Therefore, we employ k-NN-DTW to evaluate the features (i.e. data streams) individually, based on its prediction capability of the reach distance.

The results in terms of Mean Absolute Percentage Error (MAPE) are shown for each of the 12 time-series in Fig. 5. The results show that the Z-axis of the accelerometer proves to be most informative, with Y and Z-axes of the magnetometer being the next best features. A magnetometer is sensitive to external disturbances such as the earth's magnetic field, the location of the experiment, or electrical systems present in proximity, therefore 'Accel Z' is selected as the single best feature because it is robust to such interference.

Next we use a Forward Sequential Search strategy [1] to see if adding other features (i.e. time-series) will improve accuracy. It is clear from the results shown in Fig. 6 that the impact is minimal. Adding two features (Mag Y and Mag X) reduce the MAPE from 6.24 cm to 6.04 cm.

---

[2]   http://www.shimmersensing.com.

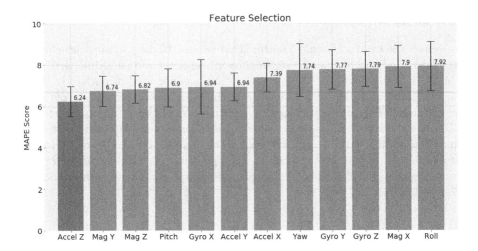

Fig. 5. Feature selection over the performance of k-NN-DTW in predicting reach distance using only one dimension.

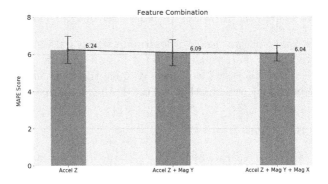

Fig. 6. Performance of kNN-DTW model when a combination of features is given.

4.2   Comparing Similarity Measures

We move on now to consider the performance of the other similarity measures (SAX and SFA) compared with DTW[3]. Given the results of the feature selection analysis we consider the Accel Z time series only.

As explained in Sect. 2, both SAX and SFA turn time-series matching into a sequence matching problem. So we have some choices on how we measure sequence similarity. Here we consider two options, standard Edit Distance (Levenshtein Distance) [3] and the Wagner-Fischer algorithm [26][4] (Fig. 7).

---

[3] Similarity computation with DTW between two time-series of unequal length is handled by padding the shorter time-series with zeroes.

[4] Edit Distance and Wagner-Fischer measures requires no size matching as it handles the unequal length of the sequences.

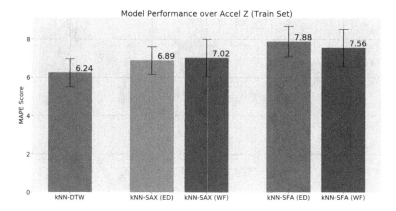

Fig. 7. Evaluation of models with 10-fold cross-validation over 'Accel Z' in the train set, and MAPE score as its evaluation metric.

The Wagner-Fischer algorithm is more nuanced than standard Edit Distance as it allows for different penalties for insertion, deletion and substitution and for distances within the alphabet to be included in the penalty score. For example, the underlying implementation of Edit distance measures the distance between "boat" and "coat" as 1, and the distance between "coat" and "goat" is also computed as 1, because these strings are only one edit away. Whereas, the Wagner-Fischer algorithm measures the distance between "coat" and "goat" as four because 'g' is 4-steps away from 'c' in the alphabet series.

We compare the three methods (DTW, SAX and SFA) used in time series regression in a k-NN model. k-NN-SAX and k-NN-SFA were evaluated on both vanilla Edit distance and Wagner-Fischer version with custom penalties. We report two sets of results, results on the training data (21 subjects) which we are using for model and parameter selection and results on the test data (8 subjects) which gives us an estimate of generalisation accuracy.

Figure 8 illustrates the performance of the kNN-models on the reach estimation task. Our conclusions are as follows:

1. k-NN-DTW beats SAX and SFA on this reach estimation task. This is consistent with earlier work that shows that DTW will beat SFA and SAX when similarity depends on the overall signal rather than local features [17].
2. Edit Distance performs better than Wagner-Fischer when used with SAX and SFA. This may be because of overfitting in the parameter setting process.

4.3 Insights

Our next objective is to see if the sensor data offers any insight into recovery from concussion. The second dataset (Sect. 3.1) contains sensor readings from six athletes who suffered concussions. There are readings, Pre-, Post-injury and on Return-to-Play (RtP) with three readings (i.e. trials) for each category. The

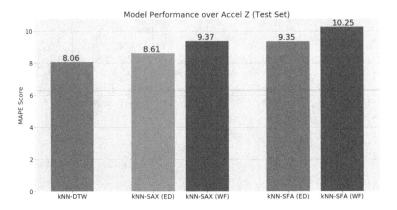

Fig. 8. Performance of the models over 'Accel Z' in the test set, with MAPE score as its evaluation metric.

Post-injury measurements were taken immediately after injury when athletes were excluded from playing due to concussion. Thus the RtP signals should be similar to the Pre- data and not the Post-.

As a baseline we have data on 'healthy' subjects which shows us what normal variations between test sessions should look like. The first two plots in Fig. 9 show data on two such subjects. The data shows two sets of three repetitions measured one week apart. It is clear from the plots that the strategies are reasonably consistent, Subject 6 has a steady acceleration while subject 11 increases acceleration through the movement and then slows sharply.

The next two plots in the Figure present the picture for two of the concussed athletes. The plots for the four other concussed athletes are shown in the Appendix. The signals for the concussed athletes were selected as follows:

- We take each of the three RtP examples and calculate the similarity to the three Pre and three Post examples.
- We present the RtP signal and its closest match (Pre- or Post-) and all the non matching signals for comparison.

We would like to see RtP signals that are similar to the Pre-injury signals. We have this for Athlete 517 and not for Athlete 400. Athlete 517 has an RtP signal similar to his Pre-injury performance and different to his Post- signal (shown in red). By contrast the signal for Athlete 400 looks like his Post-injury signal. While it would be expected that both athletes dynamic balance performance should have returned to baseline levels at the point of 'clinical recovery' (return to play), concussion presentation is multi-factorial and variable in nature, where no two injured athletes present the same. Furthermore, there is increasing evidence that neuromuscular control deficits may persist beyond resolution of symptoms, increasing their risk of future injury [8]. This may help explain why athlete 400s RtP signal looked most similar to their post-injury signal, while athlete 517 appeared to have returned to pre-injury levels of performance.

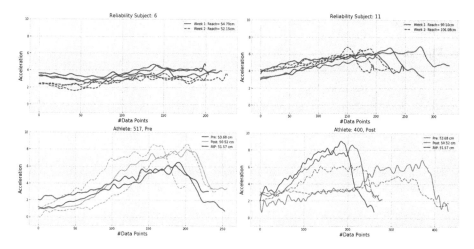

Fig. 9. A comparison between 'healthy' subjects and two subjects who suffered concussion (standing on non-dominant leg, anterior reach, Accel Z data).

5   Conclusion

Our analysis shows that using k-NN with DTW on the Accelerometer Z-axis data is effective for predicting performance on the YBT.

While the analysis relating to concussion is preliminary, we feel that the sensor data can have a role is assessing recovery from concussion. The strategy would be to gather 'healthy' baseline data during the pre-season training period, and use this to help determine when an athlete may have fully recovered post-injury. This approach may help health care professionals identify players who have not fully recovered post-concussion, facilitating the implementation of additional rehabilitation strategies to aid in the reduction of future re-injury. Due to the large degree of inter-subject variability within the data, our analysis suggests that this pre/post injury comparison needs to be player specific, and cannot be generalised between subjects.

Acknowledgments. This publication has resulted from research supported in part by a grant from Science Foundation Ireland (SFI) under Grant Number 16/RC/3872 and is co-funded under the European Regional Development Fund.

Appendix

See Fig. 10.

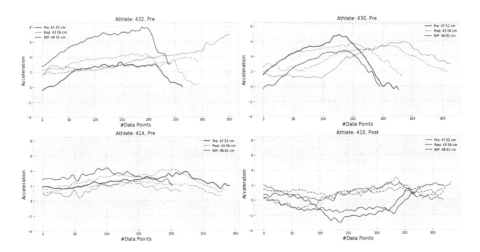

Fig. 10. A comparison between the rest four athletes in the dataset who suffered concussion. (Standing on non-dominant leg, anterior reach, Accel Z data.)

References

1. Aha, D.W., Bankert, R.L.: A comparative evaluation of sequential feature selection algorithms. In: Fisher, D., Lenz, H.J. (eds.) Learning from Data. LNS, vol. 112, pp. 199–206. Springer, New York (1996). https://doi.org/10.1007/978-1-4612-2404-4_19

2. Bregón, A., Simón, M.A., Rodríguez, J.J., Alonso, C., Pulido, B., Moro, I.: Early fault classification in dynamic systems using case-based reasoning. In: Marín, R., Onaindía, E., Bugarín, A., Santos, J. (eds.) CAEPIA 2005. LNCS (LNAI), vol. 4177, pp. 211–220. Springer, Heidelberg (2006). https://doi.org/10.1007/11881216_23

3. Ding, H., Trajcevski, G., Scheuermann, P., Wang, X., Keogh, E.: Querying and mining of time series data. Proc. VLDB Endow. **1**(2), 1542–1552 (2008)

4. Doherty, C., Bleakley, C.M., Hertel, J., Caulfield, B., Ryan, J., Delahunt, E.: Laboratory measures of postural control during the star excursion balance test after acute first-time lateral ankle sprain. J. Athl. Train. **50**(6), 651–664 (2015)

5. Elsayed, A., Hijazi, M.H.A., Coenen, F., García-Fiñana, M., Sluming, V., Zheng, Y.: Time series case based reasoning for image categorisation. In: Ram, A., Wiratunga, N. (eds.) ICCBR 2011. LNCS (LNAI), vol. 6880, pp. 423–436. Springer, Heidelberg (2011). https://doi.org/10.1007/978-3-642-23291-6_31

6. Gribble, P.A., Hertel, J., Plisky, P.: Using the star excursion balance test to assess dynamic postural-control deficits and outcomes in lower extremity injury: a literature and systematic review. J. Athl. Train. **47**(3), 339–357 (2012)

7. Herrington, L., Hatcher, J., Hatcher, A., McNicholas, M.: A comparison of star excursion balance test reach distances between ACL deficient patients and asymptomatic controls. Knee **16**(2), 149–152 (2009)
8. Howell, D.R., Lynall, R.C., Buckley, T.A., Herman, D.C.: Neuromuscular control deficits and the risk of subsequent injury after a concussion: a scoping review. Sports Med. **48**(5), 1097–1115 (2018)
9. Johnston, W., O'Reilly, M., Dolan, K., Reid, N., Coughlan, G., Caulfield, B.: Objective classification of dynamic balance using a single wearable sensor. In: 4th International Congress on Sport Sciences Research and Technology Support 2016, Porto, Portugal, 7–9 November 2016, pp. 15–24. SCITEPRESS-Science and Technology Publications (2016)
10. Johnston, W., O'Reilly, M., Argent, R., Caulfield, B.: Reliability, validity and utility of inertial sensor systems for postural control assessment in sport science and medicine applications: a systematic review. Sports Med. **49**, 783–818 (2019)
11. Johnston, W., O'Reilly, M., Coughlan, G.F., Caulfield, B.: Inter-session test-retest reliability of the quantified Y balance test. In: 6th International Congress on Sports Sciences Research and Technology Support, pp. 63–70 (2018)
12. Johnston, W., et al.: Association of dynamic balance with sports-related concussion: a prospective cohort study. Am. J. Sports Med. **47**(1), 197–205 (2019)
13. Keogh, E., Kasetty, S.: On the need for time series data mining benchmarks. In: Proceedings of the Eighth ACM SIGKDD International Conference on Knowledge Discovery and Data Mining - KDD 2002 (2002)
14. Keogh, E.J., Pazzani, M.J.: Scaling up dynamic time warping for datamining applications. In: Proceedings of the Sixth ACM SIGKDD International Conference on Knowledge Discovery and Data Mining - KDD 2000 (2000)
15. Keogh, E.J., Pazzani, M.J.: Derivative dynamic time warping. In: Proceedings of the 2001 SIAM International Conference on Data Mining, pp. 1–11. SIAM (2001)
16. Lin, J., Keogh, E., Lonardi, S., Chiu, B.: A symbolic representation of time series, with implications for streaming algorithms. In: Proceedings of the 8th ACM SIGMOD Workshop on Research Issues in Data Mining and Knowledge Discovery - DMKD 2003 (2003)
17. Mahato, V., O'Reilly, M., Cunningham, P.: A comparison of k-NN methods for time series classification and regression. In: Brennan, R., Beel, J., Byrne, R., Debattista, J., Junior, A.C. (eds.) 2018 Proceedings for the 26th AIAI Irish Conference on Artificial Intelligence and Cognitive Science. CEUR Workshop Proceedings, vol. 2259, pp. 102–113. CEUR-WS.org (2018). http://ceur-ws.org/Vol-2259/aics_11.pdf
18. Mitsa, T.: Temporal Data Mining, pp. 99–102. Chapman & Hall, London (2010)
19. Montani, S., Bottrighi, A., Leonardi, G., Portinale, L.: A CBR-based, closed-loop architecture for temporal abstractions configuration. Comput. Intell. **25**(3), 235–249 (2009)
20. Montani, S., Leonardi, G., Bottrighi, A., Portinale, L., Terenziani, P.: Supporting flexible, efficient, and user-interpretable retrieval of similar time series. IEEE Trans. Knowl. Data Eng. **25**(3), 677–689 (2013)
21. Newell, A., et al.: The knowledge level. Artif. Intell. **18**(1), 87–127 (1982)
22. Penta, K.K., Khemani, D.: Satellite health monitoring using CBR framework. In: Funk, P., González Calero, P.A. (eds.) ECCBR 2004. LNCS (LNAI), vol. 3155, pp. 732–747. Springer, Heidelberg (2004). https://doi.org/10.1007/978-3-540-28631-8_53
23. Sakoe, H., Chiba, S.: Dynamic programming algorithm optimization for spoken word recognition. IEEE Trans. Acoust. Speech Signal Process. **26**(1), 43–49 (1978)

24. Schäfer, P., Högqvist, M.: SFA: a symbolic Fourier approximation and index for similarity search in high dimensional datasets. In: Proceedings of the 15th International Conference on Extending Database Technology, pp. 516–527. ACM (2012)
25. Shahar, Y.: A framework for knowledge-based temporal abstraction. Artif. Intell. **90**(1–2), 79–133 (1997)
26. Wagner, R.A., Fischer, M.J.: The string-to-string correction problem. J. ACM **21**(1), 168–173 (1974)
27. Yujian, L., Bo, L.: A normalized Levenshtein distance metric. IEEE Trans. Pattern Anal. Mach. Intell. **29**(6), 1091–1095 (2007)

An Optimal Case-Base Maintenance Method for Compositional Adaptation Applications

Ditty Mathew$^{(\boxtimes)}$ and Sutanu Chakraborti

Department of Computer Science and Engineering,
Indian Institute of Technology Madras, Chennai, India
{ditty,sutanuc}@cse.iitm.ac.in

Abstract. Case-base maintenance method aims at maintaining a compressed case-base which is useful for solving future problems effectively. In this paper, we propose an optimization formulation to arrive at a compressed case-base that can find a solution for the rest of the cases in the case-base that involves compositional adaptation process. The objective of the optimization problem is to minimize the footprint set size and maximize the quality of solutions that can be adapted from the footprint set. We empirically studied the proposed formulation on four different datasets and the results show that the proposed model is effective and overcomes the limitation of the existing optimal footprint method in compositional adaptation applications.

Keywords: Case-base maintenance · Optimal method · Footprint-based competence model · Compositional adaptation

1 Introduction

Case-Based Reasoning (CBR) [1] is an artificial intelligence based problem solving paradigm where new problems are solved based on past experiences. Case-base maintenance is a sub-field in CBR which aims at maintaining a compressed case-base by retaining cases which are useful for solving future problems effectively. The initial works on case-base maintenance concentrated on removing noisy and redundant cases. The first approach towards this kind of data reduction is the Condensed Nearest Neighbor algorithm [13]. This method does not guarantee a minimal set but it is robust to noise. In [2], Smyth et al. proposed a case competence model which identifies a compressed case-base called footprint set. While the algorithm proposed in [2] to arrive at the footprint set attempts at maximizing the competence of the compressed case-base, it does not guarantee an optimal footprint set. In [12], Mathew et al. proposed an optimal method to identify the footprint set which aims at maximizing the effectiveness and minimizing the size of the footprint set. However, all these approaches assume that the adaptation process involves the adaptation of the solution of a single

© Springer Nature Switzerland AG 2019
K. Bach and C. Marling (Eds.): ICCBR 2019, LNAI 11680, pp. 297–313, 2019.
https://doi.org/10.1007/978-3-030-29249-2_20

case to arrive at a solution of its target case. This kind of adaptation is called single case adaptation. In certain applications, the adaptation process is more involved and is referred to as compositional adaptation [3] which means that the combined solution of multiple cases can be adapted to obtain the solution of the target case. The first approach towards case-base maintenance of compositional adaptation applications is proposed in [9]. This work proposed a case competence model to identify useful cases that can be retained when the adaptation process involves either single case adaptation or compositional adaptation. However, this approach does not guarantee a minimal size footprint set. In this paper, we propose an optimal method to identify a footprint set that can be applied in compositional adaptation applications.

We propose a convex optimization formulation to identify a footprint set from a case-base that involves compositional adaptation. The objective of the optimization function is to minimize the cardinality of the footprint set and maximize the quality of solutions that can be adapted from the footprint set to solve the rest of the cases in the case-base. To the best of our knowledge, this is the first-of-its-kind approach to identify optimal compression of case-base during case-base maintenance of compositional adaptation applications. The structure of this paper is as follows. Section 2 provides an overview of single case and compositional adaptation process and reviews the literature on case-base maintenance models. Section 3 presents the proposed optimization formulation for case-base maintenance of compositional adaptation applications. We discuss the experimental results in Sect. 4 and conclude in Sect. 5.

2   Background

This section provides an overview of single case and compositional adaptation process, literature in case-base maintenance, previous approaches in case-base maintenance of compositional adaptation applications and optimal footprint method for case-base maintenance in single case adaptation.

2.1   Single Case and Compositional Adaptation

In single case adaptation, the solution of a single case is retrieved and adapted to solve a target problem [14,15]. For example, in a cooking recipe application, suppose the case-base contains a *recipe for potato fry*. Let the new query be a *recipe for fish fry* and this new problem is similar to the problem *recipe for potato fry*. Then the solution of potato fry recipe can be adapted by substituting *potato* with *fish* for the *recipe for fish fry*. Here a single case is adapted to arrive at a solution for the query and this kind of adaptation is called single case adaptation.

In compositional adaptation, the solution for a query is obtained by adapting the combined solution of solutions of multiple cases that are similar to the query. For example, in a recipe recommendation application, suppose the case-base contains the *recipe for potato fry* and *cauliflower fry*. Let the new query be a *recipe for elephant yam fry* and this query is similar to the cases *recipe for potato fry* and *recipe for cauliflower fry*. Then the solution of elephant yam fry recipe can be adapted from the combined solution of cases in the case-base such as *recipe of potato fry* and *recipe of cauliflower fry*. In [16], Müller et al. propose a CBR system which decomposes the cooking recipes in the case-base into reusable streams and the reusable streams of multiple recipes that are similar to the target query are adapted to retrieve the solution. Another compositional adaptation application is the CBR system for predicting pollution levels [19], which predicts the pollution level of the target problem as the mean of the pollution levels of the top-k similar cases. In medical diagnosis, [18] studied the compositional adaptation approach during the multiple disorder situation. Arshadi et al. [17] applied compositional adaptation for designing tutoring library which adapts from the multiple solutions retrieved in the past for similar queries for the user's search topic.

2.2   Case-Base Maintenance

The impact of utility of a case-base in Case-Based Reasoning depends on its size and growth. The efficiency of a case-base is adversely affected when it contains a large number of cases that are not useful to solve new problems. Hence, it is desirable to weed out such cases. The goal of case-base maintenance process is to retain only the useful cases and improve the efficiency of the case-base. On reducing the size of case-base, the initial works are concentrated on removing noisy and redundant cases. The noise reduction in the case-base increases the accuracy and the elimination of redundant cases improves the retrieval efficiency [20]. The first approach towards this kind of data reduction is the Condensed Nearest Neighbor algorithm [13]. This method does not guarantee a minimal set but it is robust to noise. A case competence model is proposed by Smyth et al. in [4] to direct the learning and deletion of cases in the case-base. Here, the authors proposed case deletion policies which maximizes the competence of the case-base while minimizing its size. In [5], Smyth et al. proposed a case competence model which identifies the competent cases for the competence directed case-base maintenance. This model measures the case competence based on the range of target problems that each case solves. Footprint-based retrieval [2] is an efficient retrieval approach in Case-Based Reasoning, which guides the search procedure using a case competence measure called *relative coverage* [2]. This approach identifies a competent subset of a case-base called the footprint set for case-base maintenance. However, the *relative coverage* measure used in this approach covers only the situation where a single retrieved case is adapted to solve a problem. The *relative coverage* is used to order cases based on their individual contribution in arriving at a solution for other cases.

Case-Base Maintenance of Compositional Adaptation Applications.
The footprint-CA algorithm proposed by Mathew et al. in [9] reports the first approach towards case-base maintenance of compositional adaptation applications. This approach uses a measure called *retention score* [9,10] to estimate the

retention quality of cases in the case-base when the adaptation process can be either single case or compositional adaptation. A generalized version of retention score called *weighted retention score* is proposed in [11], which considers the problem solving ability of cases that are involved in arriving at a solution for a target problem. The retention score is a special case of weighted retention score. We first discuss the formulation of scores such as retention score and weighted retention score, and then the footprint-CA algorithm.

The retention quality of a case depends on the set of problems that it solves and the number of other cases that are required to solve this set of problems. As we want to reduce the size of case-base, we would like to retain fewer good retention quality cases that cover more useful cases. If the solutions of a set of cases can be adapted to arrive at a solution for the target problem, then this target problem is said to be *solved* by this set of cases. The formulation of retention score uses the following terminologies such as *covered cases* and *support cases*. The covered cases of a case c (i.e. *CoveredCases(c)*) include all cases that c can be used to solve either on its own, or in conjunction with other cases. The support cases of a case c_i to solve the problem c_j (*SupportCases(c_i, c_j)*) is a set of cases that the case c_i requires to solve c_j. Reachability$_{CA}$ of a case c is the set of all cases that are involved in arriving at a solution of c. Let the combined solution of cases c_1 and c_2 can arrive at a solution for the case c_3, and also the solution for c_3 can be obtained from the case c_4. Then Reachability$_{CA}(c_3) = \{\{c_1, c_2\}, \{c_4\}\}$. The *retention score* of a case c is high if

1. it can solve several cases that have high retention scores
2. the number of support cases needed to solve these covered cases is few
3. the retention scores of these support cases are high

The first criterion captures the coverage of a case and it also ensures that the covered cases should have high retention score for the corresponding case to have a high retention score. Thus, this criterion assigns a high retention score to a case if its covered cases are very important. For example, let *CoveredCases(c_1)* = $\{c_3, c_4\}$, then the importance to retain c_1 in the case-base depends on the importance of c_3 and c_4. The second criterion is based on the idea that if a case requires many support cases to solve its covered case then its support cases should also be retained in the case-base. The third criterion ensures that the effective cost of retaining the support cases in conjunction with c is low. The formulation of retention score (RS) is given in Eq. 1.

$$RS_{k+1}(c) = \sum_{c_i \in CoveredCases(c)} \left(RS_k(c_i) \sum_{\substack{\mathbb{C}' \in Reachability_{CA}(c_i) \\ c \in \mathbb{C}'}} \frac{\min\limits_{c_j \in \mathbb{C}' \text{ and } c_j \neq c} (RS_k(c_j))}{|\mathbb{C}'|} \right) \quad (1)$$

where $RS_{k+1}(c)$ is the retention score of a case c at $(k+1)^{th}$ iteration. This formulation is a recursive formulation similar to PageRank [21] where the notion of circularity of PageRank is substituted by the revised notion of circularity as discussed in the criteria of retention score. The set $\mathbb{C}'$ is a set of cases which contain c such that $\mathbb{C}'$ can solve c_i. All cases in $\mathbb{C}'$ except c are the support cases of c. The denominator handles the situation when a case does not require any support cases to solve the corresponding covered case.

For the first iteration of the retention score estimation, the retention score (RS_0) of a case c can be estimated as,

$$RS_0(c) = \sum_{c_i \in CoveredCases(c)} \left(CoverageScore(c, c_i) \sum_{\substack{\mathbb{C}' \in Reachability_{CA}(c_i) \\ c \in \mathbb{C}'}} \frac{1}{|\mathbb{C}'|} \right) \quad (2)$$

where $CoverageScore$ is defined as,

$$CoverageScore(c, c_i) = \frac{1}{1 + |\{\mathbb{C}_j : \mathbb{C}_j \in Reachability_{CA}(c_i) \text{ and } c \notin \mathbb{C}_j\}|} \quad (3)$$

For each covered case c_i in Eq. 2, the $CoverageScore$ of c captures the individual contribution of c in solving c_i. The contribution of c in solving c_i is high if c is involved in all solutions of c_i, and the individual contribution of c to solve c_i is less when c_i can be solved without using c also. The denominator of Eq. 2 ensures that the retention score increases with decrease in the number of support cases that c requires to solve c_i and vice versa. As cardinality of $\mathbb{C}'$ decreases, the number of support cases decreases.

The weighted retention score measure [11] captures the *problem solving ability* of each set of cases that solves a target problem. The $Reachability_{CA}$ set of a case c contains sets of cases where the cases in each set can arrive a solution for c. The problem solving ability is defined for each set of cases in the $Reachability_{CA}$ set and it is the extent to which the adapted solution obtained from each set of cases in the $Reachability_{CA}$ set is close to its desired solution. The idea of weighted retention score is that "A case has high weighted retention score if it covers many cases that have high retention score with high problem solving ability, and it solves each covered case with the support of less number of cases, and the minimum of the retention scores of the support cases is high". The weighted retention score formulation is given below.

$$WRS_{k+1}(c) =$$

$$\sum_{c_i \in CoveredCases(c)} \left(WRS_k(c_i) \sum_{\substack{\mathbb{C}' \in Reachability_{CA}(c_i)c \in \mathbb{C}'}} w(\mathbb{C}', c_i) \frac{\min\limits_{c_j \in \mathbb{C}' \text{ and } c_j \neq c}(WRS_k(c_j))}{|\mathbb{C}'|} \right) \quad (4)$$

where the weight $w(\mathbb{C}', t)$ for a target problem t is the problem solving ability of $\mathbb{C}'$ in solving t where $\mathbb{C}' \in Reachability_{CA}(t)$. This formulation is the same as the retention score measure when the problem solving ability of all set of cases that are involved in arriving at its solution are one. Hence this measure is a more generalized formulation.

In footprint-CA algorithm, the cases are ordered in the decreasing order based on retention score or weighted retention score and the footprint-CA set is initialized to an empty set. Each case is added to the footprint-CA set in the sorted order if its solution cannot be adapted using the subset of cases in the footprint-CA set. This algorithm is given in Algorithm 1. This algorithm is a greedy algorithm and it ensures that the cases with high retention score or weighted retention score are added first to the footprint set.

Algorithm 1. Footprint$_{CA}$ algorithm

Input: Cases sorted based on retention score, **Output:** Footprint$_{CA}$
Cases ← Sorted cases according to their retention score
Footprint$_{CA}$ ← {}
Changes ← true
while *Changes* **do**
    Changes ← false
    for *each c ∈ Cases* **do**
        if *none of the composite solution of c is a subset of Footprint$_{CA}$* **then**
            Changes ← true
            Add *c* to Footprint$_{CA}$

Optimal Footprint Method for Case-Base Maintenance. The optimal footprint method proposed by Mathew et al. [12] for case-base maintenance aims at identifying a minimal size footprint set with maximum ability to obtain solutions that are close to the desired solution of target problems in single case adaptation settings. This method addresses the limitation of the footprint approach in [2] that this method (i) does not guarantee to obtain a minimum size footprint set, (ii) does not consider the problem solving ability of cases in the footprint set to arrive at solution of rest of the cases in the case-base. The goal of optimal footprint method is to estimate an optimal footprint set (Footprint$_{opt}$ ⊆ ℂ) as representatives such that

1. Footprint$_{opt}$ can solve all cases in ℂ with high solution quality
2. Footprint$_{opt}$ size is minimal

In this approach, a problem solving ability matrix is constructed in which the ij^{th} value indicates the extent to which the case c_i is able to arrive at a solution for the case c_j. Let it be denoted as $P(c_i, c_j)$. A loss function of a case c_i is defined with respect to the optimal footprint set (*Footprint$_{opt}$*) as $1 - P(c_j, c_i)$ where c_j can obtain solution for c_i. If a case c_i can be solved by more than one case in *Footprint$_{opt}$*, the maximum problem solving ability is taken. The loss function of a case c_i is defined mathematically as,

$$loss_{c_i} = 1 - \max_{c_j \in ℂ} P(c_j, c_i) * x_{c_j} \tag{5}$$

where x_{c_j} indicates whether the case c_j is an element of *Footprint$_{opt}$* or not.

$$x_{c_i} = \begin{cases} 1 & \text{if } c_i \in Footprint_{opt}, \\ 0 & \text{otherwise.} \end{cases} \tag{6}$$

The objective of the proposed optimization formulation by Mathew et al. [12] is to minimize the footprint size and the overall loss of all cases in the case-base, and the footprint set is constrained to obtain solution for all cases in the case-base. The objective function and constraints are defined as.

$$\min \qquad \sum_{c_i \in \mathbb{C}} (loss_{c_i} + x_{c_i}) \tag{7a}$$

$$\text{subject to} \qquad \sum_{c_i \in \mathbb{C}} Cov_{c_i} = |\mathbb{C}|, \tag{7b}$$

$$x_{c_i} \in \{0, 1\} \qquad \qquad \forall 1 \le i \le n. \tag{7c}$$

where Cov is a vector, which indicates whether a case c_i can be solved by the footprint $Footprint_{opt}$ or not. This formulation is defined only for the case when the solution of a single case is adapted to arrive at a solution of target case (i.e. the single case adaptation process). The definition of loss function does not capture compositional adaptation process where the solutions of multiple cases are adapted to arrive the solution of the target case.

3   Proposed Optimization Formulation

In this section, we discuss the proposed optimization formulation to identify the optimal footprint set for compositional adaptation applications. First we introduce the preliminary definitions and then discuss the optimization formulation.

3.1   Preliminary Definitions

Let $\mathbb{C}$ be a set of cases in the case-base. We define a binary vector x such that $x = < x_{c_i} >_{c_i \in \mathbb{C}}$, where each element x_{c_i} corresponds to a case $c_i \in \mathbb{C}$ and it indicates that whether the case c_i is present in the optimal footprint set ($Footprint\text{-}CA_{opt}$) or not, i.e.,

$$x_{c_i} = \begin{cases} 1 & \text{if } c_i \in Footprint\text{-}CA_{opt}, \\ 0 & \text{otherwise.} \end{cases} \tag{8}$$

Consider the set of cases C' such that $C' \in Reachability_{CA}(c_i)$. The solutions of a set of cases in C' can be used to arrive at a solution for the case c_i. The extent to which each set C' in the $Reachability_{CA}(c_i)$ arrives at a solution for c_i is captured by the problem solving ability. It is denoted as $P(C', c_i)$ and its properties are given below.

1. $0 \le P(C', c_i) \le 1$
2. $P(C', c_i) = 1$ if $C' = \{c_i\}$

The objective of the proposed formulation is to minimize the footprint set and maximize the problem solving ability of footprint set. The vector x captures the presence of each case in the footprint set. In order to represent the second criterion, we define the problem solving ability of footprint cases to arrive at a solution for each case in the case-base. We denote the maximum problem solving ability of the footprint set to arrive at a solution for the case c_i as $MAX\_PSA_{FP}(c_i)$. This term considers the maximum problem solving ability due to the reason that the solution for a case can be obtained by multiple sets of cases in the footprint set. For example: Let the footprint cases be c_1, c_2, c_3 and the $Reachability_{CA}(c_4) = \{\{c_1, c_3\}, \{c_2\}\}$. The cases $\{c_1, c_3\}$ together solve the case c_4 with the problem solving ability 0.8 and the case $\{c2\}$ solves c_4 with problem solving ability 0.5. Then the $MAX\_PSA_{FP}(c_4) = 0.8$. The properties of (MAX_PSA_{FP}) are,

1. For each case $c_i \in \mathbb{C}$, $MAX\_PSA_{FP}(c_i)$ ranges between 0 and 1
2. If a case c_i is present in the footprint set, then $MAX\_PSA_{FP}(c_i)$ is 1.
3. If the cases in the footprint set cannot obtain a solution for a case c_i, then $MAX\_PSA_{FP}(c_i)$ is 0.

Mathematically, the MAX_PSA_{FP} of a case c_i is defined as

$$MAX\_PSA_{FP}(c_i) = \max_{\mathbb{C}' \in Reachability_{CA}(c_i)} P(\mathbb{C}', c_i) * I(\mathbb{C}') \qquad (9)$$

where $P(\mathbb{C}', c_i)$ is the problem solving ability of the set of cases $\mathbb{C}'$ to arrive at a solution for c_i and $I(\mathbb{C}')$ is an indicator function which returns 1 if all cases in $\mathbb{C}'$ are elements of footprint set otherwise it returns 0. More precisely, $I(\mathbb{C}')$ is defined as

$$I(\mathbb{C}') = \begin{cases} 1 & \text{if all cases in } \mathbb{C}' \text{ are elements of footprint set} \\ 0 & \text{Otherwise} \end{cases} \qquad (10)$$

In order to capture the above indicator function by the optimization solver, the indicator function $I(\mathbb{C}')$ can be re-written as

$$I(\mathbb{C}') = \max\left(0, \left(\sum_{c_j \in \mathbb{C}'} x_{c_j}\right) - |\mathbb{C}'| + 1\right) \qquad (11)$$

In Eq. 11, if all cases in $\mathbb{C}'$ are not elements of footprint set, the second term returns a negative value and due to the max term this function will return zero. If all cases in $\mathbb{C}'$ are elements of footprint set, then the second term will be one. Using the formulation of $I(\mathbb{C}')$ in Eq. 11, $MAX\_PSA_{FP}(c_i)$ is re-written as

$$MAX\_PSA_{FP}(c_i) = \max_{\mathbb{C}' \in Reachability_{CA}(c_i)}\left(0, \left(P(\mathbb{C}', c_i)*\left(\sum_{c_j \in \mathbb{C}'} x_{c_j} - |\mathbb{C}'| + 1\right)\right)\right) \qquad (12)$$

Let $Cov = <Cov_{c_i}>_{c_i \in \mathbb{C}}$ be a vector, which indicates whether a case c_i can be solved by the footprint $Footprint\text{-}CA_{opt}$ or not. Mathematically, it is defined as,

$$Cov_{c_i} = \begin{cases} 1 & \text{if } c_i \text{ can be solved by } Footprint\text{-}CA_{opt} \\ 0 & \text{otherwise} \end{cases} \qquad (13)$$

3.2   Optimization Formulation

The objective of our proposed method is to arrive at an optimal footprint set ($Footprint\text{-}CA_{opt}$) for compositional adaptation applications which satisfies the following criteria.

1. minimize the size of $Footprint\text{-}CA_{opt}$
2. maximize the problem solving ability of the footprint set $Footprint\text{-}CA_{opt}$
3. it should be possible to arrive at solutions for the rest of the cases in the case-base by adapting the solutions of cases in the $Footprint\text{-}CA_{opt}$

The first criterion can be recorded by the binary vector x and the summation of this vector results in the footprint set size. The second criterion is captured by the term $MAX\_PSA_{FP}(c_i)$, i.e., the maximum problem solving ability of the footprint set to arrive a solution of the case c_i. These two factors are represented in the objective function of the following optimization formulation and the third criterion is captured by the constraint 14b. We define the optimization formulation as

$$\min \qquad \sum_{c_i \in \mathbb{C}} \alpha * \left(1 - MAX\_PSA_{FP}(c_i)\right) + (1 - \alpha) * x_{c_i} \qquad (14a)$$

$$\text{subject to} \qquad \sum_{c_i \in \mathbb{C}} Cov_{c_i} = |\mathbb{C}|, \qquad (14b)$$

$$x_{c_i} \in \{0, 1\} \forall 1 \le i \le n. \qquad (14c)$$

Here, α is the trade-off parameter between footprint set size and problem solving ability and $0 \le \alpha \le 1$. The constraints of the above formulation are linear and convex. However, the formulation is concave due to the max term in the calculation of $MAX\_PSA_{FP}(c_i)$. A similar concave formulation was encountered in the optimization formulation [12] discussed in Sect. 2.2 due to the max term. In [12], Mathew et al. uses an equivalent linear function of max term by using a binary variable. The equivalent linear function to find the value of y which is equal to $\max(v_1, v_2, \ldots, v_n)$ is given below where v_i is an integer for all $1 \le i \le n$.

$$y \ge v_i \qquad\qquad \forall 1 \le i \le n, \qquad (15a)$$

$$y \le v_i + 1 - d_i \qquad\qquad \forall 1 \le i \le n, \qquad (15b)$$

$$\sum_{i=1}^{n} d_i = 1, \qquad\qquad (15c)$$

$$d_i \in \{0, 1\} \qquad\qquad \forall 1 \le i \le n, \qquad (15d)$$

where the constraint 15a assigns a lower bound for y which will be equal to the $\max(v_1, v_2, \ldots, v_n)$. In constraint 15b, the binary variable d finds an upper bound for y. The constraints 15c and 15d ensure that only one value of d is 1 and remaining are 0. The v_i value corresponds to $d_i = 1$ acts as the upper bound for y variable as per constraint 15b. All these constraints together find a feasible solution only when $d_i = 1$ for the v_i with maximum value.

Using this idea, the equivalent formulation of $MAX\_PSA_{FP}(c_i)$ in Eq. 12 can be written as,

$$MAX\_PSA_{FP}(c_i) \geq \left(P(C', c_i) * \left(\sum_{c_j \in C'} x_{c_j} - |C'| + 1 \right) \right),$$
$$\forall C' \in Reachability_{CA}(c_i), 1 \leq i \leq n \qquad (16a)$$

$$MAX\_PSA_{FP}(c_i) \leq \left(P(C', c_i) * \left(\sum_{c_j \in C'} x_{c_j} - |C'| + 1 \right) \right) + 1 - d_j,$$
$$\forall C' \in Reachability_{CA}(c_i), 1 \leq i \leq n, \qquad (16b)$$

$$MAX\_PSA_{FP}(c_i) \geq 0 \quad \forall 1 \leq i \leq n \qquad (16c)$$

$$\sum_{i=1}^{n} d_i = 1, \qquad (16d)$$

$$d_i \in \{0, 1\} \quad \forall 1 \leq i, \leq n. \qquad (16e)$$

Using this equivalent formulation of $MAX\_PSA_{FP}(c_i)$, the convex optimization formulation of our proposed solution is given below.

$$\min \sum_{c_i \in \mathbb{C}} \alpha * \left(1 - MAX\_PSA_{FP}(c_i) \right) + (1 - \alpha) * x_{c_i} \qquad (17a)$$

$$\text{subject to } \sum_{c_i \in \mathbb{C}} Cov_{c_i} = |\mathbb{C}|, \qquad (17b)$$

$$x_{c_i} \in \{0, 1\} \quad \forall 1 \leq i \leq n, \qquad (17c)$$

$$MAX\_PSA_{FP}(c_i) \geq \left(P(C', c_i) * \left(\sum_{c_j \in C'} x_{c_j} - |C'| + 1 \right) \right),$$
$$\forall C' \in Reachability_{CA}(c_i), 1 \leq i \leq n, \qquad (17d)$$

$$MAX\_PSA_{FP}(c_i) \leq \left(P(C', c_i) * \left(\sum_{c_j \in C'} x_{c_j} - |C'| + 1 \right) \right) + 1 - d_j,$$
$$\forall C' \in Reachability_{CA}(c_i), 1 \leq i \leq n, \qquad (17e)$$

$$MAX\_PSA_{FP}(c_i) \geq 0 \quad \forall 1 \leq i \leq n \qquad (17f)$$

$$\sum_{i=1}^{n} d_i = 1, \qquad (17g)$$

$$d_i \in \{0, 1\} \quad \forall 1 \leq i, \leq n. \qquad (17h)$$

This is a mixed integer programming problem, and the objective function and constraints are linear and convex.

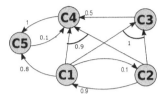

Fig. 1. A sample case-base network

Table 1. $Reachbility_{CA}$ and problem solving ability of cases in the $Reachbility_{CA}$ set of each case to arrive a its solution

Case (c)	$Reachability_{CA}(c)$	Problem solving ability
c1	$\{\{c2\}\}$	$P(\{c1\}, c1) = 1,$ $P(\{c2\}, c1) = 0.9$
c2	$\{\{c1\}\}$	$P(\{c1\}, c2) = 0.1,$ $P(\{c2\}, c2) = 1$
c3	$\{\{c1, c2\}\}$	$P(\{c1, c2\}, c3) = 1,$ $P(\{c3\}, c3) = 1$
c4	$\{\{c1, c2\}, \{c3\}, \{c5\}\}$	$P(\{c1, c2\}, c4) = 0.9,$ $P(\{c3\}, c4) = 0.5,$ $P(\{c4\}, c4) = 1,$ $P(\{c5\}, c4) = 0.1$
c5	$\{\{c1\}, \{c1\}\}$	$P(\{c1\}, c5) = 0.8,$ $P(\{c4\}, c5) = 1,$ $P(\{c5\}, c5) = 1$

For example, consider a sample case-base which contains cases $\mathbb{C} = \{c1, c2, c3, c4, c5\}$. We draw a case-base network in Fig. 1 where an edge from case $c1$ to case $c2$ means the solution of $c1$ can be adapted to arrive at a solution for the case $c2$ and the edge weight indicates the problem solving ability of $c1$ to find solution for $c2$. In this network the solution for the case $c3$ can be obtained by the combined solution of the cases $c1$ and $c2$. This is indicated by the arc joining the edges $(c1, c3)$ and $(c2, c3)$ and it is an example of compositional adaptation. The $Reachability_{CA}$ of all cases in this network and the problem solving ability of the elements in the $Reachability_{CA}$ of each case to arrive at its solution are listed in Table 1. In this table, we have included the problem solving ability of each case to arrive at its own solution which is marked as 1, i.e., $P(\{c_i\}, c_i) = 1 \ \forall \ c_i \in \mathbb{C}$.

Using the details in Table 1, we obtain the optimal footprint set ($Footprint\text{-}CA_{opt}$) based on the proposed formulation. This footprint set is compared with the $Footprint\text{-}CA$ sets obtained using retention score and weighted retention score. The footprint set comparison is listed in Table 2. In this table, the column $Total \ MAX\_PSA_{FP}$ stands for the summation of MAX_PSA_{FP} of all cases in the case-base and the column $Objective \ value$ stands for the value of the objective function in the proposed formulation with respect to the corresponding footprint set. In this example, we can observe that the footprint set size obtained using optimal footprint method is minimum with maximum $Total \ MAX\_PSA_{FP}$ and this results in minimum objective value compared to other methods.

4   Experimental Evaluation

We evaluate the proposed methods on prediction datasets which are available in UCI Repository [6] such as housing, auto MPG, hardware, and automobile. The goal of these datasets is to predict the housing price, fuel consumption, estimated relative performance, and automobile price respectively. These datasets are pre-processed by removing the data instances with unknown values, and the feature

Table 2. Comparison of footprint size and total MAX_PSA_{FP} for the example in Fig. 1

Method	Footprint set	Total MAX_PSA_{FP}	Objective value
$Footprint\text{-}CA_{opt}$	$\{c_1, c_2\}$	4.7	2.3
$Footprint\text{-}CA_{WRS}$	$\{c_1, c_4, c_3\}$	4.1	3.9
$Footprint\text{-}CA_{RS}$	$\{c_1, c_3\}$	3.4	3.6

values are normalized between 0 and 1. We considered only numeric features in all these datasets. The dataset statistics are summarized in Table 3.

Table 3. Dataset statistics

Dataset	#instances	#features
Housing	506	13
Auto MPG	392	7
Hardware	209	7
Automobile	194	12

We construct a case-base out of each dataset and identify the footprint set. Each instance in the dataset is considered as case; the feature values of each data instance are part of the problem component of the case and the target value is considered as its solution component. We say a case (or data instance) c_1 solves another case c_2 when the problem components of c_1 and c_2 are similar and the solution adapted using c_1's solution is close to the solution of c_2. The similarity of problem components are estimated using the k-nearest neighbor algorithm [8]. The solution for the target case is adapted from the solutions of k-nearest cases by taking average of its target value. The closeness of the predicted solution to its actual solution is validated by keeping an acceptable prediction error (APE) [7] fixed at 5% where APE is defined as,

$$APE = \frac{|y_{actual} - y_{predict}|}{y_{actual}} \tag{18}$$

where y_{actual} is the actual solution of c_i and $y_{predict}$ is the estimated solution predicted by the set of cases in $\mathbb{C}' \in Reachability_{CA}(c_i)$. For each case in the case-base, we obtain the $Reachability_{CA}$ set and the problem solving ability of elements in its $Reachability_{CA}$ to arrive at its solution. The k-nearest cases of a case c_i are considered as a set of cases (say $\mathbb{C}'$). This set is an element of c_i's $Reachability_{CA}$ set if the value predicted using the k-nearest cases is within the acceptable prediction error. The problem solving ability of each element in the $Reachability_{CA}(c_i)$ set to find solution for the case c_i ($P(\mathbb{C}', c_i)$ where $\mathbb{C}' \in Reachability_{CA}(c_i)$) is measured as,

$$P(\mathbb{C}', c_i) = \frac{1}{1 + (y_{actual} - y_{predict})^2} \tag{19}$$

We estimate the $Footprint\text{-}CA_{opt}$ set using the proposed optimization formulation. $Footprint\text{-}CA_{opt}$ is compared with the footprint sets obtained using greedy algorithms such as retention score based footprint ($Footprint\text{-}CA_{RS}$) and weighted retention score based footprint ($Footprint\text{-}CA_{WRS}$). For each dataset, we constructed case-base using 2-nn and 3-nn. We report the footprint set size and sum of MAX_PSA_{FP} values of all cases in the case-base for all three footprint sets. The results obtained for the case-base constructed using 2-nn and 3-nn are given in Tables 4 and 5 respectively. The objective of the proposed formulation is obtain a footprint set with minimum size and maximum ability to find solution for the rest of the cases in case-base (total MAX_PSA_{FP}) that is close to its desired solution. In Table 4, we can observe that the total MAX_PSA_{FP} is highest for the $Footprint\text{-}CA_{opt}$ in all datasets. In housing dataset, the optimal footprint has the minimum size and maximum total MAX_PSA_{FP} value. For the remaining datasets, retention score based footprint set size is the minimum. However, its total MAX_PSA_{FP} value is much smaller than that of optimal footprint set.

Table 4. Comparison of footprint size and Total MAX_PSA_{FP} over different datasets for $k = 2$

Dataset	$Footprint\text{-}CA_{opt}$		$Footprint\text{-}CA_{WRS}$		$Footprint\text{-}CA_{RS}$	
	Footprint size	Total MAX_PSA_{FP}	Footprint size	Total MAX_PSA_{FP}	Footprint size	Total MAX_PSA_{FP}
Housing	**362**	**465.75**	365	455.62	366	444.27
Auto MPG	271	**353.32**	270	348.77	**268**	335.35
Hardware	143	**177.82**	143	176.39	**142**	172.36
Automobile	129	**184.1**	128	181.82	**127**	177.94

Table 5. Comparison of footprint size and total MAX_PSA_{FP} over different datasets for $k = 3$

Dataset	$Footprint - CA_{opt}$		$Footprint - CA_{WRS}$		$Footprint - CA_{RS}$	
	Footprint size	Total MAX_PSA_{FP}	Footprint size	Total MAX_PSA_{FP}	Footprint size	Total MAX_PSA_{FP}
Housing	**393**	**467.26**	394	464.39	399	458.03
Auto MPG	**296**	**365.51**	296	359.49	299	348.82
Hardware	159	**186.36**	159	183.27	**154**	181.13
Automobile	139	**186.36**	138	184.45	**137**	179.21

In Table 5, we report the footprint set size and total MAX_PSA_{FP} values obtained for all datasets when the case-bases are constructed using 3-nn. The optimal footprint sets score the highest total MAX_PSA_{FP} values for all datasets and lowest footprint set size for housing and auto MPG datasets.

We analyze the trade-off between the size and the problem solving ability of footprint set. The α parameter in Eq. 17a of the optimization formulation is varied between 0 to 1 for this analysis. At $\alpha = 0$, we obtain a footprint set which is optimized only based on size and at $\alpha = 1$, the footprint set will be optimized only based its problem solving ability. We illustrate the trade-off analysis in

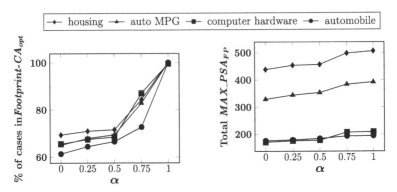

Fig. 2. Trade-off between footprint size and MAX_PSA_{FP} for all datasets when case-bases are constructed using 2-nn

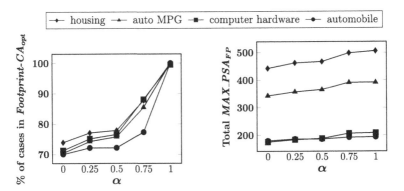

Fig. 3. Trade-off between footprint size and MAX_PSA_{FP} for all datasets when case-bases are constructed using 3-nn

Figs. 2 and 3 for all four datasets when case-bases are constructed using 2-nn and 3-nn respectively. We report the footprint set size as the percentage of cases identified as cases in the footprint set. We can observe that the percentage of cases in the footprint set increases with an increase in α. For $\alpha > 0.5$, footprint set size increased substantially and at $\alpha = 1$ all the cases in the case-base are marked as footprint cases in all datasets for both 2-nn and 3-nn. The total MAX_PSA_{FP} represents the overall problem solving ability of footprint cases. This value increases with an increase in the value of α. The overall problem solving ability of footprint cases is highest at $\alpha = 1$ for all datasets and at this α value all cases in the case-base are footprint cases.

The performance generalization of the proposed method is studied by analyzing the prediction performance of the footprint set. We split each dataset into 5-fold train-test split and obtained the footprint sets from the train data. Each footprint set is used for training and the prediction performance is evaluated over the test data using mean squared error (MSE). The average footprint set size and the mean squared error over the test data when trained with footprint sets

Table 6. Average footprint size and MSE analysis over test data when trained with footprint sets- $Footprint\text{-}CA_{opt}$, $Footprint\text{-}CA_{WRS}$ and $Footprint\text{-}CA_{RS}$

Dataset	Footprint size			Mean squared error		
	$Footprint\text{-}CA_{opt}$	$Footprint\text{-}CA_{WRS}$	$Footprint\text{-}CA_{RS}$	$Footprint\text{-}CA_{opt}$	$Footprint\text{-}CA_{WRS}$	$Footprint\text{-}CA_{RS}$
Housing	**337.6**	339.0	340.0	**22.3**	23.27	24.82
Auto MPG	**233.2**	**234.8**	238.0	**9.0**	9.18	9.22
Hardware	**154.6**	154.8	154.8	**363.04**	365.17	367.12
Automobile	**123.2**	123.6	124.4	**6.89**	7.38	7.42

such as $Footprint\text{-}CA_{opt}$, $Footprint\text{-}CA_{WRS}$ and $Footprint\text{-}CA_{RS}$ are reported in Table 6. We can observe the size and mean squared error of $Footprint\text{-}CA_{opt}$ is the lowest for all datasets. The improvements of $Footprint\text{-}CA_{opt}$ are statistically significant in paired t-test for all datasets with p-value less than 0.1.

Table 7. Run time analysis

Dataset	Time in seconds		
	$Footprint\text{-}CA_{opt}$	$Footprint\text{-}CA_{WRS}$	$Footprint\text{-}CA_{RS}$
Housing	5.79	0.24	0.18
Auto-MPG	4.6	0.15	0.39
Hardware	2.63	0.1	0.08
Automobile	2.47	0.11	0.09

In Table 7, we report the time taken to arrive at all footprint sets such as $Footprint\text{-}CA_{opt}$, $Footprint\text{-}CA_{WRS}$ and $Footprint\text{-}CA_{RS}$. The machine in which we run the program is configured with i7 processor, 1.9 GHZ, 16 GB RAM and SSD harddisk. In this analysis, we can observe as expected that the optimal footprint is taking more time compared to other greedy algorithms. However, the optimal footprint set is obtained within a few seconds for all datasets.

5 Conclusion

In this paper, we propose an optimization formulation to arrive at an optimal footprint set for the compositional adaptation applications of which single case adaptation is a special case. The proposed formulation minimizes the size of the footprint set as well as maximizes the ability of footprint set to arrive at a solution close to the desired solution for the rest of the cases in the case-base. The trade-off analysis between the footprint set size and its performance are illustrated using the proposed formulation. In the future, we would like to extend the evaluation to more complex domains with more complex compositional adaptation approaches.

References

1. Kolodner, J.L.: An introduction to case-based reasoning. Artif. Intell. Rev. **6**, 3–34 (1992)
2. Smyt, B., McKenna, E.: Footprint-based retrieval. In: Althoff, K.-D., Bergmann, R., Branting, L.K. (eds.) ICCBR 1999. LNCS, vol. 1650, pp. 343–357. Springer, Heidelberg (1999). https://doi.org/10.1007/3-540-48508-2_25
3. Wilke, W., Bergmann, R.: Techniques and knowledge used for adaptation during case-based problem solving. In: Pasqual del Pobil, A., Mira, J., Ali, M. (eds.) IEA/AIE 1998. LNCS, vol. 1416, pp. 497–506. Springer, Heidelberg (1998). https://doi.org/10.1007/3-540-64574-8_435
4. Smyth, B., Keane, M.T.: Remembering to forget. In: Proceedings of the 14th International Joint Conference on Artificial Intelligence (IJCAI), pp. 377–382 (1995)
5. Smyth, B., McKenna, E.: Modelling the competence of case-bases. In: Smyth, B., Cunningham, P. (eds.) EWCBR 1998. LNCS, vol. 1488, pp. 208–220. Springer, Heidelberg (1998). https://doi.org/10.1007/BFb0056334
6. Asuncion, A., Newman, D.: UCI Machine Learning Repository (2007)
7. Bennett, J.O., Briggs, W.L., Badalamenti, A.: Using and Understanding Mathematics: A Quantitative Reasoning Approach. Pearson Addison Wesley, Reading (2008)
8. Cover, T.M., Hart, P.E.: Nearest neighbor pattern classification. IEEE Trans. Inf. Theory **13**(1), 21–27 (1967)
9. Mathew, D., Chakraborti, S.: Competence guided casebase maintenance for compositional adaptation applications. In: Goel, A., Díaz-Agudo, M.B., Roth-Berghofer, T. (eds.) ICCBR 2016. LNCS (LNAI), vol. 9969, pp. 265–280. Springer, Cham (2016). https://doi.org/10.1007/978-3-319-47096-2_18
10. Mathew, D., Chakraborti, S.: Competence guided model for casebase maintenance. In: International Joint Conference on Artificial Intelligence, pp. 4904–4908 (2017)
11. Mathew, D., Chakraborti, S.: A generalized case competence model for casebase maintenance. AI Commun. **30**(3–4), 295–309 (2017)
12. Mathew, D., Chakraborti, S.: An optimal footprint method for case-base maintenance. In: The Thirty-First International Flairs Conference, pp. 383–388 (2018)
13. Hart, P.: The condensed nearest neighbor rule (Corresp.). IEEE Trans. Inf. Theory **14**(3), 515–516 (1968)
14. Gebhardt, F., Voß, A., Gräther, W., Schmidt-Belz, B.: Reuse of a single case: adaptation. In: Gebhardt, F., Voß, A., Gräther, W., Schmidt-Belz, B. (eds.) Reasoning with Complex Cases. SECS, vol. 393, pp. 131–152. Springer, Boston (1997). https://doi.org/10.1007/978-1-4615-6233-7_13
15. Voss, A., Bartsch-Spörl, B., Oxman, R.: A study of case adaptation systems. In: Gero, J.S., Sudweeks, F. (eds.) Artificial Intelligence in Design '96, pp. 173–189. Springer, Dordrecht (1996). https://doi.org/10.1007/978-94-009-0279-4_10
16. Müller, G., Bergmann, R.: Compositional adaptation of cooking recipes using workflow streams. In: Computer Cooking Contest, Workshop Proceedings of International Conference on Case-Based Reasoning (2014)
17. Arshadi, N., Badie, K.: A compositional approach to solution adaptation in case-based reasoning and its application to tutoring library. In: Proceedings of 8th German Workshop on Case-Based Reasoning (2000)
18. Atzmueller, M., Baumeister, J., Puppe, F., Shi, W., Barnden, J.A.: Case-based approaches for diagnosing multiple disorders. In: FLAIRS, pp. 154–159 (2004)

19. Lekkas, G.P., Avouris, N.M., Viras, L.G.: Case-based reasoning in environmental monitoring applications. Appl. Artif. Intell. Int. J. **8**(3), 359–376 (1994)
20. Cummins, L., Bridge, D.: Choosing a case base maintenance algorithm using a meta-case base. In: Bramer, M., Petridis, M., Nolle, L. (eds.) SGAI 2011, pp. 167–180. Springer, London (2011). https://doi.org/10.1007/978-1-4471-2318-7_12
21. Page, L., Brin, S., Motwani, R., Winograd, T.: The PageRank citation ranking: bringing order to the web. Stanford InfoLab (1999)

Towards Human-Like Bots Using Online Interactive Case-Based Reasoning

Maximiliano Miranda$^{(\boxtimes)}$, Antonio A. Sánchez-Ruiz, and Federico Peinado

Departamento de Ingeniería del Software e Inteligencia Artificial,
Universidad Complutense de Madrid,
c/ Profesor José García Santesmases 9, 28040 Madrid, Spain
{m.miranda,antsanch}@ucm.es, email@federicopeinado.com
http://www.narratech.com

Abstract. The imitation of human playing style has been gaining relevance in both the Artificial Intelligence for Games research community and the Digital Game industry over the last decade, achieving a special importance in recent years. The goal of these virtual players is to deceive real players and be perceived just as another human player. Although this challenge can be addressed using different Imitation Learning techniques, classic supervised learning approaches do not usually work well due to the violation of the independent and identically distributed assumption for random variables. No regret algorithms in online learning settings seem to outperform previous approaches. In this work we describe an interactive and online case-based reasoning system in which the bot gives control to the human player when it reaches game states that are not well represented by cases in its case base, and regains control when the game states are known again. Results show that (1) the amount of human intervention decreases rapidly, (2) the case base needed to achieve reasonable imitation is considerable smaller than that used in a non-interactive approach (3) the resulting agent outperforms other agents using non-interactive CBR.

Keywords: Interactive online learning ·
Learning from demonstration · Human behavior imitation ·
Case-based reasoning · Interactive entertainment

1 Introduction

One of the main challenges in Artificial Intelligence (AI) has always been to build agents that mimic human behavior. Researchers are always looking for problems that are demanding but feasible at the same time, in order to advance in their mission of resembling human intelligence in a computer. For this reason, digital games are popular testbeds and several competitions on developing believable characters have emerged during the last decade [8].

There is a widespread assumption in the Digital Game industry that wherever there is a machine-controlled character, the game experience will benefit

© Springer Nature Switzerland AG 2019
K. Bach and C. Marling (Eds.): ICCBR 2019, LNAI 11680, pp. 314–328, 2019.
https://doi.org/10.1007/978-3-030-29249-2_21

if it behaves in a less "robotic" way. In fact, if virtual players (bots) play in a more "human" way, real players perceive the game to be less predictable, more replayable, and more challenging than if the bots were hand-coded [24]. For this reason, player modeling in video games has been an increasingly important field of study, not only for academics but for professional developers as well [29].

These human-like computer bots can come in handy in a wide variety of scenarios, from enriching the game experience to helping video game developers in the production stage. Regarding the game experience, these agents can be used to create more believable enemies in action games to confront the human player, but also to collaborate with him or to illustrate how to succeed in a particular game level, helping the players who get stuck. It is reasonable to think that these computer-played sequences will be more meaningful if the bots imitate the playing style of other human players.

Virtual players can also be used during the testing stage of the video game development process, not only to check whether the game crashes or not, but to verify if the levels have the right difficulty or to find new ways to solve puzzles. This application could be especially useful in games with a strong focus in the procedural content generation, where it is virtually impossible for human testers to cover all the content than can be generated. Finally, the creation of behaviors for non-player characters (NPCs) is a complex task that requires the collaboration among programmers and game designers. Different approaches have been proposed to create these behaviors without technical knowledge using program by demonstration [23] or temporal difference reinforcement learning [7].

Despite the popularity of the Turing test, there is no rigorous standard to determine how human is an artificial agent in a video game [6]. Furthermore, there is not a clear concept about what a believable AI should achieve, and its expected behavior will vary strongly depending on what it is supposed to imitate: to emulate the behavior of other players or to create lifelike characters [12].

Previous works addressed the problem of imitating human players using different supervised learning approaches. Neuroevolution [14] has been used in Ms. Pac-Man as it had been used successfully in other game domains [17], and more recently case-based reasoning [15] has been proposed due to its capacity for imitating spatially-aware autonomous agents in a real-time setting [1], obtaining promising results when imitating human players with different playing styles. In all these approaches the problem of violating the independent and identically distributed (i.i.d.) assumption for random variables in supervised learning is faced, because the "training set" (the traces of examples used in the training) and the "test set" (how the agent actually plays) does not come from the same distribution. These agents apparently achieved very good results (high accuracy, recall, and f-score) but they performed far worse when acting on their own.

In this work we describe an interactive and online case-based reasoning (CBR) system in which the bot gives control of the character to the human player when the bot reaches game states that are not well represented by cases in its case base, and regains control when the game states are known again. This way, the human player acts like an instructor that teaches the CBR system how to correct its bad decisions and come back to well-known game states. Results show that the amount of human intervention decreases rapidly, the case base needed to achieve reasonable imitation is considerable smaller than that used in a non-interactive approach and the resulting agent outperforms a non-interactive one. In order to evaluate the "humanness" of the bot, standard low level metrics such as accuracy or recall have been proven to be ineffective to determine whether two behaviors are similar or not when these are stochastic or require memory of past states [16], so we use a set of high level metrics [15] that are able to capture to some extend different styles of play and skill levels of different players.

The rest of the paper is structured as follows. Next section summarizes the related work in the field. Section 3 describes the internals of the game framework that we use in our experiments. The following two sections describe the interactive online CBR agent and the role of the human expert during the training process. Sections 6 and 7 detail the setup of the experiments and the promising results that we have obtained. Finally, we close the paper with some conclusions and future lines of research.

2 Related Work

Several works regarding the imitation of behavior in video games can be found in the scientific literature, for imitating human players and also script-driven characters. The behavior of an agent can be characterized by studying all its proactive actions and its reactions to sequences of events and inputs over a period of time, but achieving that involves a significant amount of effort and technical knowledge [28] in the best case. Machine Learning (ML) techniques can be used to automate the problem of learning how to play a video game either progressively using players' game traces as input, in direct imitation approaches, or using some form of optimization technique such as Evolutionary Computation or Reinforcement Learning to develop a fitness function that, for instance, "measures" the human likeness of an agent's playing style [25].

Traditionally, several ML algorithms, like ANNs and Naive Bayes classifiers, have been used for modeling human-like players in first-person shooter games by using sets of examples [4]. Other techniques based on indirect imitation like dynamic scripting and Neuroevolution achieved better results in Super Mario Bros than direct (ad hoc) imitation techniques [17].

Case-based reasoning has been used successfully for training RoboCup soccer players, observing the behavior of other players and using traces taken from the game, without requiring much human intervention [3]. Related to CBR and Robotic Soccer, Floyd *et al.* [1] also noted that when working in a setting with time constraints, it is very important to study what characteristics of the cases really impact the precision of the system and when it is better to increase the size of the case base while simplifying the cases. Furthermore, they described how applying preprocessing techniques to a case base can increase the performance of a CBR system by increasing the diversity of the case base.

About how the cases for the case base are obtained, we follow a similar approach as the described by Lam *et al.* [10], as the cases are generated in an automated manner by recording traces of the player that will be imitated as pairs of *scene* state (scene as a representation of the player's point of view) - player's outputs. Floyd and Esfandiari [2] incorporated active learning with learning by observation studying how to create sequences of problems to show to the expert. Finally, Lamontagne *et al.* [11] also studied how the cases could be built from sequential traces during game demostrations in Pac-Man.

Concerning human-like agents, there have been several AI competitions including special tracks for testing the human likeness of agents using Turing-like tests. One of these competitions is the Mario AI Championship[1], which included a "Turing test" track where submitted AI controllers compete with each other for being the most human-like player, judged by human spectators [26]. The BotPrize competition [8] focuses on developing human-like agents for Unreal Tournament, encouraging AI programmers to create bots which cannot be distinguished from human players. Finally, Ms Pac-Man vs Ghosts, the framework that we use for this work, has been used in different bot competitions during the last years [13]. After some years discontinued, it returned in 2016[2] to be celebrated yearly [27].

The problem of violating the i.i.d. assumption in imitation learning has been addressed before with no regret algorithms in online learning settings resulting in algorithms like SMILe and DAGGER proposed by Ross *et al.* [21,22] which outperform previous approaches like SEARN [9] in the Super Tux Kart and Super Mario Bros video games. However, these methods have limitations when the demonstrator is a human player. Because of this, more recently, Packard and Ontañón presented *SALT*, which main idea is to let the learning agent play until it has moved out of the space for which it has training data, giving the control to the expert at this point to show the agent how to get back into this space, turning the learning process into an i.i.d. task allowing the use of supervised learning algorithms [18]. Further on, they extended this approach studying its efficiency in environments where the amount of training data the learning agent is allowed to request from the expert is limited [19]. These methods inspired the interactive-CBR system described in this work.

Finally, about the use of Pac-Man as test bed for imitation learning, it should be noticed that, during decades, it has been considered a promising platform for research due to its many characteristics that make it stand out from other games. Thus, there have been nearly 100 different approaches covering a wide selection of techniques used to develop controllers for Pac-Man or the ghosts, including Rule-based and finite state machines, tree search and Monte Carlo, evolutionary algorithms, neural networks, neuro-evolution and reinforcement learning [20].

---

[1] http://www.marioai.org/.
[2] http://www.pacmanvghosts.co.uk/.

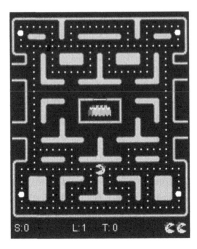

Fig. 1. A screenshot of Ms. Pac-Man vs. Ghosts

3   Ms. Pac-Man vs. Ghosts

Pac-Man is an arcade video game produced by Namco and created by Toru Iwatani and Shigeo Fukani in 1980. Since its launch it has been considered as an icon, not only for the video game industry, but for the 20th century popular culture [5]. In this game, the player has direct control over Pac-Man (a small yellow character), pointing the direction it will follow in the next turn. The level is a simple maze full of white pills, called Pac-Dots, that Pac-Man eats gaining points. There are four ghosts (named Blinky, Inky, Pinky and Sue) with different behaviors trying to capture Pac-Man, causing it to lose one life. Pac-Man initially has three lives and the game ends when the player looses all of them. In the maze there are also four special pills, bigger than the normal ones and named Power Pellets or Energizers, which make the ghosts "edible" during a short period of time. Every time Pac-Man eats one of the ghosts during this period, the player is rewarded with several points.

Ms. Pac-Man vs Ghosts (see Fig. 1) is an implementation of Pac-Man's sequel Ms. Pac-Man in Java designed to develop bots to control both the protagonist and the antagonists of the game. This framework has been used in several academic competitions during the recent years [27] to compare different AI techniques. Some of these bots are able to obtain very high scores but their behavior is usually not very human. For example, they are able to compute optimal routes and pass very close to the ghosts while human players tend to keep more distance and avoid potential dangerous situations.

The Ms. Pac-Man vs. Ghosts API represents the state of the game as a graph in which each node corresponds to a walkable region of the maze (which visually is a 2×2 pixels square). Each node can be connected to up to other four nodes, one in each direction (north, east, south and west), and can contain a Pac-Dot, a Power Pellet, one or more ghosts, Ms. Pac-Man herself or nothing at all (i.e.

the node represents an empty portion of the maze). The full graph representing the state of the first level of the game contains 1293 nodes.

The framework provides a few examples of simple bots that can be used as entry points for developing new ones. Among the controllers included for the ghosts there is the *StarterGhosts* controller with which the ghosts have a simple behavior: if the ghost is edible or Ms. Pac-Man is near a Power Pellet, they escape in the opposite direction. Otherwise, they try to follow Ms. Pac-Man with a probability of 0.9, or make random movements with a probability of 0.1. Visually, this makes the controller to appear like having 2 different states: they try to reach Ms. Pac-Man unless they are edible (or Ms. Pac-Man is very close to a Power Pellet) in which case they will try to escape. We use this controller for the ghosts because it has a behavior that is neither too complex nor too simple, enriching the game space states with slight variations depending on Ms. Pac-Man's situation and some minor random decisions.

4   The CBR Agent

It is interesting to note that even a classic arcade game such as Pac-Man hides a very high dimensional feature space that is a challenge for ML algorithms. The full state representation of the game contains 256 different parameters and the player can perform 5 possible actions at each game step (move left, right, up, down or neutral). Typically, an averaged-skill human player needs between 1200 to 1800 game steps to complete one level of the game, and the trace of a game contains thousand of pairs (state, action).

Next, we summarize the CBR system that we use as the basis to implement the interactive system that we describe in the following section (see [15] for more details). Cases describe pairs (state, action) where the state is represented using the following abstract features:

- distances to the closest Pac-Dot in each direction (p).
- distances to the closest Power Pellet in each direction (pp).
- distances to the closest ghosts in each direction (g).
- time (game steps) the nearest ghost will remain edible in each direction (eg).
- direction chosen by the player in the previous game step (la).
- direction chosen by the player in the current game step (a).

The first 4 features are four-dimensional vectors containing scalar values (distance or time) for each direction. The last 2 features are categorical variables with 5 possible values: left, right, up, down, neutral (neutral means to maintain a direction than makes unable to transition to another node of the maze, i.e. the agent is stuck in a corner without selecting a really possible direction). The remaining edible time in each direction is a scalar value ranging from 0 (the closest ghost via that direction is not edible) and 200 (Ms. Pac-Man has just eaten a Power Pellet). The direction chosen in the previous game step is important so the bot is not purely *reactive* (decisions based only on the current game state) and contains a simple internal state describing where it comes from. Although limited and short-term, this *memory-based* representation has an important impact in the performance of the bot.

The similarity between cases is computed as a linear combination of the similarities between the vectors which, in turn, is computed as the inverse of the euclidean distance. This way, we can weight the contribution of each feature to the final similarity value. The action chosen in each game state is decided by a majority vote using the 5 most similar cases in the case base, and in case of ties the nearest neighbor is chosen.

5   The Interactive-CBR Agent

The CBR virtual agent described in the previous section suffers a problem that, in fact, is common to other agents trained from human traces using supervised learning algorithms. When the agent makes a mistake, i.e. selects a wrong action, it is likely to reach a new game state that is more different from the ones represented in its case base. This is so because we consider an action to be wrong if it was never selected in that game state by the human player the CBR agent learned from. This way, the CBR agent is more likely to make another mistake that will take it to another game state even more different from the ones in its memory, increasing the probability of making new mistakes again.

Since we cannot prevent the CBR agent from making mistakes from time to time, we introduce the role of a human expert or demonstrator that will teach the CBR agent how to recover from those mistakes in real time and come back to familiar situations. Once the game state is a situation that the CBR agent recognizes (because its case base contains similar game states), the agent can regain control of the game and keep playing. The intuition behind this approach is that the more strategies the CBR agent learns to recover from mistakes, the less long-time and therefore fatal mistakes will make.

A simplified flow diagram of the system is presented in the Fig. 2. Note that both the CBR agent and the human expert can start playing the game. Besides, the CBR agent can begin with an empty case base or using cases extracted offline from previous games.

This approach is similar to *SALT* but introduces different policies to those presented by Packard and Ontañón [18] based on the average similarity of the case retrieval and the coefficient of a linear regression as detailed below.

In order to prevent continuous changes between the CBR agent and the human expert we consider some minimum time thresholds. In particular, when the human expert receives control of the game will play for at least 4 s, and when the CBR agent takes control of the game will play for at least 2 s. Those thresholds can be adjusted for each game and expert. In fact, performing our experiments we learned that 2–4 s is probably too short, especially at the beginning of training, where the CBR agent gives up control often, creating a stressful situation for the expert. It is also important to stop the game until the human expert is ready to take control. And provide visual feedback to let the human player who is in control at all times. In order to do it, we change the color of Ms. Pac-Man depending on who is currently playing.

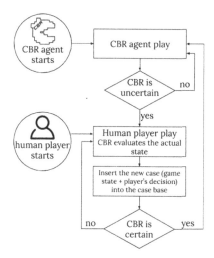

Fig. 2. Simplified flow diagram of the Interactive-CBR system.

When the CBR agent is playing, it retrieves the most similar cases to the current game state from the case base and selects the following actions according to a majority vote. It also stores the similarity values of the most similar retrieved cases for the last 5 s of the game (125 similarity values). Using this window of similarities, the agent computes 2 different values: the average similarity and the coefficient of a linear regression. This last coefficient can be used to know whether the similarities are growing (positive) or decreasing (negative) during the last seconds of the game. That is, if the CBR agent is going towards a better known set of game states or, on the contrary, is getting lost in less familiar situations. Whenever the average similarity is too low or the slope of the linear regression is too negative, the CBR agent gives control of the game to the human expert. It is important to mention that there is no learning when the agent plays.

When the human expert plays, the CBR agent incorporates new cases to the case base. It also retrieves the most similar case in the case base to the current situation, even though the next action will be performed by the human expert. This is because it is necessary to update constantly the window of similarities of the last seconds of the game to decide when to take back control. When the average similarity value exceeds a threshold, the game states are familiar again and the CBR agent can play once again.

This way, the CBR system learns using an interactive approach and we expect the human player to intervene less frequently and for shorter periods of time.

6   Experimental Setup

We have performed 2 different experiments to test the system. In the first one, the interactive CBR agent begins with an empty case base and the human player starts playing the game. The interactive CBR agent learns how to play switching control of the game with the human expert from time to time for 20 full games. In the second experiment we study whether our interactive learning approach can improve the imitative capacity of a CBR agent that has been previously trained offline with the traces of 100 full games played by the same human player.

To evaluate the performance of the CBR agent we use both low level standard classification measures such as accuracy, recall and f-score, and high level measures characterizing the style of play. To compute the accuracy, recall and f-score we compare in how many game states the bot chooses the same action the human player chose in the original game. This is equivalent to relocating Ms. Pac-Man every time the bot makes a mistake, and it is a standard approach in learning by imitation [17] because just one different decision can produce a completely different game in a few game cycles.

On the other hand, the high level measures describe different ways to play the game and were detailed in a previous work [15], although we have added two new ones for these experiments (craving and hungry):

- *Score*: the final game score.
- *Time*: the duration of the game in game steps.
- *Restlessness*: number of direction changes per second.
- *Recklessness*: average distance to the closest ghost.
- *Aggressiveness*: number of ghosts eaten.
- *Clumsiness*: number of game steps in which the player is stuck against a wall.
- *Survival*: the number of lives left when the player completes the first level or 0 if the player dies before.
- *Craving*: average time elapsed between a Power Pellet is eaten and the first edible ghost is eaten.
- *Hungry*: average time between two eaten Power Pellets.

We compute the average values of these high level metrics for each player after playing 100 new games: the human player, the CBR agent that learns offline from the traces of the human player, and the Interactive CBR agent that learns online with the human player. This way we can compare the performance of the interactive CBR agent and the offline CBR agent with respect to the actual human player we aspire to imitate.

7   Results and Discussion

In the first experiment, after 20 games, the case base is filled up with 11,313 cases. Figure 3 shows that, as we expected, the amount of human intervention rapidly decreases during the experiment. Blue and red bars represent, respectively, the amount of time played by the human player and the CBR agent. Different games are marked using white and grey vertical columns.

During the first games, the human player plays most of the time but after 6 games the situation starts to change. From the ninth game, where the case base of the agent already has 7,588 cases, the games are mainly played by the agent. At this point, the CBR agent is capable of covering the most part of the maze

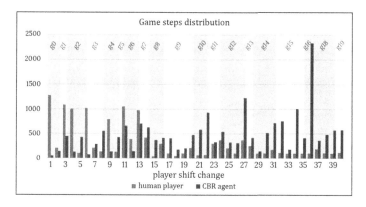

Fig. 3. Game steps distribution between human and the CBR agent play. (Color figure online)

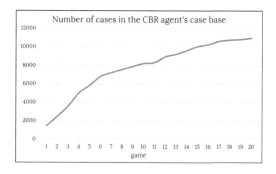

Fig. 4. CBR agent's case base size evolution during the test session.

and the similarities of the cases retrieved only fall in very specific situations (e.g. a corner with more than one ghost near, or ghosts in the middle of a corridor Ms. Pac-Man is facing straight). In particular the eighteenth game (*g17* in the Fig. 3) is entirely played by the CBR agent.

Figure 4 shows the number of cases in the case base during the experiment. As the amount of human intervention decreases over time, so does the velocity of gathering new cases.

Concerning the improvement in the ability of the CBR agent to imitate the playing style of the human player, we set the agent to play 100 new games by itself and measure the high level features explained in Sect. 6. Figure 5 shows the average percentage difference between the values obtained by the human player in 100 games and the values obtained by the CBR agent after each training game. The difference suffers a significant fall until the ninth game, where the average difference is bellow 30%. From there, the decrease is more discrete until reaching an average difference of 25.72% after the twentieth game.

The temporary increase after the third game is related with the capacity of the bot at this point to gain points eating pills, but it never tries to eat

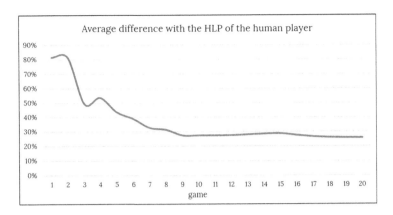

Fig. 5. Evolution of the average percentage difference between the high level parameters obtained by a human player and the ones obtained by the CBR agent every game finished (less is better).

any ghost. The capacity of going towards edible ghosts seems to be replicated after the fourth game, but this specific behavior is not good enough until the sixth game (after games 4 and 5, many times the agent goes towards the ghosts without considering if they are edible or not).

From a simple phenomenological evaluation we can highlight some clear landmarks: after one game (with 1,450 cases in the case base) the CBR agent seems to have learned up to three different game openings depending on the closest pills distribution but it also gets stuck easily and doesn't evade any ghost. At the end of the third game, with 3,583 cases in the case base, the agent has learned to clean the top region of the maze but still does not have knowledge enough to face the ghosts (they still eat Ms. Pac-Man easily). Advancing until the end of the fifth game, with 5,849 cases in the case base, the agent starts to run away from the closest non-edible ghost, and sometimes it goes towards the closest edible ghosts and eats it. It also seems to replicate little micro-errors typical of human players. From this point on, the impact of the learning seems to focus on solving particular situations like escaping from more than one ghost in corners, or learning to avoid a ghost to eat a pill that is behind it.

Comparing the performance of this online interactive approach with our non-interactive approach, we can see an important improvement in terms of the behavior of the CBR agent. Table 1 shows the percentage differences between the high level measures obtained by the human player, and the ones obtained by the interactive CBR after some training games. The closer to 0% is the value, the more similar is the behavior of the bot with respect to the human player (e.g. the interactive CBR agent after the first game obtained a *Survival* value of 100.0% because it is unable to complete the first level).

The last column of the table shows the values obtained by a not interactive CBR agent that has been trained offline with the traces of 100 complete games played by the human player. The not interactive CBR agent achieves an average

Table 1. Percentage differences (less is better) of the high level metrics for the interactive CBR agent (*ICBR n* stands for the interactive CBR agent after game *n*). The last column shows the values for the not interactive CBR agent trained offline with the traces of 100 games.

HL parameter	ICBR 1	ICBR 5	ICBR 10	ICBR 15	ICBR 20	Offline CBR 100
Score	96.15%	51.28%	3.77%	13.20%	5.41%	40.95%
Time	66.36%	30.42%	2.51%	14.63%	3.18%	18.30%
Restlessness	75.96%	6.11%	14.02%	22.95%	22.14%	13.01%
Recklessness	3.01%	12.36%	16.74%	3.56%	5.35%	24.92%
Aggressiveness	100.00%	65.90%	29.27%	17.14%	25.21%	47.13%
Clumsiness	91.21%	40.05%	42.28%	69.00%	63.82%	55.51%
Survival	100.00%	100.00%	99.56%	91.70%	79.48%	100.00%
Craving	100.00%	25.35%	16.60%	0.07%	6.02%	23.72%
Hungry	100.00%	59.77%	21.71%	26.12%	20.90%	55.85%
Avg	**81.41%**	**43.47%**	**27.38%**	**28.71%**	**25.72%**	**42.15%**

percentage difference of 42.15%, while this value is obtained with the interactive approach after only 5 games (43.47%).

Regarding the size of the case bases, the not interactive CBR agent has 127.310 cases extracted from the traces of 100 games, although the agent obtains similar values with the traces of 60 games (≈75,000 cases). Yet, the interactive CBR agent achieves a similar performance after only 5 games with a case base of 5,849 cases. Moreover, after 20 cases, the case base of the interactive agent contains only 11,313 cases and, however, it is able to replicate much better the style of play of the human player.

Finally, the evaluation using the low level metrics is not very interesting, as we showed in [15], because we obtain very high values of accuracy, recall and f-score (greater than 0.96), even after the first training game when the CBR agent's capacity to mimic the human style of play is very poor (1,450 cases in the case base). This is probably because the levels are mazes and most of the time Ms. Pac-Man just goes through corridors following the same direction.

Relative to the second experiment, the interactive CBR agent starts with a case base containing 127,310 cases extracted from the traces of 100 games previously played by the human player. As we expected, the system gives control to the human expert in very specific moments and the segments played by the human expert only last 4 s (the minimum amount of time the system forces the human player to intervene). The human expert intervenes an average of 2 times per game in this experiment and this number does not seem to evolve in contrast to the first experiment.

Table 2 shows the percentage differences of the high level metrics in this second experiment, and the impact of adding the new cases interactively is clearly visible. After 5 games, the case base contains 1000 new cases (a 0.78% of the total) and the agent improves from 42.15% to 34.05%. After 12 games, the case

Table 2. Percentage differences (less is better) of the high level metrics of the CBR agent trained offline with the traces of 100 games and then trained online interactively for another 25 games.

	127,310 cases	1000 new cases	2500 new cases	5000 new cases
Avg diff	42.15%	34.05%	27.32%	24.64%

base contains 2500 new cases (1.92% of the total) and the agent improves to 27.32%. After 25 training games, the case base only contains 5000 new cases (3.78% of the total) and the style of play is only 24.64% different from the style of the human player.

8   Conclusions

The creation of believable non-player characters in video games is a very important research topic. There is a widespread assumption that agents that behave in a more human way create richer game experiences for the players and increase their engagement. As part of our research on imitation of human playing style in video games, we propose a new online and interactive case-based reasoning agent that improves to a great extend our previous not interactive approaches.

The interactive CBR agent achieves a much higher level of "humanity" with less games and much smaller case bases. This is due to the quality of the cases which are incorporated using the interactive approach. Since the CBR bot only acquires new cases when it reaches game states that are not well known, the cases represent more relevant and interesting experiences. Moreover, when the human expert takes control, the CBR agent learns how to get back to familiar game states from which the agent can play again. That is, in some way, that the CBR agent is learning strategies to correct mistakes and we believe this is very important feature of our approach.

On the other hand, the interactive learning approach forces the human expert to be present and intervene during the whole training session. And the continuous switches between the agent and the human expert can be an stressful experience for the human. We have identified some important visual helps in the game to make more evident who is controlling Ms. Pac-Man and we plan to adjust the minimum time intervals so the training process will be easier for the expert.

Moreover, we have showed that the interactive approach can also be used in improve the behavior of other agents trained offline from the traces of games played by human players. This result makes us optimistic about the potential of mixed approaches in which the human expert only needs to intervene during part of the training process.

In the future work, we will study how to allow the expert to rectify behaviors learned by the agent by mistake. Currently, it is the CBR bot who decides when to give up control of the game to the human, but the expert should also be able to take back control of the game when the agent is not behaving as expected.

Acknowledgements. This work has been partially supported by the Spanish Committee of Economy and Competitiveness (TIN2017-87330-R) and the UCM (Group 921330), project *ComunicArte: Comunicación Efectiva a través de la Realidad Virtual y las Tecnologías Educativas*, funded by *Ayudas Fundación BBVA a Equipos de Investigación Científica 2017*, and project *NarraKit VR: Interfaces de Comunicación Narrativa para Aplicaciones de Realidad Virtual* (PR41/17-21016), funded by *Ayudas para la Financiación de Proyectos de Investigación Santander-UCM 2017*.

References

1. Floyd, M.W., Davoust, A., Esfandiari, B.: Considerations for real-time spatially-aware case-based reasoning: a case study in robotic soccer imitation. In: Althoff, K.-D., Bergmann, R., Minor, M., Hanft, A. (eds.) ECCBR 2008. LNCS (LNAI), vol. 5239, pp. 195–209. Springer, Heidelberg (2008). https://doi.org/10.1007/978-3-540-85502-6_13
2. Floyd, M.W., Esfandiari, B.: An active approach to automatic case generation. In: McGinty, L., Wilson, D.C. (eds.) ICCBR 2009. LNCS (LNAI), vol. 5650, pp. 150–164. Springer, Heidelberg (2009). https://doi.org/10.1007/978-3-642-02998-1_12
3. Floyd, M.W., Esfandiari, B., Lam, K.: A case-based reasoning approach to imitating RoboCup players. In: Proceedings of the Twenty-First International Florida Artificial Intelligence Research Society Conference, 15–17 May 2008, Coconut Grove, Florida, USA, pp. 251–256 (2008)
4. Geisler, B.: An empirical study of machine learning algorithms applied to modeling player behavior in a "first person shooter" video game. Ph.D. thesis, Citeseer (2002)
5. Goldberg, H.: All Your Base are Belong to Us: How 50 Years of Videogames Conquered Pop Culture. Three Rivers Press, New York (2011)
6. Gorman, B., Thurau, C., Bauckhage, C., Humphrys, M.: Believability testing and Bayesian imitation in interactive computer games. In: SAB (2006)
7. Harmer, J., Gisslén, L., Holst, H., Bergdahl, J., Olsson, T., Sjöö, K., Nordin, M.: Imitation learning with concurrent actions in 3D games. CoRR abs/1803.05402 (2018)
8. Hingston, P.: A new design for a turing test for bots. In: Proceedings of the 2010 IEEE Conference on Computational Intelligence and Games, CIG 2010, Copenhagen, Denmark, 18–21 August 2010, pp. 345–350 (2010)
9. Iii, H.D., Langford, J., Marcu, D.: Search-based structured prediction. Mach. Learn. **75**(3), 297–325 (2009)
10. Lam, K., Esfandiari, B., Tudino, D.: A scene-based imitation framework for RoboCup clients. In: MOO-Modeling Other Agents from Observations (2006)
11. Lamontagne, L., Rugamba, F., Mineau, G.: Acquisition of cases in sequential games using conditional entropy. In: ICCBR 2012 Workshop on TRUE: Traces for Reusing Users' Experience (2012)
12. Livingstone, D.: Turing's test and believable AI in games. Comput. Entertain. **4**, 6 (2006)
13. Lucas, S.M.: Ms Pac-Man versus ghost-team competition. In: Proceedings of the 2009 IEEE Symposium on Computational Intelligence and Games, CIG 2009, 7–10 September 2009, Milano, Italy (2009)
14. Miranda, M., Sánchez-Ruiz, A.A., Peinado, F.: A neuroevolution approach to imitating human-like play in ms. pac-man video game. In: Proceedings of the 3rd

Congreso de la Sociedad Española para las Ciencias del Videojuego, 29 June 2016, Barcelona, Spain, pp. 113–124 (2016)

15. Miranda, M., Sánchez-Ruiz, A.A., Peinado, F.: A CBR approach for imitating human playing style in Ms. Pac-Man video game. In: Cox, M.T., Funk, P., Begum, S. (eds.) ICCBR 2018. LNCS (LNAI), vol. 11156, pp. 292–308. Springer, Cham (2018). https://doi.org/10.1007/978-3-030-01081-2_20

16. Ontañón, S., Montaña, J.L., Gonzalez, A.J.: A dynamic-Bayesian network framework for modeling and evaluating learning from observation. Expert Syst. Appl. **41**(11), 5212–5226 (2014)

17. Ortega, J., Shaker, N., Togelius, J., Yannakakis, G.N.: Imitating human playing styles in super mario bros. Entertain. Comput. **4**(2), 93–104 (2013)

18. Packard, B., Ontañón, S.: Policies for active learning from demonstration. In: 2017 AAAI Spring Symposia, Stanford University, 27–29 March 2017, Palo Alto, California, USA (2017)

19. Packard, B., Ontañón, S.: Learning behavior from limited demonstrations in the context of games. In: Proceedings of the Thirty-First International Florida Artificial Intelligence Research Society Conference, FLAIRS 2018, 21–23 May 2018, Melbourne, Florida, USA, pp. 86–91 (2018)

20. Rohlfshagen, P., Liu, J., Pérez-Liébana, D., Lucas, S.M.: Pac-Man conquers academia: two decades of research using a classic arcade game. IEEE Trans. Games **10**, 233–256 (2018)

21. Ross, S., Bagnell, D.: Efficient reductions for imitation learning. In: Proceedings of the Thirteenth International Conference on Artificial Intelligence and Statistics, AISTATS 2010, 13–15 May 2010, Chia Laguna Resort, Sardinia, Italy, pp. 661–668 (2010)

22. Ross, S., Gordon, G.J., Bagnell, D.: A reduction of imitation learning and structured prediction to no-regret online learning. In: Proceedings of the Fourteenth International Conference on Artificial Intelligence and Statistics, AISTATS 2011, 11–13 April 2011, Fort Lauderdale, USA, pp. 627–635 (2011)

23. Sagredo-Olivenza, I., Gómez-Martín, P.P., Gómez-Martín, M.A., González-Calero, P.A.: Using program by demonstration and visual scripting to supporting game design. In: Benferhat, S., Tabia, K., Ali, M. (eds.) IEA/AIE 2017. LNCS (LNAI), vol. 10351, pp. 33–39. Springer, Cham (2017). https://doi.org/10.1007/978-3-319-60045-1_5

24. Soni, B., Hingston, P.: Bots trained to play like a human are more fun. In: 2008 IEEE International Joint Conference on Neural Networks (IEEE World Congress on Computational Intelligence), pp. 363–369 (2008)

25. Togelius, J., Nardi, R.D., Lucas, S.M.: Towards automatic personalised content creation for racing games. In: 2007 IEEE Symposium on Computational Intelligence and Games, pp. 252–259 (2007)

26. Togelius, J., Yannakakis, G.N., Karakovskiy, S., Shaker, N.: Assessing believability. In: Hingston, P. (ed.) Believable Bots, pp. 215–230. Springer, Berlin (2013). https://doi.org/10.1007/978-3-642-32323-2_9

27. Williams, P.R., Liebana, D.P., Lucas, S.M.: Ms. Pac-Man versus ghost team CIG 2016 competition. In: IEEE Conference on Computational Intelligence and Games, CIG 2016, 20–23 September 2016, Santorini, Greece, pp. 1–8 (2016)

28. Wooldridge, M.: Introduction to multiagent systems. Cell **757**(239), 8573 (2002)

29. Yannakakis, G.N., Maragoudakis, M.: Player modeling impact on player's entertainment in computer games. In: Ardissono, L., Brna, P., Mitrovic, A. (eds.) UM 2005. LNCS (LNAI), vol. 3538, pp. 74–78. Springer, Heidelberg (2005). https://doi.org/10.1007/11527886_11

Show Me Your Friends, I'll Tell You Who You Are: Recommending Products Based on Hidden Evidence

Anbarasu Sekar$^{(\boxtimes)}$ and Sutanu Chakraborti

Indian Institute of Technology Madras, Chennai 600036, India
anbuforever@gmail.com, sutanuc@cse.iitm.ac.in

Abstract. One of the goals of a recommender system is to minimize the cognitive load on the user and hence we cannot expect the users to give extensive feedback. The lesser the feedback, the lesser we know about the preferences of the user to make useful recommendations. This work aims to make the most use of user-provided feedback with the help of the hidden evidence in the casebase. The evidence for each product is acquired based on the relation among the products in the domain. The effectiveness of our approach is demonstrated through evaluation on three product domains.

Keywords: Conversational Case-Based Recommender Systems ·
Preference-based feedback · Evidence-Based Retrieval ·
Dominance relations · Trade-offs

1 Introduction

The advent of mobile and internet technologies has led to a vast amount of information readily available making collaborative filtering approaches to recommender systems effective. Even with such proliferation of data, there are still domains where details about a user are not adequate. Say for example, in a real estate business, it is highly unlikely that a user would make enough purchases to capture the preferences of the user beforehand. In domains such as these, even highly sophisticated methods are bound to fail due to the lack of data about the user and her preferences, this condition is a well-known issue in data-driven approaches, called the cold-start problem [1]. Case-based reasoning approaches to recommender systems come to rescue when there is no sufficient data about the user. They can be used to provide support in initial stages of collaborative filtering based approaches. When the user has some idea about what to buy, case-based recommender systems provide a way to specify their preferences and proceed with the shopping.

It is unrealistic to expect the customer/user to know everything about the product she wishes to buy. The user may have some criteria regarding the purchase but she may not be in a position to express it exactly. When a customer

© Springer Nature Switzerland AG 2019
K. Bach and C. Marling (Eds.): ICCBR 2019, LNAI 11680, pp. 329–342, 2019.
https://doi.org/10.1007/978-3-030-29249-2_22

goes shopping from an off-line store, the sales representative converses with the user and tries to understand the needs of the user and helps the user identify the various products available in the store that could satisfy the needs of the customer. *Conversational Case-Based Recommender Systems* (CCBR-RSs) aims to simulate the job of a sales representative in an online setting. A CCBR-RS engages the user in a conversation by which the user preferences are captured and it facilitates the user to iteratively modify the query such that she can identify the product of her interest.

The idea of utilizing the relationship among the products in the recommendation has been used in literature [7,9,11]. All these works build on the trade-off relation that one product has over the other. The empirical results from these works suggest the need for considering the relationship between the products in the process of recommendation. In this work, we motivate the need for looking at the conversation with the user as a process of aggregating evidence against each product. The products that have accumulated higher positive evidence will be recommended at each interaction cycle. It can be compared to the act of the sales representative who observes the preference choices of the user and when he gets enough evidence that the user would be interested in a particular product he would recommend that product. We use leave one out methodology [12] to measure the efficiency of the CCBR-RS in terms of the average number of cycles taken for a successful recommendation. The lesser the cycle lengths the efficient the system. The paper is structured as follows. First, we provide the background and works related to this work from literature in Sect. 2. Our contribution is detailed in Sect. 3. The empirical evaluation of our idea and the results are given in Sect. 4 followed by discussion in Sect. 5 and conclusion in Sect. 6.

2   Background and Related Works

The two major approaches in CCBR-RSs are Navigation by asking and Navigation by proposing [2]. Since the process of conversation with the user can be seen as the user navigating through the product space, hence the name. In the systems based on navigating by asking, the user is asked about the specifics of the feature values at each interaction. In navigation by proposing, the system proposes a set of products to the user and expects feedback from the user in the form of critiques, ratings and preference-based feedback. In preference-based feedback, the user is presented with the set of products and the user just has to pick one product as her preferred product. Among the feedback mechanisms, preference-based feedback poses lesser cognitive load on the user as pointed out in the work by Smyth [4].

We restrict our focus to navigation by proposing with preference-based feedback as it poses less cognitive load on the user in terms of the feedback the user had to provide. The interaction with the system is two-phased; the recommendation phase and the feedback phase. The interaction with the system starts by the user giving a query. The query is mostly a set of feature values. For example, in the camera domain, the query could look like {500$, 5MP}. The user need not give her preferences on all of the features in the domain. We alternate between (a) Recommendation phase, where the system recommends a set of products based on the query/feedback, and (b) Feedback phase, where the user selects one of the recommended products(preference feedback). The preference feedback in the feedback phase becomes the query for the next recommendation phase. This interaction cycle stops either when the user accepts one of the product as her product of desire or when the user gives up on her search.

In some domains like the camera domain, the features of the products can be classified as More is Better (MIB) and Less is Better (LIB). For example, when two products are similar to each other across all features except the price feature, people tend to choose the one with the lesser price, hence it is an LIB feature. Similarly, among two products with the same feature values except for the feature Zoom, people tend to prefer the product with more Zoom, making it an MIB feature. The usefulness of a product to the given query is computed using the similarity formulation detailed out in the work by Sekar et al. [11]. The utility for each feature is computed separately and combined together based on Multi-Attribute Utility Theory (MAUT) [10]. We have categorized the works from literature into three categories, we do realise that the categories are not mutually exclusive.

2.1 Query Expansion

The earliest of the works in CCBR-RSs focused on helping a user articulate her preferences in the form of a query. The idea from the work by McGinty et al. [3] in More Like This (MLT) is to use the preference-based feedback (user preferred product from a set of recommended products) as the query for the next interaction cycle. The authors point out the drawback with such an approach as every feature value in the preference feedback may not be representative of the user's preferences. The drawback in MLT is overcome in weighted MLT (wMLT) by assigning weights to the features. Higher preference is given to the feature value that is preferred over the other feature values. For example, if the user prefers a product that is of a specific manufacturer 'X' and rejects products of other manufacturers, then we can assume that the user prefers 'X' over the other manufacturers.

The work by Mouli et al. [9] points out that preferences to feature values cannot be decided without taking other feature values into account as the features of a domain are not independent of each other. For the example considered before, the reason why the user preferred the product manufactured by 'X' maybe because it is the cheapest among all products. They propose a way of learning the feature weights by posing the task as a constrained optimization problem.

2.2 Profiling Users and Products

Learning user preferences is a common task in all the CCBR-RSs. The work by McSherry [7] emphasise on the different compromises a user would be willing to make in choosing a product. The author proposes to maximize the success rate by proposing products with different compromise assumptions to the user given query. The assumption is that each user may have different preferences for compromises, hence if we can cover all the different options by recommending appropriate products, then we can maximise the success of the recommender system.

Anbarasu et al. in their works ([13] and [11]) bring out the idea of profiling the products in the casebase. The task of profiling the products can be done as a preprocessing step. In the work [13] MLT Trade-off Matching (MLT TM), the authors argue that compromises are dynamic as it deals with the gap between what the user wants and what the product could provide, but trade-offs are inherent to casebase. Given a pair of products, the relationship between them is expressed in terms of the trade-offs. They argue that given a pair of products, it may never be the case that one product completely dominates the other product. There would be a set of features in one product which would dominate the same set of features in the other product and vice versa. The action of choosing one product over the other is seen as the task of choosing a set of feature values in one product over the set of feature values in other product. The trade-offs the user is willing to make is identified and products that make trade-offs that are in line with the user's preference of trade-offs are recommended to the user. They propose a representation for trade-offs and define a similarity measure over a pair of trade-offs, we make use of their representation scheme and the similarity measure in our work, the details of which are explained in our Approach Section.

2.3 Coverage and Diversity

The principle of similar problems have similar solutions in a recommendation setting may prove to be ineffective. For example, consider the case when a user searches for a product and if all the products recommended have same specifications except for the colour of the product. The recommendation may be useful to the user in the later stages of the recommendation process but may not be helpful in the initial stages as in the initial stages it is highly likely that the users search for all alternatives before settling for a product. If the only difference in the products recommended together is the colour than any one of the products could be representative of all the recommended products. McSherry in Coverage-Optimized Retrieval [8] details several criteria under which the set the recommended products could be used to maximally cover the casebase so as to improve the success rate of the recommender system.

Diversity has been long considered as one of the ways to improve the efficiency of the system [5,6]. It has been noted in the work by McGinty et al. [6] MLT Adaptive Selection (MLT AS) that users tend to be involved in two phases in a shopping exercise. The two phases are Refine phase and Refocus phase. In the Refine phase, the user tends to exploit a small neighbourhood of the search space to identify the products of their desire. In refocus phase, the users tend to explore the options that are diverse to the products that they have been considering. The authors show empirically that introducing diversity methodically in the process of recommendation can result in better efficiency of the CCBR-RS. The authors use a greedy approach to select products that are maximally similar to the query and at the same time maximally diverse to each other.

2.4   Comparison to Related Works

The central theme of our work is that we consider the conversation process as accumulating evidence from the user against each product in the domain. Every product preferred or rejected by the user has an effect on each product in the domain either in a positive or a negative way. The work by McGinty et al. [3] in MLT and wMLT is equivalent to accumulating evidence on the level of features. The preference feedback is used to aggregate the evidence that a particular feature (or the feature value) may be important to the given user. The feature weights can be seen as the strength of evidence for the features. In our work we have aggregated the evidence of the products rather than the features, so the drawbacks as discussed in the work by Mouli et al. [9] due to feature dependency will not affect our system. The work by Anbarasu et al. [13] is equivalent to the task of collecting evidence for a product in terms of the trade-offs that it make with the other products, the idea which we have incorporated in our system.

3   Approach

The feedback provided by the user in each interaction cycle gives us direct evidence for the products viewed by the user. We formulate ways in which the feedback provided on a minimal set of products is propagated to the rest of the products in the domain. The aggregated evidence against each product is then used to predict the usefulness of the product.

3.1   Evidence

Let us assume we have transaction data, where a list of products bought together is called a transaction. If from all the transactions we come to know that, in 50% of the times bread is brought along with butter then in a new transaction if we come to know that bread is purchased then we can say that the evidence that butter would be purchased is 50%. Unlike the case of working with a transactional dataset, we work on a setting where we have no data on the transactions of any users. When a user is given a choice between bright and pastel coloured dress if the user prefers a dark one we assume it as evidence for the desirability of bright coloured dress in general. In predicting the usefulness of an item, the evidence that we have collected based on the user's preference choice will be used.

Dominant and dominated products: When a user is given a choice between a pair of products, we term the one she chooses as the dominant product. The product that is chosen over is called the dominated product.

Given a product domain, we can safely assume that the product with the highest utility to the user dominates every other product in the domain. For

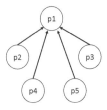

Fig. 1. Precedence graph of product p1

example consider a small product domain with 5 products *p1, p2, p3, p4* and *p5*. Let *p1* be the product that has the highest utility to the user, then we can represent it as a precedence graph as shown in Fig. 1. Based on the precedence graph we can form the pairs $(p1, p2), (p1, p3), (p1, p4)$ and $(p1, p5)$. In each of these pairs, *p1* is the dominant product. Each pair is a relation from the dominant/preferred product to the dominated/rejected product.[1] We term this relation dominance relation. The set of all dominance relations of a product p is termed Product's Dominance Pairs (PDR_p). The set of products that dominate in any of the pairs from PDR_p is denoted by PP_p (Preferred Products in dominance relation pairs of p). The set of products that are dominated in any of the pairs from PDR_p is denoted by RP_p (Rejected Products in dominance relation pairs of p). For the precedence graph shown in Fig. 1, $PDR_{p1} = \{(p1, p2), (p1, p3), (p1, p4), (p1, p5)\}$, $PP_{p1} = \{p1\}$ and $RP_{p1} = \{p2, p3, p4, p5\}$. Similarly, for each product, given the precedence graph for that product, we can form dominance pairs where the dominance relation among the products is fixed accordingly. These pairs are obtained off-line and are used in similar ways like the transaction dataset for assigning evidence for the products, the details of which we explain in the coming Section.

3.2 Predicting the Usefulness of a Product

Given a CCBR-RS, in each interaction cycle, the preference feedback (preferred product) from the user is used to form relation pairs. Each pair consists of the preferred product and one of the rejected products. These pairs are aggregated over interaction cycles. In a given interaction cycle, we try to predict the usefulness of a product based on the dominance relations that we have aggregated so far. The set of aggregated dominance pairs are termed User's Dominance Relations (UDR). UDR is aggregated during runtime. The set of products that dominate in any of the pairs from UDR is denoted by UPP (User Preferred Products). The set of products that are dominated in any of the pairs from UDR is denoted by URP (User Rejected Products). Consider the products *p1, p2, p3* and *p4*. Say the products *p1* and *p2* are shown to user and the user prefers product *p1*. Product *p1* is dominant and product *p2* is dominated. For the example taken, $UDR = \{(p1, p2)\}$, $UPP = \{p1\}$ and $URP = \{p2\}$. Having

---

[1] We use the terms dominant and preferred; dominated and rejected interchangeably.

known this information, can we predict what product the user will prefer if she is given the choice between product $p3$ and $p4$?

Coarse Hypothesis: *The dominance relation between two pairs of products are similar if they involve similar dominant and dominated products.*

For the example that we have taken, based on our hypothesis, if products $p3$ and $p1$ are similar; and the products $p4$ and $p2$ are similar, then the dominance relation between $p3$ and $p4$ will be similar to the dominance relation between $p1$ and $p2$ ($p3$ will dominate $p4$). Now that we have the user's dominance relations and the dominance relations for each product, we can predict the usefulness of each product based on our hypothesis.

The similarity between two dominance relations R and S, denoted by *domSim* is given in Eq. 1, where *sim* is the similarity between two products. R_d, R_r and S_d, S_r denotes dominant and rejected products of R and S respectively (The subscripts d and r stands for dominant and rejected respectively). Let $DSim$ be the similarity between two sets of dominance relations. The similarity between the sets of dominance relations is as given in Eq. 2.

$$domSim(R,S) = sim(R_d, S_d) + sim(R_r, S_r) \qquad (1)$$

$$DSim(PDR_p, UDR) = \frac{\sum\limits_{R_1 \in PDR_p} \sum\limits_{R_2 \in UDR} domSim(R_1, R_2)}{|PDR_p| * |UDR|} \qquad (2)$$

Let us assume that the aggregated pairs for a given user are $UDR = \{(p2, p4), (p2, p5)\}$ where $UPP = \{p2\}$ and $URP = \{p4, p5\}$. If the user's target product is $p1$, then her set of dominance relations would be similar to the set of dominance relations in PDR_{p1}. The $DSim(PDR_{p1}, UDR)$ for the example would be $(sim(p1,p2)+sim(p2,p4)+sim(p1,p2)+sim(p2,p5)+...+sim(p1,p2)+sim(p5,p4)+sim(p1,p2)+sim(p5,p5))/(4*2)$. It has to be noted that the preferred product in the dominance relations belonging to the set PDR_p is always p. $PP_p = \{p\}$ and $|PDR_p| = 1$. The rejected products in the dominance relations belonging to the set PDR_p are all the products in the domain except the product p itself. So for a product p to be more useful to the given user, its similarity to all the products that are preferred by the user should be high and the similarity of all the products rejected by the user with products other than p should be high. Based on this observation, we can see that the formulation simplifies to the formulation given in Eq. 3. We term our method **Evidence-Based Retrieval (EBR)**.

The more a product is similar to the user preferred product and the lesser it is similar to the user rejected products, the more evident it is that it will be preferred by the user. The positive part of the formulation adds up to the evidence for the product. The negative part can be seen as negative evidence for the product. The α and β controls the preference given to positive negative evidence respectively. The simplified formulation can also be directly derived from the main formulation(see Appendix A for details).

$$DScore(p, UPP, URP) = \alpha * \frac{\sum\limits_{a \in UPP} sim(p, a)}{|UPP|} - \beta * \frac{\sum\limits_{b \in URP} sim(p, b)}{|URP|} \qquad (3)$$

3.3   Evidence from Trade-Offs

In the work by Anbarasu et al. [13], the trade-offs a user makes in choosing one product over the other is captured as the preference of the user and is used to predict if a given product would be useful to that user. A product may not dominate the other product on all the features. When a user selects one product over the other, it is equivalent to choosing the set of features on which that product is better over the set of features on which the dominated product is better. This information is captured in the form of trade-offs. When comparing two products, the one with a higher value for an MIB feature is considered to dominate the one with lesser value, it is the other way around for an LIB feature. In our previous section we considered dominance at the product level, trade-offs captures dominance at the feature level.

Consider two pairs of products that are similar to each other. Let us say, we have two pairs of cameras{(C1, C2),(C3, C4)}, assume all of them are similar to each other and let C1, C3 be the preferred products. Say, in the first pair C1 is preferred over C2 because of higher zoom over lower cost and in the second pair C3 is preferred over C4 because of higher resolution over lower price. Even though the similarity between C1 and C3; and the similarity between C2 and C4 may be high, the dominance relations (C1, C2) and (C3, C4) are not similar as one may be willing to pay more for extra zoom but may not be willing to pay more for the extra resolution. This is because the hypothesis we assumed in Sect. 3.2 does not take the trade-off information into consideration.

Revised Hypothesis: *The dominance relation between two pairs of products are similar if they involve similar dominant and dominated products, provided the trade-off relations between the pairs are also similar.*

In the revised hypothesis, we consider the contribution of the trade-off information as evidence for selecting a product. We use the same representation scheme for trade-offs and the similarity between the trade-offs as in [13]. The representation used in [13] uses symbols to indicate this choice. In their work, "1" is used to indicate dominant features of user selected product, "−1" for dominated features and "0" for features where both the products have the same value. The trade-off choice of choosing a product 'a' over 'b' is indicated as T_{ab}. Given two trade-offs, the ratio of the number of matching symbols to the total number of features is taken as the similarity between the trade-offs. The similarity between trade-offs is as given in Eq. 4, where T1 and T2 are the trade-offs and $\mathbb{1}(T1_a = T2_a)$ is an indicator function that gives a value 1 if both the values are same, 0 otherwise. Table 1 shows the representation of trade-offs, T_{AB}: Camera A is preferred over Camera B; T_{CB}: Camera C is preferred over Camera B. The last row shows the similarity computations between T_{AB} and T_{CB}.

Table 1. Trade-off representation and similarity between trade-offs

	Price($)	Resolution(MP)	Zoom(X)
Camera A	500	10	12
Camera B	500	12	10
Camera C	700	12	12
T_{AB}	0	-1	1
T_{CB}	-1	0	1
tradSim(T_{AB}, T_{CB})	$(0+0+1)/3 = 0.33$		

$$tradSim(T1, T2) = \frac{\sum\limits_{a \in Attributes} \mathbb{1}(T1_a = T2_a)}{|Attributes|} \tag{4}$$

Each pair in UDR has the information on what product is preferred over what other product. We can thus arrive at a trade-off for each dominance pair, these are the trade-offs the user has made. If a product p is compared with the rejected product in each dominance pair, we can arrive at a set of trade-offs, these are the trade-offs the prospective buyer of product p will make. We can compare these trade-offs to arrive at the evidence for product p through the trade-offs, we term it trade-off evidence. The trade-off evidence score for a product p is given in Eq. 5. The revised hypothesis is used to combine the dominance evidence scores and trade-off evidence scores for each product, as given in Eq. 6. The combined score is used to rank the products in the domain. The γ value decides the preference given to dominance evidence and trade-off evidence scores.

$$Tscore(p, UDR) = \frac{\sum\limits_{R \in UDR} tradSim(T_{R_d R_r}, T_{p R_r})}{|UDR|} \tag{5}$$

$$Score(p, UPP, URP, UDR) = \gamma * Dscore(p, UPP, URP) + (1 - \gamma) * Tscore(p, UDR) \tag{6}$$

3.4   Recommendation Process

The following are the steps in the recommendation process

Step 0: For each product p in the domain arrive at the product dominance pairs PDR_p based on dominance assumption. The ways in which these dominance pairs could be arrived at may be different. Based on domain knowledge one can come with appropriate ways to identify the dominance pairs. In our experiments we assumed that if a product is deemed to be the most useful to a customer then it would dominate every other product in the domain, which is a very general assumption

Step 1: The preferences of the user are taken as the initial query. The preferences could be on one or many of the features.

Step 2: The k top similar products are recommended to the user. The preference feedback of the user is collected.

Step 3: The user dominance pairs UDR is arrived by assuming the preferred product as the dominant product and each of the rejected product as the dominated product. The UDR pairs are compared against PDR_p to get the dominance evidence for each product.

Step 4: The list of trade-offs the user makes on the rejected products is compared against the list of trade-offs a particular product p would make on the rejected products to get the trade-off evidence for the product p. We have used the dominance and trade-off evidence in our experiments, one may come up with several ways in which the evidence for each product can be accumulated.

Step 5: The products in the domain are ranked based on the evidence collected against each product and the top k products are listed to the user.

Step 6: The preference feedback of the user becomes the new query and the UDR and the trade-offs lists are updated appropriately in each interaction cycle. The system alternates between the recommendation phase and feedback phase until the user is satisfied with a recommended product or gives up the search.

4   Evaluation and Results

Our goal is to improve the efficiency of the CCBR-RSs so that the cognitive load on the user is reduced. The measure we used to evaluate is the total number of interaction cycles before success. We followed the evaluation scheme used in previous works [3,9,11–13]. We simulate the user with an artificial agent that selects the product at each interaction cycle as the preference feedback. Before a conversation instance, a product is selected randomly and is left out from the case base. The product that is most similar to the left out product is fixed as the target. The aim is to start with a subset of the feature values of the left our product as the initial query and converse until the target product appears in the recommendation set. In each interaction cycle, out of the recommended products the agent selects the product that is most similar to the left out product as the preference feedback. The number of cycles taken to reach the product of desire is used as the measure of the efficiency of the system. We simulate easy, medium and hard queries by forming queries with a varying number of features namely 5, 3 and 1 respectively. The more the feature values we know about the left out product the easier it becomes to identify the target.

We use three datasets, camera, used cars and pc datasets [14] which have 210, 956 and 120 cases. We report the average over 1000 conversation instances for each query type namely easy, medium and hard queries. We compared our work (EBR) with a method from each category we outlined in Sect. 2. Query expansion based on preference-based feedback [3] (MLT), a work that uses diversity scheme [6] (MLT AS) and a recent work that uses trade-off relationship among data [13] (MLT TM). We have reported the results for the best combination of parameter values. The parameter values that gave the best results for camera and pc datasets are $\alpha = 0.5$; $\beta = 0.5$ and $\gamma = 0.8$; the best results for Car dataset are $\alpha = 0.5$; $\beta = 0.5$ and $\gamma = 0.9$. If we recall, 1-γ is the preference given to trade-off evidence. The dominance evidence seems to be more important than the evidence from trade-offs. The information on the preferences given to the evidence may be given by the domain expert. The results are shown in Tables 2, 3 and 4.

Table 2. Efficiency in Camera dataset (the lesser the average cycle length the better)

Query size	MLT	MLT AS	MLT TM	EBR
1	11.41	6.90	6.28	**5.13**
3	9.54	5.89	5.45	**4.64**
5	6.42	4.04	3.94	**3.59**

Table 3. Efficiency in Car dataset (the lesser the average cycle length the better)

Query size	MLT	MLT AS	MLT TM	EBR
1	24.42	14.32	12.14	**9.28**
3	19.18	10.91	9.64	**7.55**
5	15.12	8.08	7.53	**5.85**

Table 4. Efficiency in PC dataset (the lesser the average cycle length the better)

Query size	MLT	MLT AS	MLT TM	EBR
1	8.29	6.09	5.50	**4.08**
3	6.14	4.22	3.96	**3.20**
5	3.67	2.19	2.19	**1.97**

The results are tested for statistical significance (paired t-test with $p < 0.05$). The results that are significantly better than the rest of the methods are highlighted. EBR has performed better than all the other methods. It can be seen that there is a significant reduction in the average cycle length across all datsets and all query sizes. The average reduction in cycle length across camera, car and pc datasets when compared against MLT TM are 13%, 22% and 18% respectively. The average reduction in cycle length across query sizes 1, 3 and 5 when compared against MLT TM are 22%, 18% and 13%. Across all datasets the trend is that harder queries require more cycle lengths but interestingly the percentage of reduction in cycle lengths by employing EBR is larger for the harder queries. Increasing cycle lengths mean more feedback from the user. Since EBR uses user's feedback to aggregate evidence, the increased feedback due to longer cycle lengths results in better efficiency.

5   Discussion

The results are significantly better than those of the works from the literature. However, the intent of this work is not to come up with the best in class recommender system but to introduce the idea of looking at the casebase as one unit that has intricate relation among the cases that can be utilized to get the most from the feedback provided to the case-based reasoning system. The evidence collected through interaction is bound to have noise. One has to find ways in which the noise can be reduced to achieve better performances. In our work, we can assume that users may not make informed decisions in the initial stages so the feedback may not be as useful as in the later stages of recommendation cycle. A decay mechanism that gives more importance to the feedback received in the later stages than to the feedback in the early stages of recommendation cycle could be employed. It should also be noted that the noise in feedback tends to propagate across interaction cycles resulting in lesser efficiency. During the evaluation, we assume that the artificial agent makes optimal choices. Optimal choices mean less noise in feedback. In reality, users tend to make sub-optimal choices, our model needs to be robust enough to accommodate such sub-optimal behaviour from the users.

6   Conclusion

In this work, we have successfully demonstrated the idea of viewing the task of conversation with the user as the task of accumulating evidence for each product. We show empirically that the relationship among the products can be utilized to build an efficient recommender system by evaluating our system on three real-world datasets. Our future work would involve filtering noise from the feedback, identifying evidence that could be of help in the process of recommendation and considering higher order relationship among products to exploit the most out of the user provided feedback.

A   Appendix

The terms in the formulations are same as mentioned before in the paper. The expansion of Eq. 2 is given below where R_{1d}, R_{2d} represents the dominant products of relations R_1 and R_2 respectively. Similarly R_{1r}, R_{2r} represents the dominated products of relations R_1 and R_2

$$DSim(PDR_p, UDR) = \frac{\sum_{R_1 \in PDR_p} \sum_{R_2 \in UDR} (sim(R_{1d}, R_{2d}) + sim(R_{1r}, R_{2r}))}{|PDR_p| * |UDR|}$$

(7)

The equation is split into two parts the first one is $\dfrac{\sum\limits_{R_1 \in PDR_p} \sum\limits_{R_2 \in UDR} sim(R_{1d}, R_{2d})}{|PDR_p| * |UDR|}$

and the second part is $\dfrac{\sum\limits_{R_1 \in PDR_p} \sum\limits_{R_2 \in UDR} sim(R_{1r}, R_{2r})}{|PDR_p| * |UDR|}$ Each of them is dealt with separately. We name the first part as Positive evidence score and the second one as Negative evidence score.

When comparing PDR_p with UDR, it can noticed that the dominant product of all the relations in PDR_p is always p. $PP_p = \{p\}$. The dominant products in the relations from UDR is from the set UPP.

Positive evidence score: $\dfrac{\sum\limits_{R_1 \in PDR_p} \sum\limits_{R_2 \in UDR} sim(R_{1d}, R_{2d})}{|PDR_p| * |UDR|}$

$\propto \sum\limits_{R_1 \in PDR_p} \sum\limits_{R_2 \in UDR} sim(R_{1d}, R_{2d})$

$\propto |PDR_p| * \sum\limits_{R_2 \in UDR} sim(p, R_{2d})$ [p is the only dominant product in PDR_p]

$\propto \sum\limits_{d \in UPP} sim(p, d)$ [UPP: Set of products preferred by the user]

The dominated products in the relations of PDR_p contains all products except the product p. If P is the set of all the products in the domain, then the dominated products in the relations of PDP_p can be represented as $P - p$.

Negative evidence score: $\dfrac{\sum\limits_{R_1 \in PDR_p} \sum\limits_{R_2 \in UDR} sim(R_{1r}, R_{2r})}{|PDR_p| * |UDR|}$

$= \dfrac{\sum\limits_{x \in P-p} \sum\limits_{R_2 \in UDR} sim(x, R_{2r})}{|PDR_p| * |UDR|}$ [P-p: Dominated products in PDR_p]

$\propto \sum\limits_{x \in P-p} \sum\limits_{y \in URP} sim(x, y)$ [URP: Set of products rejected by the user]

Let $X = \sum\limits_{x \in P} \sum\limits_{y \in URP} sim(x, y)$

$\sum\limits_{x \in P-p} \sum\limits_{y \in URP} sim(x, y) = X - \sum\limits_{y \in URP} sim(p, y)$

Negative evidence $\propto X - \sum\limits_{y \in URP} sim(p, y)$

$\propto - \sum\limits_{y \in URP} sim(p, y)$

Positive evidence $\propto \sum\limits_{d \in UPP} sim(p, d)$

Negative evidence $\propto - \sum\limits_{y \in URP} sim(p, y)$

The rank of a product is determined by its Positive and Negative evidence scores, we normalise the Positive evidence and Negative evidence scores to lie between 0 to 1 and combine them to give a combined score. The preference given to the evidence scores are controlled by α and β values

$$DScore(p, UPP, URP) = \alpha * \frac{\sum\limits_{a \in UPP} sim(p,a)}{|UPP|} - \beta * \frac{\sum\limits_{b \in URP} sim(p,b)}{|URP|}$$

References

1. Adomavicius, G., Tuzhilin, A.: Toward the next generation of recommender systems: a survey of the state-of-the-art and possible extensions. IEEE Trans. Knowl. Data Eng. **17**(6), 734–749 (2005)
2. Shimazu, H.: ExpertClerk: navigating shoppers' buying process with the combination of asking and proposing. In: IJCAI 2001: Proceedings of the Seventeenth International Joint Conference on Artificial Intelligence, vol. 2, pp. 1443–1448. Morgan Kaufmann (2001)
3. McGinty, L., Smyth, B.: Comparison-based recommendation. In: Craw, S., Preece, A. (eds.) ECCBR 2002. LNCS (LNAI), vol. 2416, pp. 575–589. Springer, Heidelberg (2002). https://doi.org/10.1007/3-540-46119-1_42
4. Smyth, B.: Case-based recommendation. In: Brusilovsky, P., Kobsa, A., Nejdl, W. (eds.) The Adaptive Web. LNCS, vol. 4321, pp. 342–376. Springer, Heidelberg (2007). https://doi.org/10.1007/978-3-540-72079-9_11
5. Smyth, B., McClave, P.: Similarity vs. diversity. In: Aha, D.W., Watson, I. (eds.) ICCBR 2001. LNCS (LNAI), vol. 2080, pp. 347–361. Springer, Heidelberg (2001). https://doi.org/10.1007/3-540-44593-5_25
6. McGinty, L., Smyth, B.: On the role of diversity in conversational recommender systems. In: Ashley, K.D., Bridge, D.G. (eds.) ICCBR 2003. LNCS (LNAI), vol. 2689, pp. 276–290. Springer, Heidelberg (2003). https://doi.org/10.1007/3-540-45006-8_23
7. McSherry, D.: Similarity and compromise. In: Ashley, K.D., Bridge, D.G. (eds.) ICCBR 2003. LNCS (LNAI), vol. 2689, pp. 291–305. Springer, Heidelberg (2003). https://doi.org/10.1007/3-540-45006-8_24
8. McSherry, D.: Coverage-optimized retrieval. In: Proceedings of the Eighteenth International Joint Conference on Artificial Intelligence, pp. 1349–1350. Morgan Kaufmann Publishers Inc. (2003)
9. Mouli, S.C., Chakraborti, S.: Making the most of preference feedback by modeling feature dependencies. In: Proceedings of the 9th ACM Conference on Recommender Systems. ACM (2015)
10. Keeney, R.L., Raiffa, H.: Decisions with Multiple Objectives: Preferences and Value Trade-Offs. Cambridge University Press, Cambridge (1993)
11. Sekar, A., Chakraborti, S.: Towards bridging the gap between manufacturer and users to facilitate better recommendation. In: The Thirty-First International Flairs Conference (2018)
12. Ginty, L.M., Smyth, B.: Evaluating preference-based feedback in recommender systems. In: O'Neill, M., Sutcliffe, R.F.E., Ryan, C., Eaton, M., Griffith, N.J.L. (eds.) AICS 2002. LNCS (LNAI), vol. 2464, pp. 209–214. Springer, Heidelberg (2002). https://doi.org/10.1007/3-540-45750-X_28
13. Sekar, A., Ganesan, D., Chakraborti, S.: Why did Naethan pick android over Apple? Exploiting trade-offs in learning user preferences. In: Cox, M.T., Funk, P., Begum, S. (eds.) ICCBR 2018. LNCS (LNAI), vol. 11156, pp. 354–368. Springer, Cham (2018). https://doi.org/10.1007/978-3-030-01081-2_24
14. http://www.mycbr-project.net/download.html

A Tale of Two Communities: An Analysis of Three Decades of Case-Based Reasoning Research

Barry Smyth[(✉)] [iD]

Insight Centre for Data Analytics, School of Computer Science,
University College Dublin, Dublin, Ireland
barry.smyth@ucd.ie

Abstract. We analyse three decades of case-based reasoning (CBR) research to better understand the health of CBR and its relationship to adjacent research fields. We identify two largely separate CBR communities, one based on the research published at mainstream CBR venues (ICCBR, ECCBR etc.), the other encompassing CBR work with no direct connection to these venues. We analyse their scale, impact, and focus, and the potential to bring them closer together in the future.

1 Introduction

This year the case-based reasoning community returns to Otzenhausen, Germany, 26 years after the first European Workshop on Case-Based Reasoning [14], which many regard as *the* formative event in the history of CBR. The 1993 Otzenhausen meeting led to a long-running series of workshops and conferences, as EWCBR became ECCBR, and later merged with ICCBR. Returning to Otzenhausen is a natural time for community *reflection* and the purpose of this paper is to support this by analysing more than 600,000 articles, including CBR papers, their referenced and citing papers, and other (non-CBR) papers by CBR authors. It echoes, and expands upon, similar analyses carried out in the past [5–7], while at the same time introducing new ideas about how we might evaluate the state of CBR today.

During ICCBR community meetings, one frequent topic for discussion concerns the existence of another CBR community without a close connection to mainstream CBR venues. This discussion often arises in the context of how we might increase the size of ICCBR, attract additional submissions, and otherwise further accelerate the development of CBR. We investigate whether such a community exists – spoiler, it does! – and we compare and contrast the scale of activity across both communities: their output and impact; the topics they

---

Supported by Science Foundation Ireland through the Insight Centre for Data Analytics under grant number SFI/12/RC/2289.

K. Bach and C. Marling (Eds.): ICCBR 2019, LNAI 11680, pp. 343–357, 2019.
https://doi.org/10.1007/978-3-030-29249-2_23

emphasise; the most influential ideas that have emerged etc.[1]. Our aim is to better understand the similarities and differences between both communities and to identify opportunities to bring them closer together in the future.

In the next section we will describe our main publication dataset and how we distinguish between the two CBR communities mentioned above. Subsequent sections examine publication output, community dynamics, and citation impact. We also describe the results of a topic modelling and citation analysis in order to identify the principal research themes, and the most influential papers, which have emerged from the last three decades of research, and more recently.

2   Datasets and Communities

This work begins with a dataset provided by Semantic Scholar (SS[2]), which provides publication meta-data and citation data for more than 46 m publications, primarily from the fields of computer science and health science.

2.1   The CBR Dataset

We select a *CBR dataset* of 675,118 papers by 1,042,490 unique authors from the following subsets of SS data:

- V_p, the set of *venue* papers; papers published at I/ECCBR and EWCBR.
- S_p, the set of papers returned in an SS *search* for CBR papers[3].
- C_p, the full set of CBR papers ($C_p = V_P \cup S_p$).
- L_p, the set of *linked* papers that cite, or are cited by, papers in C_p.
- R_p, the set of *related* (non-CBR) papers by authors of papers in C_p.

There are corresponding sets $(V_a, S_a, C_a, L_a, R_a)$ for the *authors* of these papers. As summarised in Fig. 1(a), there are 8,223 unique CBR papers in C_p, 66,941 linked papers, and 632,770 papers ($606,165 + 2,012 + 6,211 + 18,382$) authored by CBR authors. The corresponding data for authors is in Fig. 1(b).

---

[1] We will avoid the temptation to *name-check* individual researchers, on the grounds that such rankings can end up as distractions to the central argument.

[2] SS is an open, research-article search engine; see https://www.semanticscholar.org.

[3] We identified candidate papers based on a set of *strong* (e.g. case-based reasoning, derivational analogy), *moderate* (e.g. case adaptation, case based), and *weak* (CBR, case retrieval, case learning) search terms, and a scoring metric to identify CBR papers with a high degree of accuracy. Due to space restrictions it is not possible to provide a complete account of the terms and weightings used. The process involved considerable trial and error and validation tests were performed to ensure good precision and recall during the final dataset preparation.

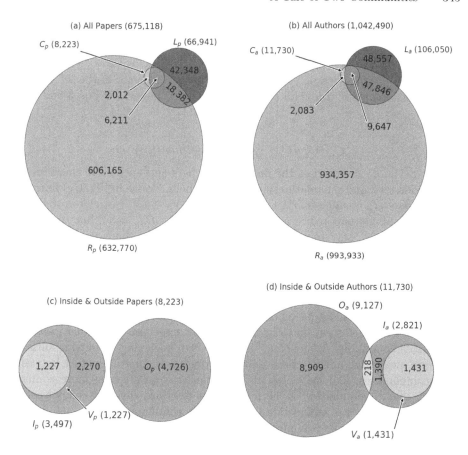

Fig. 1. Venn diagrams of the various datasets and subsets of data used in this study.

2.2 A Tale of Two Communities

A central idea in this work is that the CBR field is shared by two largely separate communities. We refer to CBR researchers connected to the *mainstream venues*[4] as the *inside community*, and to CBR researchers without a direct connection to the mainstream venues as the *outside community*.

More precisely, a CBR paper, p, is in the set of *inside papers*, I_p, if and only if p is co-authored by a *venue* author. And a CBR author, u, is in the set of *inside authors*, I_a, if and only if u is an author of an inside paper; see Eqs. 1 and 2. Notice, that $V_a \subset I_a$; an inside author does not have to be a venue author, but they must *co-author* with a venue author. Thus, inside authors are *connected* to the mainstream venues by venue authors, but they do not necessarily need to publish in the mainstream venues themselves.

---

[4] We use the term '*mainstream*' to refer to ICCBR/ECCBR/EWCBR, but only as a convenience, and without attempting to impugn the many other research venues where CBR papers appear.

$$I_p = \{p \epsilon C_p \mid \exists u \epsilon Authors(p) \wedge u \epsilon V_a\} \tag{1}$$

$$I_a = \{u \epsilon C_a \mid \exists p \epsilon I_p \wedge u \epsilon Authors(p)\} \tag{2}$$

Conversely, a CBR paper, p, is an *outside paper* if it is not in the set I_p, and u is an *outside author* if u is an author of an outside paper; see Eqs. 3 and 4.

$$O_p = C_p \setminus I_p \tag{3}$$

$$O_a = \{u \epsilon C_a \mid \exists p \epsilon O_p \wedge u \epsilon Authors(p)\} \tag{4}$$

Figure 1(c, d) summarises the number of papers and authors in these inside and outside sets, and the relationships between them. Notice in Fig. 1(c) how the inside papers are a superset of the venue papers ($V_p \subset I_p$). Notice too that the inside and outside papers are mutually exclusive ($I_p \cap O_p = \phi$), but the inside and outside *author* sets are not ($I_a \cap O_a \neq \phi$). In Fig. 1(d) there are 218 authors who are both inside and outside authors.

Each one of these 218 authors is an author of an inside paper but they are not venue authors – they have *co-authored* with a venue author, but only outside the core venues – and each is also a co-author of an outside paper. For example, an author might have been a co-author of a non-venue, inside paper, as a PhD student, then went on to bring their CBR expertise to another group in the outside community, where they also published, becoming an outside author. They are interesting authors because they represent a point of contact between both communities, and may play an important role in creating more contact between these communities in the future. But they are also rare, emphasising the conspicuous lack of connection between both communities.

The scale of the two communities is also worth noting. At the start of this study it was not clear whether the outside community would prove to be more than a limited body of CBR work. Instead we find a significant body of CBR research that is even larger than that of the inside community. The outside community spans a similar period of time but has >35% more papers, and >2x the number of unique authors, when compared to the inside community.

2.3 Inside/Outside Venues

Why are there two, mostly separate, communities? To explore this further, Fig. 2 shows the number of CBR papers at the top-25 most frequently targeted inside and outside venues; for reasons of clarity the mainstream venues have been omitted, as they tend to dominate, making other venues more difficult to compare. We can see that the most popular (non-mainstream) venue for CBR research is *Expert Systems with Applications*, with just over 150 CBR papers during the last 30 years, the vast majority of which ($\approx$80%) have come from the outside community. In contrast, *FLAIRS*, *IJCAI*, and *AAAI* are more frequent targets for the inside community. These are the types of venues – AI, ML, expert systems etc. – that one might expect for CBR publications.

More revealing is a similar plot in Fig. 3, but this time focusing on the *non-CBR* papers published by the inside and outside authors. Now there is a clearer

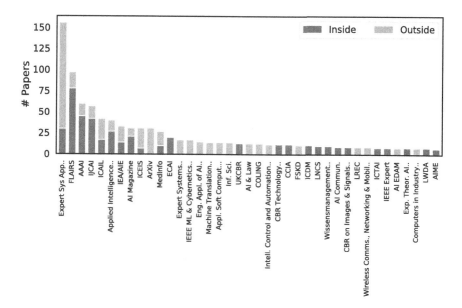

Fig. 2. Top publication venues (excluding I/ECCBR/EWCBR) for inside and outside papers.

division between the top inside and outside venues. The former, as was the case for CBR output, targets mostly AI/ML related venues (*AI Magazine*, *ECAI*, *RecSys* etc.), but the latter targets *PloS One*, *Physics Review*, the *Journal of Biochemistry*, and *Applied Materials* etc. This suggests that while inside authors are mostly AI/ML researchers, outside authors are much more likely to be biologists, physicists, material scientists, and chemists.

The inside community is a community of AI/ML researchers with a focus on CBR, whereas outside researchers come from many different areas, scientific and commercial, using CBR as a technology to solve challenging problems in their home domains. To put this another way, the inside community is *about* a related set of topics (AI, ML, CBR), whereas the outside community is *about* many different topics. As such, we might expect the former to be more coherent and less fragmented than the latter. If so, then there should be a stronger *community effect* for the inside community when it comes to citation impact.

3   Publications, Authors, and Impact

Next we examine the output and impact of these communities, by looking at the volume of publications per annum, the number of active authors, and the citation impact that their work is having.

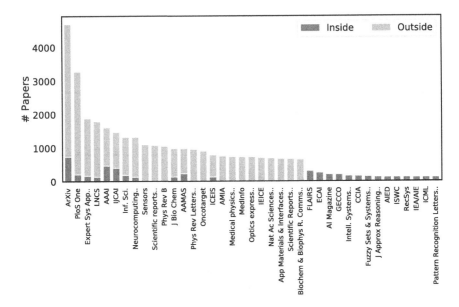

Fig. 3. Top publication venues for non-CBR papers produced by inside and outside authors.

3.1 Publication Output

Figure 4(a) shows the cumulative number of CBR publications produced; for reference, we also show the cumulative output of the mainstream venues. There has been a steady output from each community but since 2009 the total number of CBR papers produced by the outside community has surpassed that of the inside community, a trend that continues to this day.

Approximately 50% of the inside and outside output is made up of application papers; see Fig. 4(b)[5]. CBR has always been an application-oriented field and this is in contrast with a much lower, but growing, fraction of application papers among the non-CBR papers in our dataset, also shown.

3.2 New, Returning, and Churning Authors

While both communities are broadly similar in terms of their publication output, differences begin to emerge when we look at their respective author-bases. Figure 5 shows: (a) the cumulative number of active authors per year; (b) the fraction of new authors per year; (c) the fraction of returning authors; and (d) the fraction of churning/lost authors.

Compared to the inside community, the outside community is characterised by higher levels of new authors and lower levels of returning authors, suggesting

---

[5] We determine application papers based on the presence of keywords such as 'application', 'domain', 'deploy', for example, in the title or abstract.

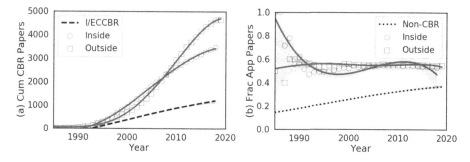

Fig. 4. The (a) cumulative number of CBR papers published per year and (b) the fraction of application-oriented papers per year.

that many outside authors are engaged in CBR research for a shorter period of time. In fact, on average inside authors publish CBR papers over a 5-year period, compared a 3-year period for outside authors; just over 20% of inside authors remain CBR-active for more than 5 years, compared to <10% for outside authors. This is consistent with the idea that the inside community is focused on advancing the fundamentals of CBR – with its researchers engaged for the long-term – while the outside community is a community of practice, with less long-term investment in CBR by its researchers.

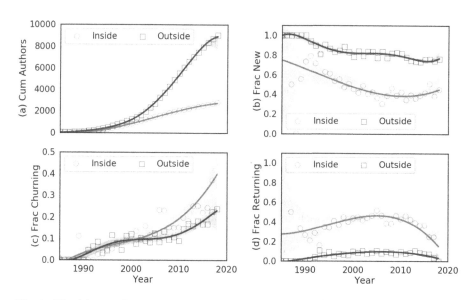

Fig. 5. The (a) cumulative number of publishing CBR authors per year; (b) the fraction of new authors per year; (c) the fraction of churning/lost authors per year; (d) the fraction of returning authors per year.

3.3   Citations and Impact

Another difference between the communities is revealed when we consider citation impact. Figure 6(a) shows a significant citation benefit for inside papers compared to outside papers. From an early point in the development of CBR, inside papers have tended to attract more citations than the outside papers.

Figure 6(b) shows that ≈50% of the citations to CBR papers by the inside community come from other CBR papers, compared to just over 25% for the outside community. Once again, this difference is consistent with the notion that CBR papers by the inside community are more likely to make a central CBR contribution, attracting CBR cites, whereas CBR papers by the outside community are more likely to use CBR in the service of some other task.

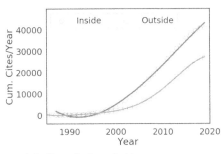
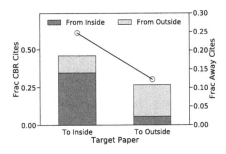

(a) Cumulative citations per year.   (b) Fraction of cites from CBR papers.

Fig. 6. A citation analysis summary for inside and outside communities.

Figure 6(b), also shows the extent to which one community cites their own work versus the work of the other. We refer to inside papers citing inside papers, and outside papers citing outside papers, as *home cites*. Conversely, inside papers citing outside papers, and outside papers citing inside papers, are *away cites*. Figure 6(b) shows that the inside community benefits from a much higher proportion of away cites (≈25%) than the outside community (≈12%). In other words, outside papers are more likely to cite inside papers than the other way around. All other things being equal, this may suggest a *discoverability* issue for the outside community, which contributes to its lower citation impact; fostering greater links between the community may help to address this.

Figure 7 looks at a number of summary impact metrics, and inside papers continue to benefit. They attract more cites per paper. The time to the first citation is shorter (they are more discoverable). Their citation half-life[6] is longer, and the number of years to peak-cites[7] is greater. Inside papers enjoy a more immediate, significant and sustained impact, compared with outside papers, indicating, as predicted, the inside community benefits from a stronger *community effect*.

---

[6] The number of years it takes to accumulate 50% of their cites.
[7] The time it takes for the paper to have its best citation year.

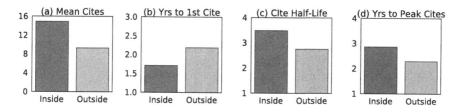

Fig. 7. Summary impact metrics for inside/outside papers.

4   The Evolution of Research Topics

To explore the themes from the last three decades of research we construct topic models for the inside and outside papers by transforming titles and abstracts into term-based representations, using tokenisation, lemmatisation, and stemming. Latent Dirichlet Allocation (LDA) [2] is applied to the resulting *document-term* matrix to produce a *document-topic* matrix (encoding the probability distribution of the topics per document) and a *topic-term* matrix (encoding the probability distributions of the terms per topic). We cluster the papers based on their dominant topics, and use t-SNE (t-distributed stochastic neighbour embedding, [10]) to produce the 2D topic maps shown in Figs. 8 and 9.

In these visualisations each paper is represented by a disc, with papers from the same topic grouped together by t-SNE, and coloured similarly. Distance denotes similarity, the radius of each disc is proportional to the number of citations attracted by the paper, and the opacity of the disc is proportional to the recency of the paper (more recent papers are *more* opaque). Finally, each topic is labeled using the top terms from the LDA probability distributions.

Fig. 8. The inside topics discovered from the inside papers. (Color figure online)

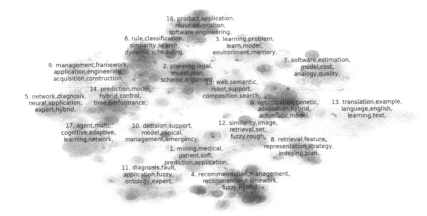

Fig. 9. The outside topics discovered from the outside papers. (Color figure online)

Although there is no direct mapping between inside and outside topics – *inside topic i* is not related to *outside topic i* – there are clearly similarities between the areas explored by both communities, reflecting common themes within CBR research (similarity and retrieval, learning and adaptation, prediction classification, recommendation etc.) But there are differences too. Planning and strategy games (*inside topic 1*), learning and analogy (*inside topic 3*), and maintenance and competence (*inside topic 9*) are important themes within the inside community, but they are less evident among the outside topics. Conversely, the outside topics exhibit a greater emphasis on certain application themes – medical data-mining (*outside topic 1*), software engineering and estimation (*outside topic 7*), and (example-based) translation (*outside topic 13*) – which are less well represented by the inside papers.

Figure 10(a–f) summarises aspects of each of these topics; remember inside topic *i* has no relationship to outside topic *i*. The inside/outside topics are similar in terms of their fractions of papers and application papers. Outside topics tend to peak sooner (≈ 8.5 years) and more recently (≈ 2013) versus ≈ 11.5 years and 2007 for inside papers, respectively. The citation benefit for inside papers persists across topics too: inside topics enjoy more cites per paper, and a higher (topic-based) h-index in almost all cases.

There is obviously more that could be explored with respect to the evolution of CBR research topics. It would be interesting, for example, to pay more attention to recent, emerging topics, by building our topic models over a subset of recent papers, or to try and predict future topics. Alternative approaches to clustering papers could also be considered, for example by using co-citation, rather than term-overlap, as a measure of inter-paper similarity. For now we will leave these as open ideas for future work.

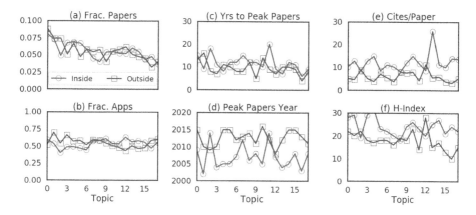

Fig. 10. Summary of various aspects of the inside and outside topics.

5   Influential Papers

In any research area there are some papers that stand out as particularly important to the evolution of the field. Sometimes they are breakthrough papers that introduce new or improved techniques, or even new research directions. Sometimes they are survey papers that bring together a body of research, perhaps reframing it, or integrating it with relevant ideas from other fields. Here we seek to identify the most influential CBR papers, over the past three decades of CBR research, as well as those that have emerged more recently.

5.1   Link Analysis and Influence Metrics

Important papers tend to stand out as being among the most cited works in a field, but citation count alone is not always sufficient to identify the *most* influential articles. In recent years, *link analysis* techniques have been used to evaluate the importance of nodes in a graph, based on various features of network topology. For example, algorithms such as *PageRank* [3] and *HITS* [8] consider a node to be important if it is connected to other important nodes. It is common to use these ideas to reveal influential papers in a *citation graph*, where the nodes are papers and the edges are the citation links between them [4].

We build a single citation graph based on all of the CBR papers (from both communities) and implement three different scoring metrics: (1) the number of cites that the paper has attracted; (2) the *PageRank* score of paper; and (3) the *HITS authority score*. Each metric generates a single score for a paper, which we convert into a rank, and then we calculate the sum of these ranks to generate an overall ranking; using ranks is a simple but effective way to combine these scores in a scale-free manner.

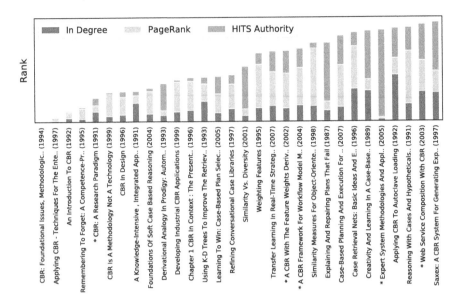

Fig. 11. Top-25 most influential CBR papers (inside & outside communities).

5.2 Seminal CBR Papers

Figure 11 shows the top-30 papers based on this overall ranking. Each bar corresponds to a single paper and shows its ranking across all 3 metrics. Outside papers are indicated with an asterisk prefixing their title on the x-axis. Over one-third of the publications are survey papers or introductory books, including 4 of the top-5 [1, 9, 11–13]. Only 5 (16%) of the most influential papers come from the outside community and most of the top-30 come from the very early years of CBR research; the mean publication year is 1997. This is not so surprising, as many of these papers established the foundations of the field, and their impact has been building over a long period of time, but it begs the question as to where future influential papers are likely to come from.

5.3 Emerging Influencers

To shed some light on this, Fig. 12 presents a similar set of ranking results, but focusing on the most influential CBR papers just from the last decade (2009–2019). There are far fewer survey papers – notwithstanding that the top ranked paper is a survey of CBR in health sciences – and there is an abundance of outside papers; 15 of the top 30 are outside papers. Perhaps the outside community will prove to be more influential over the coming years.

There is also evidence of a number of increasingly important and novel application domains among these more recent papers. For example, 7 of the papers focus on healthcare and clinical applications (from classical diagnosis and classification to duty rostering), 5 of the papers focus on financial applications

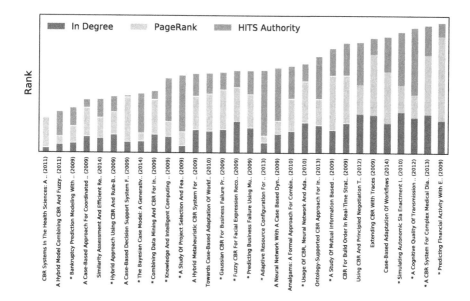

Fig. 12. Top-25 most influential recent (>2008) CBR papers (inside & outside communities).

(including bankruptcy and business closure prediction), 3 consider the application of CBR to workflows, while others explore applications in cloud computing, cost estimation, and optical networks.

Before concluding, it is worth highlighting another factor that distinguishes the inside and outside communities: *location*. To date, mainstream CBR venues have taken place in Europe and the US only, which may limit the interest of researchers from wider afield. This deserves further analysis than is possible here, but when we look at locations of the authors of these recent and influential papers we find, in a large majority of the cases, that the outside authors are based outside of Europe and the US; among the 15 outside papers, only 3 are from Europe or the US, with the rest from China, India, Korea, Pakistan, and Australia. This might be a sign that ICCBR needs to start looking further afield for future conference locations? It also points to a set of authors who may be well positioned to help organise and even host future ICCBR events.

6   Conclusions

The aim of this paper has been to examine the last three decades of CBR research. The data supports the existence of two significant but largely separate communities of CBR researchers: an *inside* community of AI/ML researchers, focusing on core techniques and applications, and an *outside* community of practitioners, focusing on a diverse tasks and applications from a variety of scientific and commercial domains.

The outside community is larger, but its members remain CBR authors for a shorter period of time. Many outside authors may be "passing through", leveraging CBR ideas in their research for a limited period of time only. The outside community's output lags behind that of the inside community in terms of citations, but this may be a consequence of the lack of connection between both communities and, the lack of citations from inside papers to outside papers, in particular. Despite this the outside community produces influential CBR papers, especially when we consider recent research.

One conclusion to draw from this is that it is worthwhile creating stronger links between both communities. Encouraging outside researchers to become involved in mainstream venues, may help to promote and sustain CBR within the outside community. Improving the flow of information between both communities will improve discoverability, especially for the work of the outside community. Furthermore, the outside community appears to be especially well positioned with respect to novel application domains for CBR, which may introduce new research challenges and themes to the inside community.

Precisely how we might bring about this increased engagement between the communities is a matter for the CBR community as a whole. There are some practical things that can be considered in the short-term, from inviting senior outside authors to present at ICCBR, to encouraging targeted sessions or workshops on emerging themes that are associated with the outside community. Longer-term actions might require other forms of outreach: involving outside researchers in ICCBR's programme and organising committees; encouraging host bids from locations that are well represented by the outside community (e.g. China, India, etc.). Whatever the approach, the good news is that, success will strengthen the field of CBR, helping to sustain the next 30 years of research.

References

1. Aamodt, A., Plaza, E.: Case-based reasoning: foundational issues, methodological variations, and system approaches. AI Commun. **7**(1), 39–59 (1994)
2. Blei, D.M., Ng, A.Y., Jordan, M.I.: Latent Dirichlet allocation. J. Mach. Learn. Res. **3**, 993–1022 (2003)
3. Brin, S., Page, L.: The anatomy of a large-scale hypertextual web search engine. Comput. Netw. ISDN Syst. **30**(1–7), 107–117 (1998)
4. Ding, Y., Yan, E., Frazho, A., Caverlee, J.: Pagerank for ranking authors in co-citation networks. J. Am. Soc. Inf. Sci. Technol. **60**(11), 2229–2243 (2009)
5. Freyne, J., Coyle, L., Smyth, B., Cunningham, P.: Relative status of journal and conference publications in computer science. Commun. ACM **53**(11), 124–132 (2010)
6. Greene, D., Freyne, J., Smyth, B., Cunningham, P.: An analysis of research themes in the CBR conference literature. In: Althoff, K.-D., Bergmann, R., Minor, M., Hanft, A. (eds.) ECCBR 2008. LNCS (LNAI), vol. 5239, pp. 18–43. Springer, Heidelberg (2008). https://doi.org/10.1007/978-3-540-85502-6_2
7. Greene, D., Freyne, J., Smyth, B., Cunningham, P.: An analysis of current trends in CBR research using multi-view clustering. AI Mag. **31**(2), 45–62 (2010)

8. Kleinberg, J.M.: Authoritative sources in a hyperlinked environment. J. ACM (JACM) **46**(5), 604–632 (1999)
9. Kolodner, J.L.: An introduction to case-based reasoning. Artif. Intell. Rev. **6**(1), 3–34 (1992)
10. van der Maaten, L., Hinton, G.: Visualizing data using t-SNE. J. Mach. Learn. Res. **9**(Nov), 2579–2605 (2008)
11. Slade, S.: Case-based reasoning: a research paradigm. AI Mag. **12**(1), 42 (1991)
12. Smyth, B., Keane, M.T.: Remembering to forget: a competence-preserving case deletion policy for case-based reasoning systems. In: Proceedings of the 14th International Joint Conference on Artificial intelligence (IJCAI), pp. 377–382. Morgan Kaufmann (1995)
13. Watson, I.: Applying Case-Based Reasoning: Techniques for Enterprise Systems. Morgan Kaufmann Publishers Inc., Burlington (1998)
14. Wess, S., Althoff, K.-D., Richter, M.M. (eds.): EWCBR 1993. LNCS, vol. 837. Springer, Heidelberg (1994). https://doi.org/10.1007/3-540-58330-0

Going Further with Cases: Using Case-Based Reasoning to Recommend Pacing Strategies for Ultra-Marathon Runners

Cathal McConnell and Barry Smyth$^{(\boxtimes)}$ (iD)

Insight Centre for Data Analytics, School of Computer Science,
University College Dublin, Dublin, Ireland
`barry.smyth@ucd.ie`

Abstract. We build on recent work on the application of case-based reasoning to help marathon runners to plan and pace their races. We apply related ideas to the domain of ultra running (typically >100 km routes across mountainous or desert terrain). This new domain introduces its own distinct challenges: distance and terrain make for a more physically demanding and less predictable event; weather can play a very significant role in how competitors perform; and, unlike road marathons, race routes and distances vary from year to year, making it more difficult to compare race records. We evaluate case-based methods for pace prediction and pacing recommendation for runners in the Ultra Trail du Mont Blanc (UTMB), one of the world's toughest ultra-marathons.

Keywords: Case-based reasoning · Marathon running ·
Race-time prediction · Pacing recommendation

1 Introduction

These days, almost everything we do generates a data record that is stored somewhere [7]. Privacy issues aside, this affords new opportunities when it comes to better understanding how people live, work, and play [16,19]. One particularly exciting opportunity relates to personal health and exercise. Certainly there are many reasons to prioritise a more active lifestyle in today's sedentary world [6,8, 15], and recently researchers have begun to turn their attention to many aspects of this challenging task. There are numerous examples of different ways in which sensors, mobile technology, and machine learning techniques have been used to encourage, enable, and otherwise support more active lifestyles, even helping elite athletes to train and compete more effectively [1,2,4,10,11,17,18,25].

Endurance sports represent an especially rich domain for this type of research for a number of reasons: large amounts of data are generated as people train and

Supported by Science Foundation Ireland through the Insight Centre for Data Analytics under grant number SFI/12/RC/2289.

K. Bach and C. Marling (Eds.): ICCBR 2019, LNAI 11680, pp. 358–372, 2019.
https://doi.org/10.1007/978-3-030-29249-2_24

compete; they are increasingly popular events, which brings many first-time competitors, who stand to benefit from more personalised advice and support.

In this work we explore whether case-based reasoning (CBR) techniques can be used to support ultra-marathon runners to better plan their racing strategies. We extend recent work on CBR in marathon running [20–23] to show how it can be adapted from fixed-length (42.2 km), mostly flat, road-races to much longer, more open-ended mountain races, involving elevation changes of tens of thousands of metres, and where weather and sheer time-on-feet (>24 h) can play a critical role in performance. In line with the work of [20–23] our aim is to support ultra-marathoners by helping them to determine a challenging but achievable finish-time, and by recommending a pacing strategy with which they can achieve this time.

In the next section we will briefly summarise related work in the area and contrast this work on ultra-marathon prediction with the recent work on prediction in road marathons [20–23] on which it is based. We use the famous UTMB (Ultra Trail de Month Blanc) as a real-world case-study, evaluating the effectiveness of our approach on race-data from more than a decade of races.

2   Related Work

This work sits at the intersection between personal sensing, machine learning, and connected health. An explosion of wearable sensors and mobile devices has created a tsunami of personal data [7], and the promise that it can be harnessed to help people to make better decisions and live healthier and more productive lives [9,16]. Indeed, within the case-based reasoning community there has been a long history of applying case-based, data-driven methods to a wide range of healthcare problems [5]. The world of sports and fitness has also embraced this data-centric vision, as teams and athletes endeavour to harness the power of data to optimise the business of sports and the training of athletes [13,14].

2.1   CBR for Marathon Races

Recently there has been a surge of interest in using mobile sensors and ML techniques to better support elite and recreational athletes as they train and compete, from classifying activities [4], fitness estimation [1,2], to supporting injury recovery [11], and even the generation of personalised training plans [10,18].

There is a body of work that uses linear models to predict future race-times based on previous race-times; e.g. [3]. Less well developed, however, is the translation of a goal-time into a specific race strategy, and a concrete set of pacing recommendations, which is relevant to this work. The most directly relevant research is [20–23], which considered goal-time prediction and pacing recommendation, but for road marathons rather than ultra races. In short, [20–23] demonstrated how to predict race-times and pacing plans for a runner by adapting the race-times and pacing plans of runners who had run similar races (with respect to finish-time and pacing) to the target runner in the past. The basic idea was presented and validated in [21] and further improved upon in [22] by using more sophisticated representations of a runner's racing history.

2.2 From Road Marathon to Ultra-marathon

We will use the aforementioned work as a starting point for the present work on ultra-marathons, but first it is worth highlighting a number of important features which distinguish ultra marathons from road marathons as follows:

1. Ultra-marathons are significantly longer and more arduous than conventional road marathons. Technically speaking an ultra marathon is any race that is greater than the conventional marathon distance (42.2 km), but in practice many of the better known races are >100 km, extending over multiple days, and covering challenging, if not down-right hostile, mountainous or dessert terrain. This makes ultra marathons a more challenging prediction task than road races; this is further compounded by the much smaller number of competitors (100's) participating in ultra-races versus the thousands or even tens of thousands who participate in marathons.
2. External factors, especially weather, play a much more important role in ultra-marathon performance, compared with conventional road-races. For example, 48 h in the Alps can see conditions varying from the full heat of the summer to torrential rain and wind, to full-on blizzard conditions over high mountain passes. Once again, this adds to the unpredictability of a race, such as the UTMB, and makes accurate prediction that bit more challenging again.
3. Road marathons tend to follow the same route year after year, with the same evenly-spaced timing stations. This greatly simplifies the representation of race records and makes it straightforward to compare race records across the years, an important factor in many machine learning approaches, and especially for case-based reasoning. In contrast, dynamic external conditions mean frequent route changes in ultra marathons so that the course and length of races such as the UTMB can vary from year to year. And this in turn can impact on the location of the timing stations so that the timed segments also vary from year to year, making it more difficult to compare like-with-like across the years.

Thus, when it comes to modeling and predicting performance in ultra-marathon races, such as the UTMB, these considerations introduce significant additional complexity compared with more stable and predictable road marathons. In what follows we will describe how we have adapted the approach described by [21] to cope with these complexities, describing how we compare non-identical race records during case similarity assessment and retrieval, and an initial attempt to introduce the effects of weather as part of the adaptation stage.

2.3   The Ultra Trail du Mont Blanc

The UTMB takes place annually in late August or early September in Chamonix in France. The UTMB course follows the route of the Tour du Mont Blanc through France, Italy and Switzerland[1]. It has a distance of approximately 171 km (106 mi), and a total elevation gain of around 10,040 m (32,940 ft). It is widely regarded as one of the most difficult foot races in the world, and one of the largest, with more than 2,500 participants. Entry is based on a points-based qualification standard, which runners meet by participating in a range of qualifying races over the previous two years. Runners range in age and ability, and each year approximately one-third of starters fail to finish within the 46.5 h time-limit; while the race winners will typically complete the course in just over 20 h, most will take 32–46 h to finish, and will have run through two nights in the mountains.

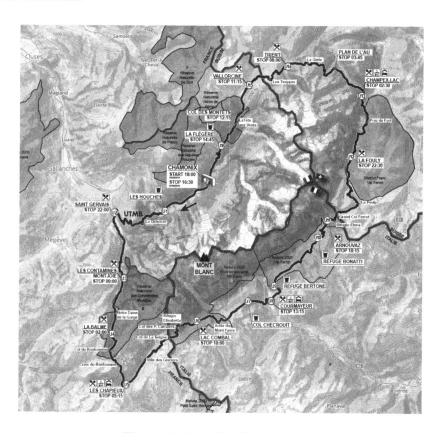

Fig. 1. The 2019 UTMB course map.

---

[1] The Tour du Mont Blanc is one of the most popular long-distance walks in Europe. It circles the Mont Blanc massif and is normally walked in a counter-clockwise direction in 11 days; https://en.wikipedia.org/wiki/Tour_du_Mont_Blanc.

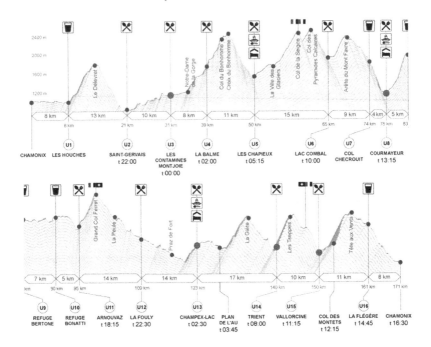

Fig. 2. The 2019 UTMB elevation profile, stages, and aid stations.

The route map and profile for the 2018 race is shown in Figs. 1 and 2, highlighting the race's significant elevation gains, and the various stages and aid stations along the way. While most of these stages remain fixed from year to year, it is not unusual for there to be some route changes based on conditions in the mountains. On average, a typical year's course will have approximately 24 stages but the exact stages will vary from year to year with between 1 and 5 route changes in a given year.

The UTMB race data used in this study covers the years 2003–2017, inclusive, but excluding 2010 and 2012; the 2010 race was canceled due to a large mudslide and the 2012 race was significantly truncated, by about 70 km, due to adverse weather conditions. Each race record includes details on the participant (gender, age category, nationality, team etc.) plus timing, distance, and elevation data for the race stages. In total there are 19,579 race records for 15,144 unique runners. However, in what follows we will focus on runners who have at least two race records and there are only 1,266 such runners, with 3,222 race records in total.

3   Pace Prediction and Planning for Ultra Races

For endurance events such as the marathon, race planning plays an important role in race outcome. Runners need to have a target finish-time in mind in order to plan how they will pace and fuel their race. Runners who are overly ambitious run the risk of *hitting the wall* or *"blowing up"* late in the race, while those that are too conservative may fail to maximise their performance. In an ultra-marathon like UTMB, such matters are even more important because poor pacing choices will be compounded by the distance. In this section we will describe how we use a case base of UTMB race cases to support runners with two important pieces of information as they plan their race:

1. *A Pacing Prediction, p* – a predicted overall, average race-pace; thus $p * d$ is their estimated finish-time (in minutes) for a race of a given distance, d. We use average race-pace rather than finish-time because the route and route distance will vary from year to year.
2. *A Pacing Plan, P* – a sequence of recommended paces, for each stage of the race, based on distance, terrain, and weather, to provide the runner with a more fine-grained pacing plan in order to achieve it.

We convert race records into a case base of race cases, retrieving and adapting similar cases in order to generate these predictions and pacing plans as discussed in the following section.

3.1   Case Representation

The work of [21] represented each marathon race as a sequence of 9 split-times based on 5 km stages/segments (plus the final 2.2 km segments), and we follow a similar approach for the UTMB data, but using a greater number of split-times that are less regularly spaced and less consistent year on year, due to stage changes. We convert these raw split-times, and the final finish-time, into mean segment paces (measured in minutes per km). In this way, each raw race record is converted in to a pacing record with one average pace per race segment.

Drawing on the work of [21], these race records are converted into cases by pairing a *personal-best (PB)* race record, for a runner r ($PB(r, u_i)$) with some previous non-PB (nPB) race ($nPB(r, u_j)$), for the same runner. Here u_i and u_j refer to particular *ultra-marathon races* and, for the avoidance of doubt, the PB race for r is that race with the fastest mean pace[2]. Thus, in general, a runner with n race records will be represented by $n - 1$ cases, each with the same PB race but different non-PB races; see Eq. 1.

$$c_{ij}(r, u_i, u_j) = \left\langle nPB_i(r, u_i), PB(r, u_j) \right\rangle \tag{1}$$

In this way, each case represents a runner's progression and we will use these cases as the basis for pacing prediction and recommendation in a manner that is similar to the approach taken by [21], but with some important differences, as described in the next section.

---

[2] Note we do not use the fastest finish-time because race length tends to vary from year to year depending on conditions and stages and hence mean race pace serves as a more realistic measure performance.

3.2   Case Retrieval

Retrieval is a three-step process, as shown in Algorithm 1; this is a version of the approach used by [21] that has been adapted for UTMB. Given a query race record (q)—that is a runner, a finish-time/average pace, and a nPB pacing profile—we first filter the available cases (CB) based on their mean paces, so that we only consider cases for retrieval if their mean paces are within p minutes/km of the query mean pace. This ensures that we are basing our reasoning on a set of cases that are comparable in terms of performance and ability.

Algorithm 1. Outline CBR Algorithm.

Data: Given: q, query race record; CB, case base; k, number of cases to be retrieved; p, mean pace threshold, w, linear weather model.
Result: pb, predicted finish-time; pn, recommended pacing profile.
begin
    $C = \{c \in CB : pace(q) - p < pace(nPB(c)) < pace(q) + p\}$
    $C = \{c \in C : c.gender == q.gender\}$
    if $len(C) \geq k$ **then**
        $shared = segs(q) \cap segs(nPB(c))$
        $R = sort_k(sim(q, c, shared) \; \forall \; c \in C)$
        $R' = adapt(R, w)$
        $pb = predict(q, R')$
        $pn = recommend(q, R')$
        return pb, pn
    else
        return $None$
    end
end

Next, we filter on the basis of gender, only considering cases for retrieval if they have the same gender as the query runner. The reason for this is that physiological differences between men and women have a material impact on marathon performance [24] but the effect is even more pronounced in ultra races [12]; the marathon gender gap is about 12% but this increases to about 20% at UTMB and similar races.

Finally, we perform a standard, distance-weighted kNN retrieval over the remaining candidate cases C, comparing q's pacing profile to their nPB profiles. These pacing profiles are real-valued vectors and we use a simple cosine-based similarity metric for similarity assessment. Importantly, and unlike the work of [21,23], this similarity is computed only over race segments that are *shared* between the query race and each of the candidate cases. Thus, in the case where there are routing differences between the query and candidate cases we simply ignore these differences on the basis that, on average, they are likely to be few in number relative to the shared segments; it is left as a matter for future work to explore alternative ways to account for such differences. In any event, once we have calculated the case similarities we select the top k most similar as the retrieved cases, R.

3.3 Adapting for Weather

Prior to making a pacing prediction from the retrieved cases we adapt the pacing of the races to account for local weather conditions. To do this we trained a linear regression model to relate mean race temperature to mean race paces. The resulting model provides a pace adjustment factor with which we can increase or decrease a given pace based on the temperature that prevailed when it was run. In effect this allows us to normalise paces for an average race temperature ($22\,^{\circ}$C), slowing paces for warmer races and improving the pacing associated with cooler races; the model operates in the range $17\,^{\circ}$C to $30\,^{\circ}$C.

3.4 Predicting a PB Pace

Given a set of adapted, similar cases, R', we need to estimate the best achievable mean pace for q. Each case in R' represents another runner with a similar nPB to q, but who has gone on to achieve a faster personal best at UTMB. The intuition is that, since these PBs were achievable by similar runners, then a similar PB should be achievable by the query runner.

For the purpose of this work, the predicted PB pace for the query runner is a function of the personal best paces of the retrieved cases using one of two simple strategies:

1. *Fastest PB* – uses the PB pace of the single retrieved case with the fastest PB pace; we can view this as an *optimistic* prediction strategy as it assumes the query runner will perform in a way that is similar to the best performing runner among our retrieved cases.
2. *Mean PB* – uses the mean PB pace of the k retrieved cases; this is a more realistic strategy as it assumes the query runner will perform in a manner that is similar to the average runner among our retrieved cases.

$$PB_{best}(q, C) = w(q, C_{best}) \bullet pace(C_{best}(PB)) \qquad (2)$$

$$PB_{mean}(q, C) = \frac{\sum_{\forall i \in 1..k} w(q, C_i) \bullet pace(C_i(PB))}{k} \qquad (3)$$

With each approach the predicted PB paces are weighted based on the relative difference between the query runner's pace and the corresponding nPB pace of a retrieved case; see Eq. 4. This adapts the PB paces based on whether they were achieved by similar runners with slightly slower or faster nPB paces. The resulting pace prediction

$$w(q, c) = \frac{q(nPB).meanpace}{c(nPB).meanpace} \qquad (4)$$

Finally, since the resulting pace prediction has been normalised for weather, it will need to be adjusted for weather conditions on the target race day, using the same weather model.

3.5   Recommending a Pacing Plan

While supporting a runner with a realistic predicted pace for their race is useful, it is just as important, and arguably more practical, to provide the runner with a granular pacing plan so that they can adjust their pacing on a stage by stage basis. Once again we use two different strategies:

1. *Fastest Profile* – recommend the segment paces, from the fastest case, for those segments that are shared with the query runner's upcoming race.
2. *Mean Profile* – recommend the average shared segment paces for the k retrieved cases.

In this way, the query runner will be recommended a pacing plan, based on the PBs of runners of similar ability (similarly paced past races). There are a couple of important caveats to note. First, as with the PB pace predictions, these paces will need to be adjusted for the target race conditions using the linear weather model. Also, the pacing plan will only contain pace recommendations for segments that are shared with the target race route, but since the retrieved cases can come from a variety of past years, in practice, the number of missing segments is kept to a minimum.

3.6   Discussion

So far we have discussed an approach to predicting PB performance (mean race-pace) and recommending tailored pacing plans for UTMB participants by reusing and adapting the PB performances of a set of similar runners who have run similar UTMB races to the target runner in the past. We have proposed two different approaches: (1) an optimistic approach that focuses on the fastest PB in the set of similar cases; an (2) a more conservative approach that considers an average of all the PBs in the similar cases. This is based closely on the approach described by [21] but has been adapted for the unique features of ultra-marathon races: we have focused on pacing rather than finish-times because races tend to vary in distance and duration; we take account of route differences by focusing on common/shared race segments; and we incorporate a simple, linear weather model to adapt pacing based on changes in local weather conditions.

4   Evaluation

In this section we evaluate these prediction and recommendation approaches, examining their ability to make accurate pacing predictions and to recommend high-quality pacing plans.

4.1   Dataset and Methodology

To evaluate these methods we use the race data from UTMB (2003–2017) as mentioned earlier. Because our case base relies on runners who have at least 2

UTMB race records, this limits our dataset to 1,266 runners, from which we can produce a case base of 3,222 cases.

We adopt the approach taken by [21], using a standard leave-one-out methodology, with each case treated as a query runner, predicting their overall race-pace and recommending a pacing plan, and then comparing the prediction and pacing plan to the actual PB race of the target runner. For predictions we will calculate an error score, based on the difference between the predicted and actual pace. For the recommended pacing plans, we will calculate their cosine similarity with the actual PB pacing profile; we are using pacing similarity as a proxy for recommendation quality, which is not unreasonable. These error rates and similarities are then averaged, across all target/query runners, to produce overall results.

4.2   Prediction Error: On the Importance of k and Weather

To begin with we will examine the relationship between the prediction error and k (the number of cases retrieved), in order to determine a suitable k for the remainder of this evaluation. We will also examine whether the weather adjustment leads to improved prediction accuracy.

The results are shown in Fig. 3(a, b) for the *Fastest PB* and *Mean PB* strategies. In Fig. 3(a) we see that the *Fastest PB* strategy is not as effective as the *Mean PB* strategy. For the former, not only are the error rates poorer for each value of k, they deteriorate as k increases. In contrast, the *Mean PB* strategy benefits from gradually improving error rates as k increases. Incidentally we chose to limit k to a maximum of 20 cases given the size of the case base available.

It is not surprising that the *Fastest PB* strategy performs poorly with increasing k. Increasing the number of retrieved cases provides a larger set of cases from which to select a fastest one, and so we can expect it to produce more and more ambitious pace predictions as k increases. In contrast, the error rate of the more realistic predictions of the *Mean PB* approach drops to about 6% for $k \geq 10$. In subsequent sections will focus on the *Mean PB* strategy with $k = 20$.

Figure 3(b) shows the average prediction error with (*Weather*) and without (*Std*) the weather adaptation. The adaptation improves prediction error for both prediction strategies, but more so for *Mean PB*, a *relative* improvement of just over 8.5%, compared to only 5% for *Fastest PB*. Although a relatively modest improvement in prediction error it must be acknowledged that this was achieved using a straightforward adjustment, based on a simple linear weather model, and using only temperature. Obviously there is an opportunity to improve this approach significantly by including other relevant factors (wind, humidity, precipitation etc.). Nevertheless the improvement we see here bodes well for a more sophisticated model to offer further prediction benefits in the future.

4.3   Prediction Error by Gender, Age, and Ability

Next we consider how prediction error varies with gender, age, and ability and the results are presented in Fig. 4, based on the age categories used by UTMB

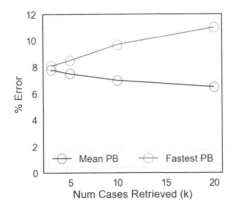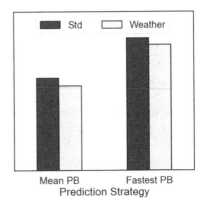

Fig. 3. The average error for predictions by *Fastest PB* and *Mean PB* strategies based on (a) different values of k and (b) the application of a weather adjustment.

(23–39, 40–49, 50–59, and 60–69) and a set of pacing ranges (<9 mins/km, 9–11 mins/km, 11–13 mins/km, 13–15 mins/km, and 15–17 mins/km) as shown.

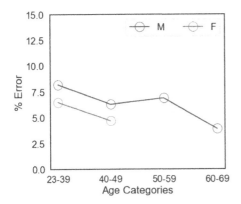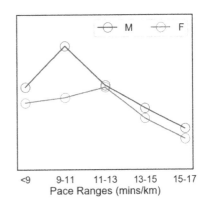

Fig. 4. The average error for pacing predictions (*MeanPB* strategy and $k = 20$) based on male & female runners with different (a) age groups and (b) ability levels.

The results show more accurate pacing predictions for women than for men, regardless or age or ability, which is consistent with similar results found by [21] for road marathons. Figure 4(a) suggests that prediction error improves with increasing age, for men and women, although there are not enough female runners in the older age-groups to demonstrate this conclusively. This suggests that older runners are more predictable, perhaps because they are less ambitious with their targets and therefore less likely to "blow up" during the race.

A similar trend is seen in Fig. 4(b) for ability: slower runners are more predictable than faster runners. This contrasts with the findings of [21] for road marathons, where faster runners were more predictable than slower runners, and is worthy of further research in order to determine the possible cause of this difference.

4.4 Recommendation Similarity by Gender, Age, and Ability

In a similar fashion we evaluate the accuracy of the pacing plans recommended to runners by comparing their recommended segment/stage paces to the actual paces run by the runner during their PB race; as mentioned earlier, we calculate an overall cosine similarity score between the recommended and actual plans. The results are presented in Fig. 5, comparing runners based on their age category and ability levels; once again note that there are not enough older female runners to generate reliable cosine similarities for older age groups.

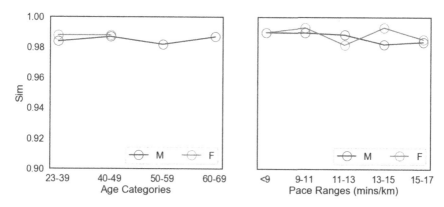

Fig. 5. The average pacing plan similarity for plans ($MeanPB$ strategy and $k = 20$) based on male & female runners with different (a) age groups and (b) ability levels.

The results indicate that recommendation quality does not depend critically on age or ability. In each case we see that pacing plans are suggested that are very similar (>0.98) to the actual pacing profile the runner achieved in their PB. A word of caution here. We do need to remember that this is not quite the complete picture because there can be missing race stages due to changes in the UTMB course from year to year. This means that the recommended plans cover most (typically >80%) but not all of the race segments in a given year.

Nevertheless the high similarity levels tell us that the recommended plans are very close to the actual PB pacing profiles, and there is no reason to expect that missing stages would be any different. It would be possible to "fill in" these gaps by recommending an average (grade-adjusted) pace to the target runner.

5   Discussion

These results suggest that the ideas proposed by [20–23] for marathon runners may also prove effective for other forms of endurance activities. The prediction errors in this work ($\approx$7% on average) are similar to, but not as good as, those reported by [21]. Follow-up work in [22] provided a number strategies for significantly reducing error-rates, which should also be applicable in this work.

Although not reported here, for reasons of space, we have also evaluated these techniques on a datset from the 100-mile Western States ultra marathon[3]. It is a similar event to UTMB, in distance and terrain, which takes place in California's Sierra Nevada every June. Using a dataset of approximately 4,000 race records (for 2008–2018), and a similar number of race stages to the UTMB, we found similar patterns of results for prediction error and pacing similarity. However the actual prediction errors ($\approx$9%) and recommendation similarities ($\approx$0.95) were not as good as those found for UTMB; once again women were more predictable than men.

A lack of data is likely the primary reason for these lower accuracy and similarity rates, highlighting one of the main shortcomings with the approach outlined here. The work of [21] leveraged large datasets of 10,000's of runners to make very accurate predictions and recommendations for marathon runners. This volume of data is just not available for ultra races, which are much more selective and do not (yet) appeal to most people, even serious runners. However, it may be possible to overcome this data sparsity challenge if we can use race records from other events as the basis of our ultra marathon predictions. For example, many ultra marathon runners also run road marathons and their PB marathon times are likely to be strong predictors of their ultra-marathon times. Indeed, as mentioned previously, in order to qualify for events such as UTMB runners must accumulate enough points across two years of mountain races and their performance in these races could also be used in our race cases.

6   Conclusions

Our aim in this work has been to examine whether the case-based marathon prediction techniques in [20–23] can be generalised for other types of endurance events and specifically ultra marathons. There are many reasons to suspect that such events will prove to be more challenging than more conventional road marathons – increased distances, challenging terrain, changing routes, the impact of weather, fewer race records to learn from – but the results presented in this paper are promising. Although not as good as the results presented in [20–23] the ability to predict ultra marathon pacing with an error rate of just $\approx$7% should prove to be useful, certainly for non-elite, recreational runners, if one can call any UTMB participant a 'recreational' runner!

Many opportunities exist for future work. Paired-race case representations could be extended to capture richer representations of runner history using the techniques outlined in [22]. Additional event-types could be incorporated to address the data-sparsity issues that are inherent in ultra marathon running, relative to more popular event-types such as marathons, half-marathons and

---

[3] https://www.wser.org.

triathlons. And, more sophisticated models of weather conditions and terrain could be developed better normalise performance across the years.

References

1. Abut, F., Akay, M.F., George, J.: Developing new VO2max prediction models from maximal, submaximal and questionnaire variables using support vector machines combined with feature selection. Comput. Biol. Med. **79**, 182–192 (2016)
2. Akay, M.F., Zayid, E.I.M., Aktürk, E., George, J.D.: Artificial neural network-based model for predicting VO2max from a submaximal exercise test. Expert Syst. Appl. **38**(3), 2007–2010 (2011)
3. Bartolucci, F., Murphy, T.B.: A finite mixture latent trajectory model for modeling ultrarunners' behavior in a 24-hour race. J. Quant. Anal. Sports **11**(4), 193–203 (2015)
4. Berlin, E., Laerhoven, K.V.: Detecting leisure activities with dense motif discovery. In: The 2012 ACM Conference on Ubiquitous Computing, Ubicomp 2012, Pittsburgh, PA, USA, 5–8 September 2012, pp. 250–259 (2012)
5. Bichindaritz, I., Montani, S., Portinale, L.: Special issue on case-based reasoning in the health sciences. Appl. Intell. **28**(3), 207–209 (2008)
6. Bramble, D.M., Lieberman, D.E.: Endurance running and the evolution of homo. Nature **432**, 345–352 (2004)
7. Campbell, A.T., et al.: The rise of people-centric sensing. IEEE Internet Comput. **12**(4), 12–21 (2008)
8. Dearden, P.: Game, set and mismatch. EMBO Rep. **8**(3), 219 (2007)
9. Ellaway, R.H., Pusic, M.V., Galbraith, R.M., Cameron, T.: Developing the role of big data and analytics in health professional education. Med. Teach. **36**(3), 216–222 (2014)
10. Fister, I., Rauter, S., Yang, X.S., Ljubič, K., Fister, I.: Planning the sports training sessions with the bat algorithm. Neurocomput. **149**(PB), 993–1002 (2015)
11. Glaros, C., Fotiadis, D.I., Likas, A., Stafylopatis, A.: A wearable intelligent system for monitoring health condition and rehabilitation of running athletes. In: 4th International IEEE EMBS Special Topic Conference on Information Technology Applications in Biomedicine, pp. 276–279 (2003)
12. Hoffman, M.D.: Performance trends in 161-km ultramarathons. Int. J. Sports Med. **31**(01), 31–37 (2010)
13. Kelly, D., Coughlan, G.F., Green, B.S., Caulfield, B.: Automatic detection of collisions in elite level rugby union using a wearable sensing device. Sports Eng. **15**(2), 81–92 (2012)
14. Lewis, M.: Moneyball: The Art of Winning an Unfair Game. WW Norton & Company, New York (2004)
15. Mattson, M.P.: Evolutionary aspects of human exercise – born to run purposefully. Ageing Res. Rev. **11**(3), 347–352 (2012)
16. Mayer-Schönberger, V., Cukier, K.: Big Data: A Revolution That Will Transform How We Live, Work, and Think. Houghton Mifflin Harcourt, Boston (2013)
17. Möller, A., et al.: Gymskill: mobile exercise skill assessment to support personal health and fitness. In: 9th International Conference on Pervasive Computing (Pervasive 2011), Video, San Francisco, CA, USA (2011)
18. Rauter, S.: New approach for planning the mountain bike training with virtual coach. Trends Sport Sci. **2**, 69–74 (2018)

19. Rooksby, J., Rost, M., Morrison, A., Chalmers, M.C.: Personal tracking as lived informatics. In: Proceedings of the 32nd Annual ACM Conference on Human Factors in Computing Systems, pp. 1163–1172. ACM (2014)

20. Smyth, B., Cunningham, P.: A novel recommender system for helping marathoners to achieve a new personal-best. In: Proceedings of the Eleventh ACM Conference on Recommender Systems, RecSys 2017, Como, Italy, 27–31 August 2017, pp. 116–120 (2017)

21. Smyth, B., Cunningham, P.: Running with cases: a CBR approach to running your best marathon. In: Aha, D.W., Lieber, J. (eds.) ICCBR 2017. LNCS (LNAI), vol. 10339, pp. 360–374. Springer, Cham (2017). https://doi.org/10.1007/978-3-319-61030-6_25

22. Smyth, B., Cunningham, P.: An analysis of case representations for marathon race prediction and planning. In: Cox, M.T., Funk, P., Begum, S. (eds.) ICCBR 2018. LNCS (LNAI), vol. 11156, pp. 369–384. Springer, Cham (2018). https://doi.org/10.1007/978-3-030-01081-2_25

23. Smyth, B., Cunningham, P.: Marathon race planning: a case-based reasoning approach. In: Proceedings of the Twenty-Seventh International Joint Conference on Artificial Intelligence, IJCAI 2018, 13–19 July 2018, Stockholm, Sweden, pp. 5364–5368 (2018)

24. Trubee, N.W.: The effects of age, sex, heat stress, and finish time on pacing in the marathon. Ph.D. thesis, University of Dayton (2011)

25. Yoganathan, D., Kajanan, S.: Persuasive technology for smartphone fitness apps. In: PACIS, p. 185. Citeseer (2013)

NOD-CC: A Hybrid CBR-CNN Architecture for Novel Object Discovery

J. T. Turner[1(✉)], Michael W. Floyd[2(✉)], Kalyan Gupta[2(✉)],
and Tim Oates[1(✉)]

[1] University of Maryland Baltimore County, Baltimore, MD, USA
{jturner1,oates}@umbc.edu
[2] Knexus Research Corporation, National Harbor, MD, USA
{michael.floyd,kalyan.gupta}@knexusresearch.com

Abstract. Deep Learning methods have shown a rapid increase in popularity due to their state-of-the-art performance on many machine learning tasks. However, these methods often rely on extremely large datasets to accurately train the underlying machine learning models. For supervised learning techniques, the human effort required to acquire, encode, and label a sufficiently large dataset may add such a high cost that deploying the algorithms is infeasible. Even if a sufficient workforce exists to create such a dataset, the human annotators may differ in the quality, consistency, and level of granularity of their labels. Any impact this has on the overall dataset quality will ultimately impact the potential performance of an algorithm trained on it. This paper partially addresses this issue by providing an approach, called **NOD-CC**, for discovering novel object types in images using a combination of Convolutional Neural Networks (CNNs) and Case-Based Reasoning (CBR). The CNN component labels instances of known object types while deferring to the CBR component to identify and label novel, or poorly understood, object types. Thus, our approach leverages the state-of-the-art performance of CNNs in situations where sufficient high-quality training data exists, while minimizing its limitations in data-poor situations. We empirically evaluate our approach on a popular computer vision dataset and show significant improvements to object classification performance when full knowledge of potential class labels is not known in advance.

Keywords: Deep learning · Novel object discovery ·
Computer vision · Convolutional Neural Networks

1 Introduction

Deep Learning has seen rapid advancement in recent years, setting benchmarks for many machine learning tasks in the areas of computer vision, natural language processing, and game AI. While these deep neural networks are fundamentally the same as the perceptrons [14] of the late 1960s, they leverage dramatic improvements in the availability of computational resources and training data

© Springer Nature Switzerland AG 2019
K. Bach and C. Marling (Eds.): ICCBR 2019, LNAI 11680, pp. 373–387, 2019.
https://doi.org/10.1007/978-3-030-29249-2_25

to significantly outperform their predecessors. In particular, the field of computer vision has benefited from the application of *Convolutional Neural Networks* (CNNs) [6] that are able to use massive image datasets to learn relevant image features rather than relying on hand-engineered feature sets. Additionally, this field has been able to utilize a seemingly endless streams of crowdsourced labeled images from sources like Facebook, Instagram, Twitter, and Reddit.

However, the ability of these deep learning architectures to learn is directly tied to the availability of high-quality, human-labeled data to use during training [16]. If training data is either rare or low-quality, deep learning systems will have difficulty accurately learning from the data. In the case of rare data, it may be possible to gather more data over time as more images become available (e.g., as a new model of mobile phone is released, when a new species is discovered and documented). The more difficult long-term problem is the quality of data, as demonstrated by the age-old idiom *"garbage in, garbage out"*. In some situations, this can be erroneous labels being given to training instances. For example, if an annotator labels an image of a *car* as a *tree*, the learning system will attempt to learn based on that erroneous data. Similarly, an annotator may only label a subset of objects in complex scenes. In a bedroom scene, the annotator may label *bedclothes*, *pillows*, and *nighstands*, but treat other objects as background, like *alarm clocks* or *lamps*. Furthermore, the system's learning will be constrained by the level of granularity used during labeling and the annotator's term preference. For example, the choice of whether to use a high-level label such as *animal* or *pet*, limit the classification granularity of a vision system compared to using lower-level labels such as *cat*, *European cat*, or *Russian Blue cat*. These issues are compounded by the fact that, given the scale of datasets used by Deep Learning systems, it is impractical for a single annotator to label an entire dataset. Instead, the annotation work is generally crowdsourced from hundreds or thousands of human annotators. It is unlikely that all of these annotators will be consistent, error-free, and complete in the labels they provide. Thus, the overall quality of the labeled datasets, and ultimately the potential performance of a machine learning system trained on the datasets, is bound by the quality of the human annotators.

We propose a method, called *Novel Object Discovery Using Convolutional Neural Networks and Case-Based Reasoning (NOD-CC)*, for object discovery and classification in images that leverages the high-end performance of CNNs while reducing their reliance on large sources of pre-labeled training data. Instead, NOD-CC attempts to classify an input image using a trained CNN, but can dynamically switch to using a case-based classification approach if the CNN is not confident in its prediction. The primary motivation of this approach is that while CNNs require a large collection of training images of each object type to learn successfully, a CBR system can be used to learn using as few as one training instance. Thus, the CBR component can be used to discover novel object types and provide classification of those types until such time as there are sufficient training examples to retrain the CNN.

In our previous work [23], we demonstrated how CBR can leverage the automated feature extraction capabilities of CNNs, and perform novel object discovery and classification. In that work, which we will refer to as *Novel Object Discovery using Case-Based Reasoning (NOD-CBR)*, the convolutional layers of a CNN (i.e., the CNN architecture excluding the fully-connected neural network layers) are used to convert input images into a feature vector representation. That feature vector representation, and optionally any detectable *object parts* that are visible, is used to retrieve similar cases and determine if an object of that type has been encountered previously. NOD-CC significantly extends NOD-CBR and provides the following key contributions:

- A hybrid architecture that includes both the NOD-CBR system as well as a fully functional CNN (i.e., a CNN that performs object classification rather than purely feature extraction).
- An architecture that provides both the high-end performance of CNNs as well as the lazy, data-poor learning capabilities of CBR.
- A series of decision algorithms that can dynamically select whether to use the CNN or CBR components of our architecture to perform object classification.
- An online method for object classification, novel object discovery, novel object labeling, and learning.
- An empirical evaluation that demonstrates the utility of NOD-CC when the full set of object types is not known in advance.

The remainder of this paper describes how NOD-CC combines Convolutional Neural Networks and Case-Based Reasoning to classify images while also performing novel object discovery. Section 2 provides an overview of similar research in both Deep Learning and CBR. Section 3 describes our hybrid architecture that combines CNNs and CBR for object classification and discovery. Our empirical evaluation is presented in Sect. 4, and provides evidence to support our claims of the utility of NOD-CC. Finally, in Sect. 5 we summarize our findings and identify important future research directions.

2 Related Work

The intersection of CBR and CNNs has been previously examined in the domains of Human Activity Recognition (HAR) and medicine. In many HAR settings, usage of high-fidelity wearable sensors for movement are used for feature extraction [20], and to further train classifiers for new users [19]. Using multi-channel medical device EEG signals, researchers have also conducted analysis on patterns of electrical signals from the brain to set a baseline for seizure detection [24], and then using a case base of seizure-like data to classify unseen patients [15]. The usage of actigraphy sensors provided large amounts of medical data that can be used to predict sleep patterns, sleep quality, and sleep activity using Deep Learning techniques such as Multilayer Perceptrons, Convolutional Neural Networks, and many variants of Recurrent Neural Networks [21]. By leveraging this high-quality sensor data, it is possible to preserve existing patient privacy in medical information while training an initial model, or fine tuning a model for new or changed data [18].

CBR has also seen a wide array of uses for image processing in medical domains [17]. Despite the wide usage and success of CBR in a variety of medical domains (e.g., [7,10,12]), most applications require hand-crafted features (e.g., [1,11,13]) generated by Subject Matter Experts (SMEs), a practice which does not scale to the Exascale-level computation and learning that Deep Learning making possible [8].

A similar application of CBR to our novel object detection system is a website classifier on sites by using image data from the websites instead of the textual data [11]. Although this work also uses the feature vector from later stages of a Convolutional Neural Network, this was used to classify the images from the website into existing categories without leaving the possibility for novelty. Also different from other previous works is that our system's novel object detection system performs unsupervised learning in an online, incremental manner, not doing offline dataset analysis to search for out of distribution classes.

3   Hybrid CNN-CBR Architecture

Our approach, Novel Object Discovery Using Convolutional Neural Networks and Case-Based Reasoning (NOD-CC), is a hybrid of two learning and classification methods (Fig. 1). The Convolutional Neural Network component (labeled as CNN) is intended to classify images of object types for which sufficient training instances are available. Additionally, it converts raw images into feature vectors for use by the Case-Based Reasoning component (labeled as CBR). The CBR component is intended to learn from and classify object types that are not classifiable by the CNN. A meta-algorithm, labeled as Controller, determines whether the classification from the CNN or CBR component is used to provide final image classification. In the following sub-sections, we will provide details about each of the three primary components: CNN, CBR, and Controller.

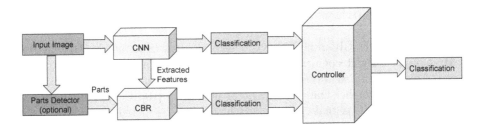

Fig. 1. Architecture of the NOD-CC image classification system. The classifications are shown in green and are produced by the three decision algorithms shown in blue. The inputs to the decision algorithms are shown in yellow, the input image in orange, and the optional parts detector in red. (Color figure online)

3.1   Convolutional Neural Network Component

When used for object classification, a Convolutional Neural Network takes as input an image and outputs a classification for what type of object is visible in the input. CNNs contain two primary stages: *convolutional and pooling layers*, and *fully-connected layers*. The convolutional and pooling layers take the raw image input and extract a feature vector containing the relevant features that were learned during training. For example, these features may include the presence (or absence) of certain edges, shapes, or complex geometric objects (composed of many shapes). This feature vector is used as input to the fully-connected layers, which then use the feature vector to determine a classification for the image. In NOD-CC, the feature vector computed by the convolutional and pooling layers is also provided to the CBR component. This is done to leverage the ability of CNNs to automatically learn and perform feature extraction, and avoids any manual feature engineering for the CBR component.

NOD-CC is agnostic to the particular CNN architecture used; since all CNNs can produce an intermediate feature vector (which can be provided to the CBR component) and a classification, any architecture can be used. In our work, we use the Inception-v3 architecture [22]. We selected this architecture based on it having been shown to achieve similar classification performance compared to more computationally expensive architectures such as ResNet [5] and ResNeXt [25]. An additional benefit of using the Inception-v3 architecture is that it allows the possibility of future extensions of NOD-CC, as part of future work, to use a hierarchical image grammar. This would allow not only novel object discovery but also hierarchical class relationships between classes (e.g., that a novel object type is a subclass of an existing object type). Inception-v3 facilitates this by using a set of auxiliary classifiers, used to combat the vanishing gradient problem during training, that could be used to facilitate predictions at multiple levels of granularity.

3.2   Case-Based Reasoning Component

Although CNNs can achieve high accuracy when classifying objects in images, their performance is dependent on the set of class labels (i.e., object types) contained in the training data. If the training data contains images labeled with the set of labels $\mathcal{L} = \{l_1, \ldots, l_n\}$, a CNN (and most other learning algorithms) will only be able to classify those n object types. Any images of objects with a label l_m (where $l_m \notin \mathcal{L}$) will either be misclassified as one of the labels in $\mathcal{L}$ or unclassified (i.e., the CNN will output a low confidence for all labels such that an *unknown* output is produced). This issue is particularly problematic for CNNs since they require a large set of example images labeled as l_m before they be accurately trained to predict that object type. CBR, on the other hand, likely does not have the same peak classification performance on massive image datasets but is capable of one-shot learning. Once a single image with label l_m is encountered, it can be stored as a case and reused to classify other instances of that object type.

For the CBR component of NOD-CC, we use our previous case-based novel object discovery approach, NOD-CBR [23]. NOD-CBR stores each training image $I_i \in \mathcal{I}$ (where $\mathcal{I}$ is the set of all images) as a case C_i in the case base CB ($C_i \in CB$). Cases are encoded as triplets containing the feature vector representation of the image F_i, a set of detectable image parts P_i, and the ground truth object label l_i: $C_i = \langle F_i, P_i, l_i \rangle$. Using case-based reasoning nomenclature, the feature vector and parts set of the image are the *problem*, and the class label is the *solution*.

Recall from the previous subsection that the convolutional and pooling layers of the CNN component convert a raw input image into a feature vector $F_i = \langle f_1, \ldots, f_v \rangle \in \mathcal{F}$ (where v is an integer value defined by the CNN architecture and $\mathcal{F}$ is the set of all feature vectors). Thus, both the CBR component and the fully-connected layers of the CNN component use an identical feature vector representation as produced by the convolutional and pooling layers mapping from images to features: $features : \mathcal{I} \to \mathcal{F}$.

Each case also contains the set of parts $P_i \subset \mathcal{P}$ that are detectable in the input image, where $\mathcal{P}$ is the set of all parts that may be detected. These parts are generic lower-level structures of an image, like *hands*, *feet*, *wheels*, or *wings*. Since parts are generic, different objects types can share parts (e.g., both *dogs* and *cats* have *legs*, *heads*, *ears*, *tails*). However, even images of the same object type may have different detectable parts based on variations in pose, occlusion, or photographic style. For example, in Fig. 2, the cats do not have an identical set of detectable parts due to different poses and image framing. Our work assumes the presence of a parts extractor that returns the set of detected parts in an image: $parts : \mathcal{I} \to \mathcal{P}$. However, as we will discuss shortly, while our approach can leverage parts information it is not necessary for classification (i.e., it can classify using only the feature vector).

The NOD-CBR object discovery and classification algorithm is shown in Algorithm 1. While full details of the algorithm are described in our previous work [23], we will provide a brief overview of its reasoning process. The algorithm takes as input an image I_{in}, a case base of training images CB, the k value to use when retrieving similar cases from the case base, the threshold λ_f used to determine if two images have similar features, and the threshold λ_p used to determine if two images have similar parts. The output is the class label for the image. Given an input image, the algorithm will extract the feature vector representation (i.e., from the CNN component) and the set of detectable parts (i.e., from the parts extractor). If the either the case base is empty (Lines 4–5), no cases are sufficiently similar to the input image's feature vector representation (Lines 7–9, based on a threshold λ_f), or there are cases with similar feature vectors but their detectable parts are not similar (Lines 11–17, based on a threshold λ_p), then NOD-CBR generates a new label for the input image. In that situation, it believes the image to be of a newly discovered object type. Otherwise (Line 19), it uses the class label from the most similar retrieved case. In all situations, a new case is retained and added to the case base (Line 20).

Fig. 2. The variation in pose of the two cats, as well as the framing of the picture can drastically effect the observable parts. The cat on the left in the so-called *catloaf* position is hiding his legs under his torso, and the way the picture is framed does not show its tail, while the cat on the right has all major parts visible.

An advantage of this approach is that it can start from a variety of initial case base configurations: an empty initial case base if no prior knowledge exists, a case base containing cases for all images used to train the CNN, or a sampling of cases of each object type if the full training set is too large. It should also be noted that while the *generateLabel*() function in Algorithm 1 will generate a unique label for a newly discovered object type, it will likely not be a meaningful class label (e.g., returning the label *object5849* rather than *lion*). However, images with newly generated labels (i.e., the newly discovered object types) could be presented to a human expert, either online or offline, to receive more meaningful object labels.

3.3 Controller Component

The CNN component and CBR component both output a classification for the input image. However, there is no guarantee that they will predict the same object type. The role of the controller is to receive as input the predictions from both components and output a final predicted class label.

In our work, we use three different Controller strategies:

– **Always CNN:** The classification output by the CNN component is used regardless of the the CBR component's classification. This is equivalent to the CNN component operating in isolation.
– **Always CBR:** The classification output by the CBR component is used regardless of the the CNN component's classification. This is equivalent to the CBR component operating in isolation.

Algorithm 1. NOD-CBR algorithm for image classification and novel object discovery

---

Function: $classify(I_{in}, CB, k, \lambda_f, \lambda_p)$ **returns** l_{in}

---

1 $F_{in} \leftarrow features(I_{in})$;
2 $P_{in} \leftarrow parts(I_{in})$;
3 $l_{in} = \emptyset$;
4 **if** $CB = \emptyset$ **then**
5 $\quad \lfloor \; l_{in} \leftarrow generateLabel()$;
6 **else**
7 $\quad topK \leftarrow retrieveTopK(F_{in}, CB, k, \lambda_f)$;
8 $\quad$ **if** $topK = \emptyset$ **then**
9 $\quad \quad \lfloor \; l_{in} \leftarrow generateLabel()$;
10 $\quad$ **else**
11 $\quad \quad nn = \emptyset; nnSim = -1$;
12 $\quad \quad$ **foreach** $C_i \in topK$ **do**
13 $\quad \quad \quad sim \leftarrow partSim(P_{in}, C_i.P_i)$;
14 $\quad \quad \quad$ **if** $sim > nnSim$ **and** $sim > \lambda_p$ **then**
15 $\quad \quad \quad \quad \lfloor \; nn = C_i; nnSim = sim$;
16 $\quad \quad$ **if** $nn = \emptyset$ **then**
17 $\quad \quad \quad \lfloor \; l_{in} \leftarrow generateLabel()$;
18 $\quad \quad$ **else**
19 $\quad \quad \quad \lfloor \; l_{in} \leftarrow nn.l_i$;
20 $CB \leftarrow CB \cup \langle F_{in}, P_{in}, l_{in} \rangle$;
21 **return** l_{in};

- **Conditional CBR:** The classification of the CNN component is used unless the CNN has low confidence in its prediction. This occurs when none of the class labels are above an abstention threshold λ_a. In situations where the CNN does not output a class label, the prediction of the CBR component is used.

4 Evaluation

Our empirical evaluation demonstrates the object discovery and classification performance of NOD-CC when the complete set of object types that will be encountered at run-time is not known in advance. More specifically, the following hypotheses are evaluated:

H1: The CNN component will be unable to correctly classify any object types not present in the training set.
H2: The CBR component, NOD-CBR, will outperform the CNN component when the training images do not contain instances of all object types that may be encountered at run-time.

H3: NOD-CC will achieve higher classification performance than the CNN component alone when the training images do not contain instances of all object types that may be encountered at run-time.

H4: NOD-CC will achieve higher classification performance than NOD-CBR alone when the training images do not contain instances of all object types that may be encountered at run-time.

4.1   Dataset

The image dataset used during our evaluation is the publicly available *PASCAL-Part* dataset [2], a subset of the images from the *Visual Object Classes Challenge 2010* dataset [3]. The dataset contains images with 20 coarse-grained ground truth object types: *aeroplane, bicycle, bird, boat, bottle, bus, car, cat, chair, cow, dining table, dog, horse, motor bike, person, potted plant, sheep, sofa, train,* and *tv monitor*. Additionally, each image has between 0 and 24 annotated detected parts. However, as discussed previously, images may differ on the number of annotated parts based on the quality and properties of each image.

Other properties of the *PASCAL-Part* dataset that make it appropriate for this evaluation include the variation between scale, orientation, pose, lighting, and ambient setting of the objects. The images include many complex, real-world environments so there is a high-degree of image clutter, object occlusion, and background scenes. For example, one image of a person is in a forested environment where less than 10% of the visible pixels are of the person, while in another a person takes up 90% of the visible pixels but is partially occluded by the water they are swimming in.

Our current work is focused on classifying a single object type in each image. To facilitate this, we filtered the *PASCAL-Part* dataset to only the images that contain a single class label, thereby reducing the dataset from 10,103 images to 4,737. While this may seem like a limitation of our approach, many computer vision applications first propose sub-regions of a cluttered image to classify (e.g., the *region proposal* stage of a Region-Based Convolutional Neural Network [4]), and then provide at most a single object label for each sub-region (i.e., a traditional CNN classification). Additionally, even though each image only contains a single labeled object, nearly all of the images contain a variety of unlabeled background objects. Since the size of the filtered *PASCAL-Part* dataset is quite small by Deep Learning standards, the CNN component used in our work, the Inception-v3 architecture, was pretrained on the much larger *Open Images v4* dataset [9] and then fine-tuned using the filtered *PASCAL-Part* dataset. It should be noted that there is no overlap between the images contained in the two datasets (i.e., pretraining on *Open Images v4* will not provide any images from the testing sets we use).

For our experiments, we used the filtered *PASCAL-Part* dataset to create 20 experimental datasets. The original dataset comes pre-partitioned into training and testing sets. For each of the 20 experimental datasets, 5 of the 20 object types were selected at random (such that no two experimental datasets used the same set of 5 object types). All images of the 5 selected object types were removed from the training set but left in the testing set. Thus, all testing sets contain images of all 20 object types, but the training sets only contained images of 15 object types. These experimental datasets were partitioned in advance, such that all experimental variations would work on an identical set of datasets.

4.2 Scoring Metrics

Our previous work [23] demonstrated the ability of NOD-CBR, when starting from an empty case base, to discover and classify novel object classes. More specifically, we evaluated its ability to maximize class purity (i.e., provide the same generated label to images of the same object type) while minimizing the divergence in the number of discovered classes from the true number of classes (i.e., not over-partitioning the data). Given that we have previously demonstrated the efficacy of NOD-CBR on these tasks, our evaluation will measure the performance of our hybrid NOD-CC architecture's classification performance when class labels from the testing set are not present in the training set (i.e., novel object types are encountered at run-time).

For each testing image provided to NOD-CC, there are four possible ways in which the classification prediction of NOD-CC can align with the image's ground truth label, ordered from best to worst:

1. **Correct:** The class label predicted by NOD-CC matches the ground truth class label. This is the ideal situation and is considered to be a 100% match. C represents the percentage of testing instances labeled correctly.
2. **Known Novel:** NOD-CC correctly predicts that the class label was not one of the class labels in its training set. Since a random guess would correctly predict a novel class 25% of the time (since 5 of 20 classes are not in the training set), we consider this to be a 25% match. KN represents the percentage of testing instances labeled as known novel.
3. **Abstention:** NOD-CC does not have enough confidence in any of its potential predications, so it abstains from making a prediction. Since guessing a class label randomly would provide the correct prediction approximately 5% of the time (since there are 20 classes), we consider an abstention to be a 5% match. Essentially, this prevents NOD-CC from being forced to provide a random guess to boost its accuracy and allows it to abstain when it is unsure. A represents the percentage of testing instances that were abstained from labeling.
4. **Incorrect:** NOD-CC predicts a known class label (i.e., a class label present in the training set) but it does not match the ground truth class label. This is incorrect and considered to be a 0% match. I represents the percentage of testing instances labeled incorrectly.

During each evaluation, each image in the testing dataset is used as input to NOD-CC and a comparison between the predicted class and ground truth class label is used to calculate our scoring metrics: accuracy (ρ_A), precision (ρ_P), recall (ρ_R), and F1 score (F_1). Although the F1 score calculation uses the well-established equation, we use modified accuracy, precision, and recall functions based on the previous discussions of the four ways NOD-CC's classification can align with the ground truth classification. These metrics have a new term $\mathcal{W}$ introduced that provides weighted credit based on the correctness of the prediction (i.e., Correct, Known Novel, Abstention, or Incorrect). Thus, more correct prediction types are preferred using these metrics.

$$\mathcal{W} = 1.00 \times C + 0.25 \times KN + 0.05 \times A + 0.00 \times I$$

$$\rho_A = \frac{\mathcal{W} \times (TP + TN)}{TP + TN + FP + FN} \qquad\qquad F_1 = 2\frac{\rho_P \times \rho_R}{\rho_P + \rho_R}$$

$$\rho_P = \frac{\mathcal{W} \times TP}{TP + FP} \qquad\qquad \rho_R = \frac{\mathcal{W} \times TP}{TP + FN}$$

For every class label in the dataset (all training classes unseen at training time are considered to be of a single class labeled as *Novel Class*), we compute the accuracy (ρ_A), precision (ρ_P), recall (ρ_R), and f-score (F_1). For each experimental run (i.e., providing the testing instances from a single experimental dataset to Algorithm 1) the mean of each of the class-level metrics is computed. We further vary our experiments by randomizing the order in which testing instances are provided to Algorithm 1. This is important since it is a learning algorithm (i.e., new cases are stored) so the order of testing instances may impact performance. For each of the 20 experimental datasets, 20 random orderings were used. This resulted in 400 total experimental runs (20 datasets × 20 orderings) and the reported results are the averages of the metrics over all 400 runs.

4.3   Always CNN Variant

As a baseline, we evaluated the **Always CNN** variant of NOD-CC (i.e., when the CBR component is ignored). The abstention parameter λ_a was determined through cross-validation on the entire dataset, such that the F_1 was maximized. Recall that the **Always CNN** variant is unable to learn online; it is only able to abstain from providing a label. Assuming a perfectly balanced set of classes, since the CNN is only trained on 15 classes with the remainder only appearing in the testing set, its maximum accuracy is bounded as: $max(\rho_A) = (\frac{15}{20} \times 100\%) + (\frac{5}{20} \times 5\%) = 76.3\%$. In reality, due to the imbalance of the datasets the true maximum accuracy was lower - 63.9% in our experiments. We report an additional metric, *Relative Mean Accuracy (RMA)*, that measures the fraction of $max(\rho_A)$ that was achieved. We also report the minimum (Min. ρ_A), maximum (Max. ρ_A), median (Med. ρ_A) and standard deviation (σ ρ_A) of the accuracy (i.e., when examining each experimental run individually). The performance of **Always CNN** is shown in Table 1.

One item of note in these baseline results is that the precision is significantly higher than the recall. This is intuitive in a system that uses a threshold to determine confidence in classifications (i.e., λ_a); the system only provides classifications when it is confident in its predictions and thereby lowers the number

Table 1. Performance of the various NOD-CC configurations

Variant	ρ_A	ρ_P	ρ_R	F_1	RMA	Min. ρ_A	Max. ρ_A	Med. ρ_A	$\sigma \rho_A$
Always CNN	42.99	61.31	37.98	44.32	67.27	25.98	61.27	41.80	9.15
Always CBR w/ Parts	58.45	54.30	59.66	56.18	81.70	43.49	68.93	61.33	6.57
Always CBR w/o Parts	49.82	49.77	49.41	48.21	69.67	37.49	63.44	59.78	7.27
Conditional CBR w/ Parts	59.90	56.52	60.98	58.17	83.77	54.00	64.12	60.35	2.66
Conditional CBR w/o Parts	53.73	51.39	53.75	52.44	75.15	49.84	61.91	55.23	2.67

of false positives. In these results, as expected, the **Always CNN** approach is never able to correctly label unknown classes, providing evidence to support **H1**.

4.4 Always CBR Variant

As an additional control, we use the *Always CBR* variant of NOD-CC (i.e., the CNN always abstains, so only CBR is used). This variant was evaluated both with observable parts information (i.e., a parts detector component was available) and without. When parts are not available, Algorithm 1 only uses the image features during retrieval. For these experiments, the CBR component was initially given a case base containing all training instances.

Always CBR has a higher theoretical maximum accuracy than **Always CNN** because it has the ability to label an image as a novel class rather than abstaining: $max(\rho_A) = (\frac{15}{20} \times 100\%) + (\frac{5}{20} \times 25\%) = 81.3\%$. Based on the class imbalance of the datasets, the true maximum accuracy was determined to be 71.5%. Similar to with *Always CNN*, this was used to calculate the RMA. The results are shown in Table 1.

Although the availability of detectable object part information is beneficial, **Always CBR** is able to outperform **Always CNN** even without parts. The only metric **Always CNN** performs better on is precision. As we mentioned previously, this is a result of the CNN algorithm being able to abstain, thereby lowering its false positive rate. Overall, the results demonstrate the benefits CBR can provide when the full set of object classes is not known in advance. Even considering the performance of these approaches relative to their maximum accuracy (i.e., RMA), **Always CBR** still outperforms **Always CNN**. These results provide evidence to support **H2**.

4.5   Conditional CBR Variant

In this variant, we use both the CBR and CNN components (i.e., our full architecture). As described previously, the classification from the CNN is used unless the CNN abstains. If the CNN does abstain, the CBR component is used for classification. We use the same configurations (i.e., λ_a threshold and initial case base) for the CNN and CBR components as described in the previous experiments. Similar to the **Always CBR** variant, we evaluate the **Conditional CBR** both with and without parts information. The results are shown in Table 1. Across all metrics, except precision, both variants of **Conditional CBR** outperform **Always CNN**. This demonstrates that the ability of CBR to dynamically detect and learn from previously unseen class types provides significant benefit to the CNN component. In situations where the CNN abstains, the CBR component is able to provide assistance. This provides support for **H3**.

When comparing **Always CBR** to **Conditional CBR**, the **Conditional CBR** variants outperform across all five core metrics (accuracy, precision, recall, f-score, and RMA). This includes both the variants that use parts information as well as those that do not. The results show that the **Conditional CBR** performance has fewer extreme results (i.e., minimums and maximums closer to the mean) and significantly lower standard deviation. This is beneficial because it provides both improved performance as well as less uncertainty about the potential performance on an unknown dataset. Additionally, these results demonstrate the combination of both the CNN and CBR components are necessary for maximum performance; neither module is sufficient for novel object discovery on their own. These results provide support for **H4**.

5   Conclusions and Future Work

In this work we described NOD-CC, a hybrid architecture that uses Case-Based Reasoning and Convolutional Neural Networks to discover novel object types during the image classification process. NOD-CC leverages the automated feature extraction and image classification performance of CNNs while minimizing their requirement for large, pre-labeled training datasets by using CBR's instance-based learning capabilities. NOD-CC can be used with any CNN implementation so it is not tied to a specific CNN architecture, training methodology, or parameter selection. This is particularly important given the rapid advancement in the field of CNNs.

Additionally, NOD-CC can use detected object parts to further improve its performance, although it performs well even if such additional information is unavailable. We evaluated our approach on a publicly available image dataset and showed NOD-CC had improved performance over a CNN or CBR module alone. Our results demonstrated that NOD-CC was able to discover previously unknown classes of objects (i.e., not represented in training data), learn from a single instance of the novel object type, and classify future instances of those objects. Additionally, NOD-CC performed these tasks without compromising the discriminatory power of the CNN.

Future work will involve using the WordNet hierarchy in conjunction with the hierarchical multi-class capabilities afforded by Inception-style architectures in order to perform hierarchical clustering of classes. Thus, a novel class could be placed in a hierarchy relative to known classes, possibly revealing a parent-child relationship. For example, if an image dataset contained labeled images of *balloons* and *baskets*, it could be learned that they are related to a newly discovered object type, an image of a *hot air balloon*. Similarly, textual relations between the known class labels could be used to generate a more semantically meaningful label for the novel object type (e.g., *balloon basket*). We also wish to investigate additional methods for using CBR for classification. Even in our dynamic approach described in this paper, we set an abstaining threshold λ_a for detection to be used unilaterally across all classes. There is an intuitive reason to believe that a CBR system (i.e., a meta-algorithm) for determining when to deploy a second CBR system (i.e., an image classifier) may be useful in this effort.

References

1. Begum, S., Ahmed, M.U., Funk, P., Xiong, N., Folke, M.: Case-based reasoning systems in the health sciences: a survey of recent trends and developments. IEEE Trans. Syst. Man Cybern. Part C (Appl. Rev.) **41**(4), 421–434 (2011)
2. Chen, X., Mottaghi, R., Liu, X., Fidler, S., Urtasun, R., Yuille, A.: Detect what you can: detecting and representing objects using holistic models and body parts. In: Proceedings of the IEEE Conference on Computer Vision and Pattern Recognition, pp. 1971–1978 (2014)
3. Everingham, M., Van Gool, L.J., Williams, C.K.I., Winn, J.M., Zisserman, A.: The Pascal Visual Object Classes (VOC) challenge. Int. J. Comput. Vis. **88**(2), 303–338 (2010)
4. Girshick, R., Donahue, J., Darrell, T., Malik, J.: Rich feature hierarchies for accurate object detection and semantic segmentation. In: Proceedings of the IEEE Conference on Computer Vision and Pattern Recognition, pp. 580–587. IEEE Computer Society (2014)
5. He, K., Zhang, X., Ren, S., Sun, J.: Deep residual learning for image recognition. In: Proceedings of the IEEE Conference on Computer Vision and Pattern Recognition, pp. 770–778 (2016)
6. Krizhevsky, A., Sutskever, I., Hinton, G.E.: ImageNet classification with deep convolutional neural networks. In: Proceedings of the Conference on Advances in Neural Information Processing Systems, pp. 1097–1105 (2012)
7. Kumar, A., Kim, J., Cai, W., Fulham, M., Feng, D.: Content-based medical image retrieval: a survey of applications to multidimensional and multimodality data. J. Digit. Imaging **26**(6), 1025–1039 (2013)
8. Kurth, T., et al.: Exascale deep learning for climate analytics. In: Proceedings of the International Conference for High Performance Computing, Networking, Storage, and Analysis, pp. 51:1–51:12. IEEE Press (2018)
9. Kuznetsova, A., et al.: The open images dataset V4: unified image classification, object detection, and visual relationship detection at scale (2018). arXiv preprint arXiv:1811.00982

10. Lenz, M., Bartsch-Spörl, B., Burkhard, H.D., Wess, S.: Case-Based Reasoning Technology: From Foundations to Applications. Springer, Heidelberg (1998). https://doi.org/10.1007/3-540-69351-3
11. López-Sánchez, D., Corchado, J.M., González Arrieta, A.: A CBR system for efficient face recognition under partial occlusion. In: Aha, D.W., Lieber, J. (eds.) ICCBR 2017. LNCS (LNAI), vol. 10339, pp. 170–184. Springer, Cham (2017). https://doi.org/10.1007/978-3-319-61030-6_12
12. Macura, R.T., Macura, K.J.: *MacRad*: radiology image resource with a case-based retrieval system. In: Veloso, M., Aamodt, A. (eds.) ICCBR 1995. LNCS, vol. 1010, pp. 43–54. Springer, Heidelberg (1995). https://doi.org/10.1007/3-540-60598-3_5
13. Micarelli, A., Neri, A., Sansonetti, G.: A case-based approach to image recognition. In: Blanzieri, E., Portinale, L. (eds.) EWCBR 2000. LNCS, vol. 1898, pp. 443–454. Springer, Heidelberg (2000). https://doi.org/10.1007/3-540-44527-7_38
14. Minsky, M., Papert, S.: Perceptrons: An Introduction to Computational Geometry. MIT Press, Cambridge (1969)
15. Page, A., Turner, J., Mohsenin, T., Oates, T.: Comparing raw data and feature extraction for seizure detection with deep learning methods. In: Proceedings of the International Florida Artificial Intelligence Research Society Conference, pp. 284–287 (2014)
16. Patrini, G., Rozza, A., Menon, A., Nock, R., Qu, L.: Making deep neural networks robust to label noise: a loss correction approach. In: Proceedings of the IEEE Conference on Computer Vision and Pattern Recognition, pp. 2233–2241 (2017)
17. Perner, P., Holt, A., Richter, M.: Image processing in case-based reasoning. Knowl. Eng. Rev. **20**(3), 311–314 (2005)
18. Ravi, D., Wong, C., Lo, B., Yang, G.Z.: A deep learning approach to on-node sensor data analytics for mobile or wearable devices. IEEE J. Biomed. Health Inform. **21**(1), 56–64 (2017)
19. Sani, S., Massie, S., Wiratunga, N., Cooper, K.: Learning deep and shallow features for human activity recognition. In: Li, G., Ge, Y., Zhang, Z., Jin, Z., Blumenstein, M. (eds.) KSEM 2017. LNCS (LNAI), vol. 10412, pp. 469–482. Springer, Cham (2017). https://doi.org/10.1007/978-3-319-63558-3_40
20. Sani, S., Wiratunga, N., Massie, S.: Learning deep features for kNN-based human activity recognition. In: Proceedings of the ICCBR Workshops: Case-Based Reasoning and Deep Learning Workshop, pp. 95–103 (2017)
21. Sathyanarayana, A., et al.: Sleep quality prediction from wearable data using deep learning. JMIR Mhealth Uhealth **4**(4), e125 (2016)
22. Szegedy, C., Vanhoucke, V., Ioffe, S., Shlens, J., Wojna, Z.: Rethinking the inception architecture for computer vision. In: Proceedings of the IEEE Conference on Computer Vision and Pattern Recognition, pp. 2818–2826 (2016)
23. Turner, J.T., Floyd, M.W., Gupta, K.M., Aha, D.W.: Novel object discovery using case-based reasoning and convolutional neural networks. In: Cox, M.T., Funk, P., Begum, S. (eds.) ICCBR 2018. LNCS (LNAI), vol. 11156, pp. 399–414. Springer, Cham (2018). https://doi.org/10.1007/978-3-030-01081-2_27
24. Turner, J., Page, A., Mohsenin, T., Oates, T.: Deep belief networks used on high resolution multichannel electroencephalography data for seizure detection. In: Proceedings of the AAAI Spring Symposium Series: Big Data Becomes Personal: Knowledge into Meaning (2014)
25. Xie, S., Girshick, R.B., Dollár, P., Tu, Z., He, K.: Aggregated residual transformations for deep neural networks. In: Proceedings of the IEEE Conference on Computer Vision and Pattern Recognition, pp. 5987–5995 (2016)

Adaptation of Scientific Workflows by Means of Process-Oriented Case-Based Reasoning

Christian Zeyen[(✉)][iD], Lukas Malburg[(✉)][iD], and Ralph Bergmann[(✉)][iD]

Business Information Systems II, University of Trier, 54286 Trier, Germany
{zeyen,malburgl,bergmann}@uni-trier.de
http://www.wi2.uni-trier.de

Abstract. This paper investigates automatic adaptation of scientific workflows in process-oriented case-based reasoning with the goal of providing modeling assistance. With regard to our previous work on the adaptation of business workflows, we discuss the differences between the workflow types and the implications for transferring the approaches to scientific workflows. An experimental evaluation with RapidMiner workflows demonstrates that the approaches can significantly improve workflows towards a given query while mostly maintaining their executability and semantic correctness.

Keywords: Process-oriented case-based reasoning ·
Workflow adaptation · Scientific workflows

1 Introduction

In the age of big data, data mining is essential for exploiting the potentials hidden in data. However, defining a suitable data analysis procedure is a very challenging task that requires significant expertise. In e-Science, scientific workflows [23] are an established means for this purpose since they describe at a more abstract level how to address a concrete problem in terms of required data, composition of suitable processing steps, and the selection of good parameter settings. In addition, workflows are a means to document an important step of scientific discovery in a way that reproducibility and reliability of results are improved. Besides their use in natural sciences, their potentials have been recently recognized in the Digital Humanities (DH), where the data mostly consists of text documents which must be processed by natural language and text mining procedures [11,12]. A major obstacle, which currently prevents their wider usage in DH as well as in other fields, is the lack of comprehensive modeling support for scientific workflows that particularly targets the needs of researchers who are not experts in software development and workflow modeling. Scientific Workflow Management Systems (SciWFMs) such as *KEPLER* [14], *Taverna* [22], *RapidMiner* [17], or *WINGS* [7] are very helpful in this respect as they already

© Springer Nature Switzerland AG 2019
K. Bach and C. Marling (Eds.): ICCBR 2019, LNAI 11680, pp. 388–403, 2019.
https://doi.org/10.1007/978-3-030-29249-2_26

support the construction and execution of scientific workflows by an integrated visual development environment. Nevertheless, workflow construction remains a demanding and time-consuming task, in particular for complex data analysis that require combinations of many processing steps [5].

An important line of research aims at supporting the development of scientific workflows by reuse of best-practice workflows [5,8] from a repository [6]. Case-Based Reasoning (CBR) has already been proposed to support the reuse of scientific workflows [1,4,7], however previous work mainly focused on the retrieval of reusable workflows, leaving the necessary adaptation up to the user. In this paper, we now address the adaptation of scientific workflows in a case-based manner. For this purpose, we build upon our previous work in Process-Oriented CBR (POCBR) [18] in which we developed various adaptation methods for business workflows, mostly demonstrated in the field of cooking recipes [19–21]. In particular, we investigate the transferability of these approaches to the considerably more complex domain of scientific workflows. This includes the representation of scientific workflows, the learning phase for adaptation knowledge from a case base, as well as the actual adaptation methods using this learned knowledge. Further, we implemented and evaluated the methods for adapting data mining workflows in the SciWFM RapidMiner [17].

The next section reviews related work in the field and analyzes the differences between business and scientific workflows. Section 3 investigates the implications of these differences and presents the results in terms of adaptation approaches that can deal with scientific workflows. Section 4 shows the implementation of the presented approaches for RapidMiner workflows and presents the results of an experimental evaluation. Section 5 concludes with an outlook at future work.

2   Foundations and Related Work

In the following, we briefly survey existing work on modeling support for scientific workflows, we briefly summarize our own work on POCBR which we aim at transferring to scientific workflows, and we discuss the differences between business and scientific workflows.

2.1   Modeling Support for Scientific Workflows

Scientific workflows are executable descriptions of automatable scientific processes such as computational science simulations and data analyses [23]. They consist of a collection of various tasks representing data processing activities such as data import, pre-processing, or modeling together with the parameters used during the execution of these tasks. Also the data items being processed as well as the usage of the data as input or output to the tasks is modeled as part of a scientific workflow. Due to the complexity involved in the creation of scientific workflows, several methods have been proposed to support users in doing so. For instance, an approach by Jannach et al. [9] supports the user with context-sensitive recommendations of single processing steps and parameter settings for the current workflow under construction, based on the analysis of a large workflow repository. Kietz et al. [10] propose a planning approach for the automatic composition of RapidMiner workflows, correctness-checking, and quick-fixes using an ontology of available computational steps, parameters, and constraints. The SciWFM WINGS [7] also uses planning and semantic reasoning to automatically create workflows based on high-level user requests.

While these approaches require either a fully specified domain model appropriate for planning or a huge number of workflow instances to learn from by induction, CBR has also been used to support workflow development. For instance, Leake and Kendall-Morwick [13] consider execution traces of workflows as cases. The approach extracts tasks from retrieved cases for extending the current workflow under construction. The approach by Chinthaka et al. [4] uses keyword descriptions of inputs and outputs provided by the user for finding the best-matching workflow. If necessary, an adaptation process tries to extend the workflow with single tasks to better fulfill the user's requirements. In our own work, we addressed the retrieval of semantically labeled workflow graphs as a starting point for reuse [16]. In general, previous case-based approaches in the field of scientific workflows mainly focused on retrieval and did not yet address extensive adaptation in a fully automatic manner.

2.2   Process-Oriented Case-Based Reasoning

Our previous work on POCBR already dealt with the adaptation of business workflows. We developed methods for substitutional *adaptation by generalization and specialization* [20] and two methods for structural adaptation, namely *operator-based adaptation* [21] and *workflow streams adaptation* [19]. The methods are currently restricted to business workflows and have been intensively investigated only in the field of cooking recipes. In this paper, we focus on adaptation by generalization and specialization and on workflow stream adaptation.

Adaptation by Generalization and Specialization. Using generalized cases [3] is an established approach that allows to represent a set of cases by a single generalized case, leading to a reduced case base and increased coverage. When applied to workflow cases, workflow components such as task and data objects are generalized to a common object that has certain properties that hold for all subsumed objects. During the specialization process, either the original or other available specializations can be inserted to better fulfill a given query.

Adaptation with Workflow Streams. Workflow stream adaptation is a compositional adaptation approach [24] which decomposes the workflow cases of the case base into meaningful sub-workflows (referred to as workflow stream) that are stored as adaptation knowledge. A retrieved workflow is adapted with respect to a query by incrementally replacing its own workflow streams with more suitable streams from the adaptation knowledge.

Learning Adaptation Knowledge. Since the presented adaptation methods require a significant amount of domain-specific adaptation knowledge, the developed methods automatically learn required knowledge (generalized workflows and workflow streams) from the workflow repository, i.e., the case base. Hence, we distinguish between a learning phase of adaptation knowledge and a problem solving phase in which for a given query the best matching workflow is adapted such that it matches the particular problem scenario at best. By applying the learned adaptation knowledge between different cases, a larger solution space is covered.

2.3 Differences Between Business and Scientific Workflows

There are significant differences between business workflows and scientific workflows, which prevent the direct application of the developed adaptation methods.

Business workflows aim at automating organizational processes, which are mainly executed by humans, supported by certain resources including application programs. While the goal of a business workflow is already determined before the workflow is executed, the goal of a scientific workflow is to validate hypotheses of a researcher based on available data. These scientific goals are experimental and exploratory and thus vary more frequently. While business workflows are executed under the control of the involved humans, scientific workflows are executed fully automatically by a computer [15]. Consequently, they depend more strongly on proper parameter settings.

Control-flow and data-flow orientation are discussed as main differences between business and scientific workflows [1,15]. In business workflows, control-flow patterns such as *AND*, *XOR*, and *LOOPS* are commonly used. Data-flow is modeled separately or implicitly due to its subordinate significance [15]. In scientific workflows, data-flow is of primary importance since the control-flow, respectively the execution order of the computational steps, results implicitly from the given data-flow. A computational step can only be executed after the required data has been generated by the previous steps, hence the workflow is data-driven. However, some SciWFMs also enable to explicitly define control-flow between computational steps or to use control-flow patterns.

Many SciWFMs such as *KEPLER* [14], *Taverna* [22], or *RapidMiner* [17] use certain interfaces (referred to as ports) for restricting the data-flow between computational steps to ensure that only valid data is exchanged. Port connections are not used in business workflows, which is thus a further significant difference.

3   Adaptation of Scientific Workflows

This section outlines the representation of scientific workflows as semantic graphs before the substitutional and structural adaptation approaches are described. We keep those descriptions on an informal level as the full algorithmic details[1] would exceed the space limitation of this paper.

---

[1] Lukas Malburg formally described the approaches in his master thesis "Adaptation of Scientific Workflows by Means of POCBR" submitted 2019 at Trier University.

3.1   Graph Representation and Semantic Similarity

To represent workflows, we use semantic graphs [1]. Figure 1 depicts such a graph of a scientific workflow from the data mining domain. In a semantic graph, semantic descriptions and specific types are assigned to each node and edge. Semantic descriptions are based on a semantic metadata language. For scientific workflows, we use object-oriented descriptions to represent heterogeneous parameters and values of computational steps. The computational steps of a scientific workflow are represented as task nodes in the graph. Data nodes between task nodes represent the data that is exchanged between computational steps. To represent port connections, corresponding information is added to the semantic descriptions. In the example graph, the semantic descriptions of data nodes comprise information about the labels of connected input and output ports. For example, it is stated that data object *Table Data* is produced by the task *Import Data* at the output port *table data* and that the same data object is consumed by the task *Discretize* at the input port *table data*.

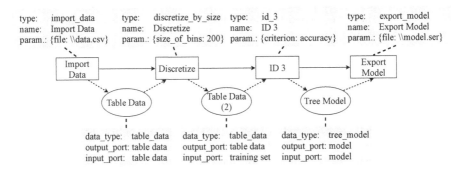

Fig. 1. Graph representation of a scientific workflow

In a scientific workflow graph, the data nodes reflect the state changes of processed data and thus have to be distinct. For instance, the task *Discretize* produces an output of the same data type but in an altered state. Hence, data nodes *Table Data* and *Table Data (2)* are represented as distinct nodes. In business workflows (cf. the cooking workflows used in our previous work [19,20]), each data node is unique, too, but state changes of data are typically not represented. Due to this difference, the usage of data-flow edges between data nodes and task nodes differs in the graph representation. While a data node can be linked with various task nodes as input or output in a business workflow graph, we restrict the connection of a data node such that it can only be connected as output of one task node and as input of another task node in a scientific workflow graph.

Task nodes are also semantically enriched by semantic descriptions that contain, for example, the type of the computational step, the name, and defined parameters. Even though control-flow is secondary for scientific workflows, we explicitly represent the control-flow between task nodes by selecting one valid execution order of tasks. By this means, we take into account SciWFMs such as RapidMiner that allow for defining control-flow.

Analogous to previous work (cf. [19,20]), we use the semantic similarity measure by Bergmann and Gil [1] to assess the similarity between two graphs. The measure uses subgraph matching and applies heuristic search to find the best possible, injective, partial mapping m between the nodes and edges of the query workflow QW and those of the case workflow CW:

$$\text{sim}(QW, CW) = \max\{\text{sim}_m(QW, CW) \mid \text{mapping } m\} \tag{1}$$

During the search, each mapping of two nodes or edges is rated by local similarity functions $\text{sim}_m \rightarrow [0,1]$. Following the local-global principle, similarities between a pair of mapped nodes or edges is assessed by comparing the attribute values of their semantic descriptions with each other. Such similarities are then aggregated to similarities of mapped nodes or edges, which, in turn, are aggregated to the global similarity value of two graphs.

3.2 Substitutional Adaptation by Generalization and Specialization

The adaptation by generalization and specialization is used to substitute task nodes in a scientific workflow graph. In the learning phase, the workflows in the case base are generalized. In the problem solving phase, a retrieval is performed in the generalized case base for the best-matching generalized workflow. Subsequently, the generalized workflow is specialized in regards to the query.

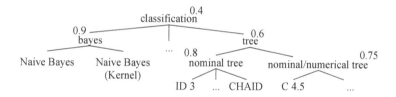

Fig. 2. Excerpt of a task taxonomy

Generalization. The generalization requires ontological information about the tasks. In particular, a taxonomy of all types of tasks is required. In such a hierarchy, the generalization considers the types along the path from a concrete type (i.e., a leaf node) towards the root as possible generalizations [20]. Figure 2 depicts an excerpt of an exemplary task taxonomy. In the taxonomy, the generalized class *bayes* subsumes the concrete tasks *Naive Bayes* and *Naive Bayes (Kernel)* and has assigned a semantic similarity value of 0.9 that holds for all subsumed tasks, i.e., for both naive bayes implementations. The values were determined manually considering the meta data of tasks. In contrast to business workflows, the generalization of scientific workflows primarily relates to tasks since data types depend on tasks and thus cannot be adapted independently. In order to generalize a task node in a workflow graph, several conditions must be satisfied. In previous work [20], two similarity thresholds have been proposed for business workflows to assess whether task nodes are suitable for generalization. The first similarity threshold $\triangle_W \in [0,1]$ prevents workflows from being generalized that are considered too heterogeneous. A second threshold is defined for a measure that takes into account the taxonomic structure of the tasks to prevent over-generalization. For scientific workflows, this parameter is not required, since port connections of tasks are considered as constraints instead. Two tasks from different workflows are considered to be generalizable, if they have equal port connections, i.e., if the specifications of connected ports are identical. Furthermore, it is required that the generalized task is able to consume and produce the same data items. For each inner node (representing a generalized task type) in the taxonomy, we determine the common port specifications of the subsumed tasks and assign them to the inner node. A task is incrementally generalized until no more common port specifications exist. The constraint of common port connections ensures that each specific task is able to consume and produce the same data as its siblings in the taxonomy. Thus, in the specialization phase, each specific child of a generalized task can be selected for substitution and it is ensured that no structural changes (e.g., adaptation of the data-flow) to the workflow are required.

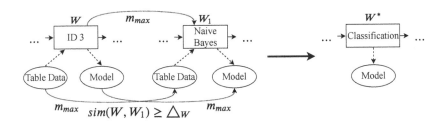

Fig. 3. Exemplary generalization of tasks

Figure 3 illustrates an exemplary generalization process. In the given example, two similar workflow graphs W and W_1 are compared with a similarity $sim(W, W_1) \geq \triangle_W$. The task node *ID 3* of workflow W is mapped to the task node *Naive Bayes* of workflow W_1. Both tasks consume and produce the same data. Assuming that the tasks have the same port connections and that the generalized task type *Classification* is a common ancestor of both tasks (see Fig. 2) that is able to consume and produce the corresponding data, the task *ID 3* can be generalized to *Classification*.

Specialization. The specialization process is similar to the process applied to business workflows (cf. [20]). For each generalized task in a retrieved generalized workflow, the most suitable specializations are inserted with respect to the given query. If a generalized task is used in the query workflow, an arbitrary specific task is selected during the specialization of business workflows. However, to ensure the executability of scientific workflows, already known specializations are preferred over unknown specializations. Hence, during generalization all original tasks are stored for each generalized task. If a known specialization exists, it is chosen as specialization. Otherwise, an arbitrary specialization is selected. In this event, the task is pre-initialized with known default parameter settings.

3.3   Structural Adaptation with Workflow Streams

The adaptation with workflow streams exploits the structure of workflows to define the borders of substitutable and meaningful sub-workflows that we call workflow streams. For this purpose, the approach poses the requirement of *block-orientation* (cf. [19]) on the representation of workflow graphs. In a nutshell, block-orientation restricts the control-flow in a graph in such a way that only a single start and end node is allowed and that all task nodes must be connected by control-flow edges. For workflow representations without an explicit definition of the control-flow, which is often the case for scientific workflows, the control-flow must be made explicit to be block-oriented.

A workflow stream is considered as a sub-workflow that produces a partial output of a workflow. A task that produces such a data object is referred to as *producer* (or creator). Due to the different representation of data objects in scientific workflow graphs (cf. Sect. 3.1), the definition of a producer differs from our previous approach. In business workflow graphs, types of data nodes are always considered different. Since data objects are stateful in scientific workflows, several data nodes may exist with an equal data type. Thus, producers are identified by looking at the input and output data types. A task is considered as producer, if it produces at least one data object of a type that is different from all of its input data types.

Workflow Stream Partitioning. A block-oriented scientific workflow can be partitioned into workflow streams similarly to a business workflow. Based on the identified producers $P \subseteq T$ from all tasks T in the workflow, the workflow is partitioned into a set of workflow streams such that each task $t \in T$ is exactly assigned to one stream. A workflow stream S is constructed for each producer $p \in P$. The producer p constitutes the end task node in the control-flow of the stream S. A task $t \in T \setminus P$ is part of the stream S,

- if t is a predecessor of p in the control-flow,
- if t is not already part of another stream,
- and if t is connected to an equal data node the producer p or another task of stream S is connected to (cf. definitions of data-flow connectedness in [19]).

In such a partitioning, each task is exactly assigned to one workflow stream. Figure 4 illustrates the partitioning of two scientific workflows. All workflow

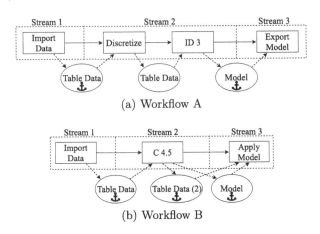

(a) Workflow A

(b) Workflow B

Fig. 4. Partitioning of scientific workflows

streams are marked with dashed rectangles. The producers in the example workflow A are *Import Data* and *ID 3*. The example demonstrates that streams do not only produce partial outputs, but also consume data (see stream 3).

Workflow Stream Substitution. For the problem-solving phase, we assume that the available workflows in the case base have been decomposed into workflow streams in a learning phase beforehand. A workflow from the case base can be adapted w.r.t. a given query workflow by replacing its workflow streams with other available streams that better fulfill the requirements of the query. For this purpose, the workflow to be adapted is decomposed into streams and for each stream a search for the best substitute stream candidate is performed. This search process applies the semantic graph similarity outlined in Sect. 3.1 to find the best workflow streams with respect to a given query.

Two streams are regarded as substitutable, if they consume and produce the same data. More precisely, a stream S_1 is substitutable by a stream S_2, if all the data objects in S_1 that are consumed or produced by a task of another stream are identical to those data objects of S_2. Such data nodes are also referred to as *data anchors* and the corresponding task is named *consumer* or *producer anchor*, respectively. In Fig. 4, data anchors are marked with ⚓ symbols.

Due to the availability of port connections in scientific workflows, a further condition is proposed to define substitutable streams. If the set of data anchors of two streams S_1 and S_2 is not identical, it is checked for each unmatched data anchor of stream S_1 whether the other stream S_2 contains a consumer or producer anchor with an unused port that is suited to consume or produce the data anchor of S_1. For instance, in the example given in Fig. 4, Stream 2 of Workflow B is substitutable by Stream 2 of Workflow A, assuming that both tasks *C 4.5* and *ID 3* have equal output port specifications. In this event, the unused port of *ID 3* is used to produce the data object *Table Data (2)*.

3.4   Combined Adaptation

Both adaptation approaches can also be used in combination. By this means, the coverage of the solution space can be increased, which has been investigated in previous work [2]. For the combined adaptation, the case base is first generalized to learn generalized workflow streams. Then, retrieval is performed and the best-matching generalized workflows are adapted. First, structural adaptation substitutes generalized workflow streams that best match the given query. Subsequently, substitutional adaptation specializes the adapted workflow to produce concrete and executable workflows.

4   Implementation and Experimental Evaluation

We fully implemented the described approaches in the POCBR component of the CAKE framework[2] such that it can adapt RapidMiner workflows. The following section outlines the implementation with a particular focus on the semantic graph representation as prerequisite of adaptation. Subsequently, we present the evaluation setup and discuss the results.

4.1   Implementation of the RapidMiner Domain

We developed a plugin for the RapidMiner environment [17] that allows for automatically exporting workflows to XML and extracting meta data about available workflow components. Essentially, the knowledge model is based on these meta data that comprise various information about each available task such as textual descriptions, parameters, value ranges, default settings, input

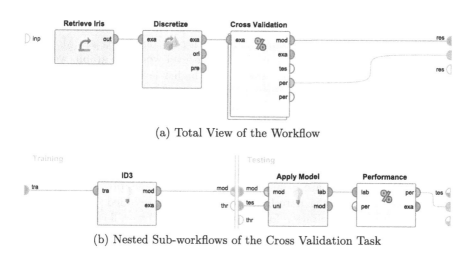

(a) Total View of the Workflow

(b) Nested Sub-workflows of the Cross Validation Task

Fig. 5. Screenshot of a RapidMiner workflow

---

[2] See http://procake.uni-trier.de.

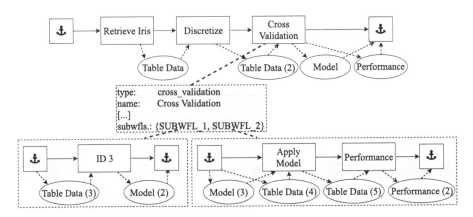

Fig. 6. Semantic graph of a RapidMiner workflow

and output ports, and data types. Figure 5 illustrates a data mining workflow that reads and discretizes a data set, learns an ID3 decision tree, and performs a cross validation to measure the model performance. Some tasks (in the following referred to as complex tasks) such as the Cross Validation enclose further tasks as sub-workflows. Figure 6 shows the semantic graph representation of the overall workflow, which is similar to the graph depicted in Fig. 1. It should be noted that only the semantic description of the complex task is shown in the figure. Sub-workflows of a complex task are represented as semantic graphs that are referenced in the corresponding semantic description. In the given example, both graphs are listed as references in the *subworkflows* attribute of the semantic description. Hence, when computing the similarity between two complex tasks, referenced sub-workflows are also compared according to their order in the *subworkflows* attribute using new instances of the graph similarity measure. By this means, they contribute to the semantic similarity of complex tasks. Moreover, during the similarity computation of the higher-level workflow graph, workflow components can also be mapped between the higher-level graph and the sub-workflow graphs. Consequently, a retrieval with a query that does not contain sub-workflows can still find similar workflows that contain similar components nested in sub-workflow structures. To represent data-flow connections to and from workflow or sub-workflow input and output ports, auxiliary tasks referred to as *IO anchors* are added to the graphs (see ⚓ symbols in Fig. 6).

The presented adaptation approaches can be applied for this graph representation. An exception are io anchors, which are not adapted by the methods to ensure that existing data-flow connections are retained. Regarding the requirement of block-orientation for the adaptation with workflow streams, sub-workflows can be adapted separately, since IO anchors ensure that their interfaces to other workflows are considered.

4.2   Hypotheses and Evaluation Setup

To evaluate the adaptation methods, we perform adaptation experiments with executable RapidMiner workflows and constructed queries. With respect to the average query fulfillments, we investigate the following hypotheses:

H1 Each adaptation method provides at least as good results as the sole retrieval.

H2 The structural adaptation method outperforms the substitutional adaptation method.

H3 The combined adaptation method provides better results as both, the structural and the substitutional adaptation method.

H4 The adaptation methods produce executable and semantically correct workflows.

As evaluation criteria, we measure the *query fulfillment*, which is determined by the semantic similarity between the query and the adapted workflow graphs (see Sect. 3.1). Furthermore, we assess the *quality*, which is determined by executability and semantic correctness of the adapted workflows.

For the experiments, we created a case base of 20 data mining workflows as follows: First, we selected three sub-trees of the task taxonomy that subsume classification and clustering tasks (e.g., ID3 or k-Means), validation tasks (e.g., cross or split validation), and performance measuring tasks (e.g., accuracy or cluster distance). Subsequently, we randomly selected one concrete task from each of the sub-trees and constructed an executable workflow containing these tasks by adding required data processing tasks. All workflows are created for the same data set to allow for a broader applicability of adaptation knowledge. Figure 5 depicts one of the created workflows. Similarly, we created 10 queries, which are different from one another and from the workflows in the case base, by randomly selecting one task from each of the three sub-trees, including generalized tasks (i.e., inner tree nodes). Such a query is a partial, non-executable workflow whose tasks are not connected with each other. Here, additionally required data processing tasks are not included.

For each query, a retrieval of the k top-ranked workflows is performed in the case base. Workflows with an identical similarity value are considered equally ranked. Subsequently, adaptation is performed on the top-k workflows and the *query fulfillment* is measured. The workflows with the highest query fulfillment are selected for further examination. We checked the *syntactic correctness* by converting the graph into the XML format, the *executability* by importing (the XML) and executing the workflow in RapidMiner, and the *semantic correctness* by manual examination after successful execution.

As baseline for the query fulfillment, a *retrieval without adaptation* is conducted. For the *structural adaptation* approach, retrieval is performed and the k top-ranked workflows are adapted w.r.t. the query. For the *substitutional adaptation* approach, the case base is generalized first, a retrieval is performed with the generalized workflows, and the k top-ranked workflows are specialized. Due to a rather narrow scope of the workflows, the threshold $\triangle W = 0$ is used for the generalization. In a third experiment, the *combined adaptation* is evaluated.

4.3 Results

Table 1 summarizes the measured query fulfillments for plain retrieval, i.e., the baseline, (row "w/o") and for the performed adaptations. An empty space in the table indicates that the measured value is equal to the baseline. For $k > 1$ the highest value is taken, since adaptability among the workflows may vary. A cross mark beside a value indicates that none of the k top-ranked workflows is executable or semantically correct. In each of these events, the error is caused by missing pre-processing tasks for converting the data set into a format required by the requested classification or clustering task (e.g., discretization). We found no case, where an adapted workflow was executable but not semantically correct. The results of structural adaptation show that adaptation of the first ranked workflows by retrieval does not necessarily lead to the highest possible similarity, which is only reached with $k \geq 3$. In the experiments with the substitutional adaptation, the generalized workflows ranked highest by retrieval are also best suited for adaptation regarding each query. The query fulfillments do not increase for higher k values. Both adaptation methods outperform the sole retrieval, which confirms Hypothesis H1. No significant differences in the results of structural and substitutional adaptation could be observed in the experiments. Thus, Hypothesis H2 cannot be confirmed. For $k = 1$, the combined approach achieved the best results on average, followed by the substitutional and structural approaches. Hence, Hypothesis H3 is confirmed. However, comparing these similarities with those obtained without adaptation, all approaches yield to significantly ($p < 0.05$) higher values on average.

Table 1. Maximum query fulfillments of k top-ranked adapted workflows

		Q1	Q2	Q3	Q4	Q5	Q6	Q7	Q8	Q9	Q10	Avg
w/o		0.82	0.60	0.85	0.85	0.91	0.81	0.78	0.85	0.71	0.72	0.79
struct.	$k = 1$			1.00			1.00 ✗	1.00			0.86 ✗	0.86
	$k = 2$	1.00		1.00	1.00		1.00 ✗	1.00		0.91	0.86 ✗	0.91
	$k \geq 3$	1.00	0.69	1.00	1.00		1.00 ✗	1.00		0.91	0.86 ✗	0.92
subst.	$k \geq 1$	1.00	0.68		1.00		1.00 ✗	1.00		0.91	0.91	0.91
comb.	$k \geq 1$	1.00	0.69	1.00	1.00		1.00 ✗	1.00		0.91	1.00 ✗	0.94

The results also show some differences between the queries. In the events, where adaptation does not exceed retrieval ($Q5$ and $Q8$) or where adapted workflows are not executable ($Q6$ and $Q10$), queries contain combinations of tasks for which no or at least no applicable adaptation knowledge is available. For example, in $Q6$, an ID3 decision tree is requested that requires discretized data, but a discretization task is not part of the query. All the adapted workflows contain the requested tasks but no discretization task since it is not requested. Consequently, Hypothesis H4 is only partially confirmed.

5 Conclusions

This work investigates the application of adaptation in POCBR for the complex domain of scientific workflows. In a first step, we determine differences between business and scientific workflows. These differences affect the adaptation methods that are already used to adapt business workflows. Thus, we present a modified approach for structural adaptation with *workflow streams* and for substitutional adaptation by *generalization and specialization*. We implement both adaptation methods in the CAKE framework for adapting RapidMiner workflows. Both adaptation methods show promising results in the experimental evaluation as they are able to significantly increase the query fulfillment.

Due to the differences between business and scientific workflows and the more complex domain of scientific workflows, further advancements in the adaptation approaches are necessary to improve executability and semantic correctness of adapted workflows. At present, adaptation only ensures the syntactic correctness and does not consider or adapt parameter settings of tasks. For future work, several ideas for improving the adaptation methods exist. The structural adaptation can be improved by adding an insert capability and by including finer-grained adaptation components such as those presented in our operator-based adaptation approach. Moreover, it can be investigated how dependencies of tasks can be determined for a given data set. For this purpose, the coverage of adaptation can be computed by applying the entire adaptation knowledge to the workflows in the case base. The resulting sets of executable and non-executable workflows can then be analyzed regarding the dependencies of tasks to derive, which adaptation components are compatible with each other.

Further, we plan to expand the evaluation to more complex, user-generated workflows and we will integrate the adaptation methods within our plugin for the RapidMiner environment in order to support their use in a convenient way. By this means, a further step towards the interactive adaptation of workflows for providing modeling assistance can be made.

Acknowledgments. This work is partly funded by the German Federal Ministry of Education and Research (BMBF, No. 01UG1606) and the German Research Foundation (DFG, No. BE 1373/3-3).

References

1. Bergmann, R., Gil, Y.: Similarity assessment and efficient retrieval of semantic workflows. Inf. Syst. **40**, 115–127 (2014)
2. Bergmann, R., Minor, M., Müller, G., Schumacher, P.: Project EVER: extraction and processing of procedural experience knowledge in workflows. In: Proceedings of ICCBR 2017 Workshops, vol. 2028, pp. 137–146 (2017). CEUR-WS.org
3. Bergmann, R., Vollrath, I.: Generalized cases: representation and steps towards efficient similarity assessment. In: Burgard, W., Cremers, A.B., Cristaller, T. (eds.) KI 1999. LNCS (LNAI), vol. 1701, pp. 195–206. Springer, Heidelberg (1999). https://doi.org/10.1007/3-540-48238-5_16

4. Chinthaka, E., Ekanayake, J., Leake, D.B., Plale, B.: CBR based workflow composition assistant. In: 2009 IEEE Congress on Services, Part I, SERVICES I, pp. 352–355. IEEE Computer Society (2009)

5. Cohen-Boulakia, S., Leser, U.: Search, adapt, and reuse: the future of scientific workflows. SIGMOD Rec. **40**(2), 6–16 (2011)

6. De Roure, D., et al.: myExperiment: defining the social virtual research environment. In: IEEE Fourth International Conference on eScience, pp. 182–189. IEEE (2008)

7. Gil, Y., et al.: Wings: intelligent workflow-based design of computational experiments. IEEE Intell. Syst. **26**(1), 62–72 (2011)

8. Goderis, A.: Workflow re-use and discovery in bioinformatics. Ph.D. thesis, University of Manchester (2008)

9. Jannach, D., Jugovac, M., Lerche, L.: Supporting the design of machine learning workflows with a recommendation system. TiiS **6**(1), 8:1–8:35 (2016)

10. Kietz, J.-U., Serban, F., Fischer, S., Bernstein, A.: "Semantics inside!" but let's not tell the data miners: intelligent support for data mining. In: Presutti, V., d'Amato, C., Gandon, F., d'Aquin, M., Staab, S., Tordai, A. (eds.) ESWC 2014. LNCS, vol. 8465, pp. 706–720. Springer, Cham (2014). https://doi.org/10.1007/978-3-319-07443-6_47

11. Kuhn, J., Reiter, N.: A plea for a method-driven agenda in the digital humanities. In: Book of Abstracts of DH 2015 (2015)

12. Kuras, C., Eckar, T.: Prozessmodellierung mittels BPMN in Forschungsinfrastrukturen der Digital Humanities. In: INFORMATIK 2017, pp. 1101–1112. GI (2017)

13. Leake, D.B., Kendall-Morwick, J.: Towards case-based support for e-science workflow generation by mining provenance. In: Althoff, K.-D., Bergmann, R., Minor, M., Hanft, A. (eds.) ECCBR 2008. LNCS (LNAI), vol. 5239, pp. 269–283. Springer, Heidelberg (2008). https://doi.org/10.1007/978-3-540-85502-6_18

14. Ludäscher, B., et al.: Scientific workflow management and the KEPLER system. Concurr. Comput. Pract. Exp. **18**(10), 1039–1065 (2006)

15. Ludäscher, B., Weske, M., McPhillips, T., Bowers, S.: Scientific workflows: business as usual? In: Dayal, U., Eder, J., Koehler, J., Reijers, H.A. (eds.) BPM 2009. LNCS, vol. 5701, pp. 31–47. Springer, Heidelberg (2009). https://doi.org/10.1007/978-3-642-03848-8_4

16. Malburg, L., Münster, N., Zeyen, C., Bergmann, R.: Query model and similarity-based retrieval for workflow reuse in the digital humanities. In: Proceedings of LWDA 2018, Mannheim, vol. 2191, pp. 251–262 (2018). CEUR-WS.org

17. Mierswa, I., Wurst, M., Klinkenberg, R., Scholz, M., Euler, T.: YALE: rapid prototyping for complex data mining tasks. In: Proceedings of the 12th ACM SIGKDD International Conference on Knowledge Discovery and Data Mining, pp. 935–940. ACM (2006)

18. Minor, M., Montani, S., Recio-García, J.A.: Process-oriented case-based reasoning. Inf. Syst. **40**, 103–105 (2014)

19. Müller, G., Bergmann, R.: Workflow streams: a means for compositional adaptation in process-oriented CBR. In: Lamontagne, L., Plaza, E. (eds.) ICCBR 2014. LNCS (LNAI), vol. 8765, pp. 315–329. Springer, Cham (2014). https://doi.org/10.1007/978-3-319-11209-1_23

20. Müller, G., Bergmann, R.: Generalization of workflows in process-oriented case-based reasoning. In: Proceedings of FLAIRS 2015, pp. 391–396. AAAI Press (2015)

21. Müller, G., Bergmann, R.: Learning and applying adaptation operators in process-oriented case-based reasoning. In: Hüllermeier, E., Minor, M. (eds.) ICCBR 2015.

LNCS (LNAI), vol. 9343, pp. 259–274. Springer, Cham (2015). https://doi.org/10.1007/978-3-319-24586-7_18

22. Oinn, T., et al.: Taverna/^{my}Grid: aligning a workflow system with the life sciences community. In: Taylor, I.J., Deelman, E., Gannon, D.B., Shields, M. (eds.) Workflows for e-Science. Scientific Workflows for Grids, pp. 300–319. Springer, London (2006). https://doi.org/10.1007/978-1-84628-757-2_19

23. Taylor, I.J., Gannon, D.B., Shields, M. (eds.): Workflows for e-Science: Scientific Workflows for Grids. Springer, London (2010)

24. Wilke, W., Bergmann, R.: Techniques and knowledge used for adaptation during case-based problem solving. In: Pasqual del Pobil, A., Mira, J., Ali, M. (eds.) IEA/AIE 1998. LNCS, vol. 1416, pp. 497–506. Springer, Heidelberg (1998). https://doi.org/10.1007/3-540-64574-8_435

Author Index

Printed in the United States
By Bookmasters